BEST
Remudas

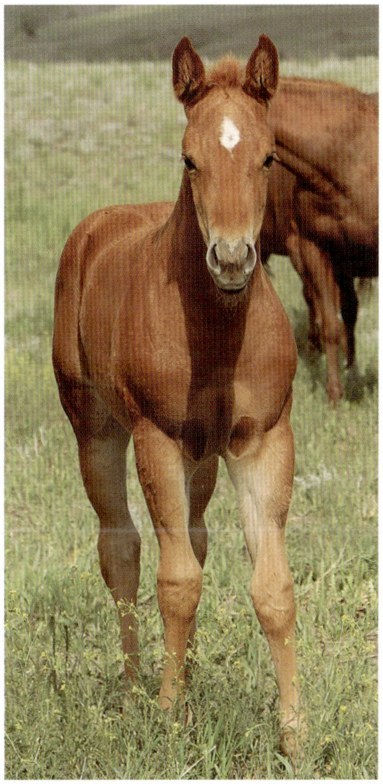

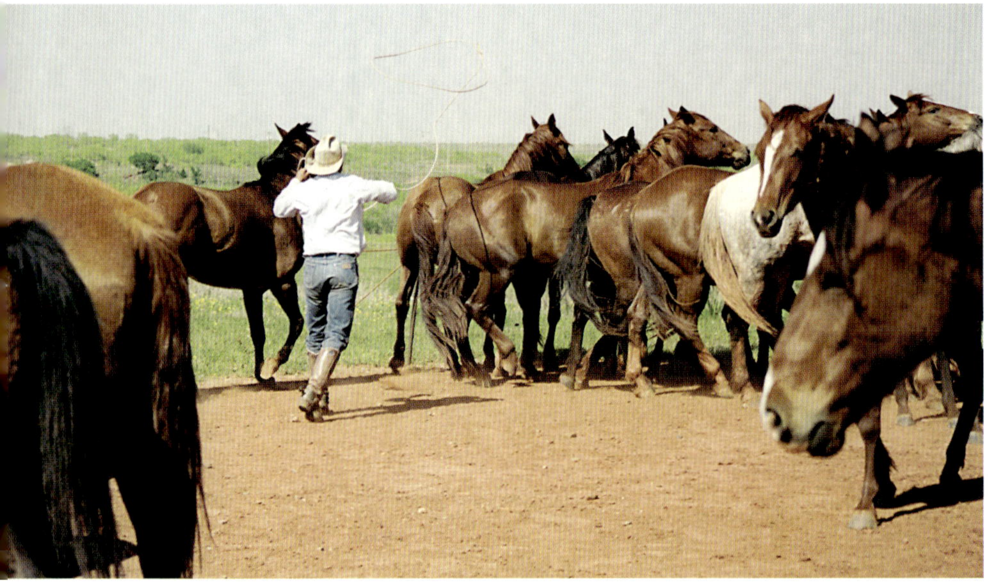

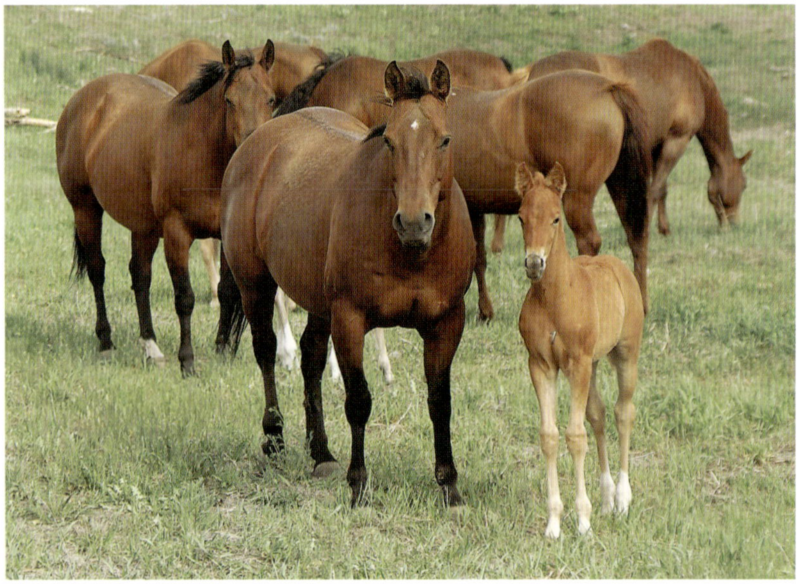

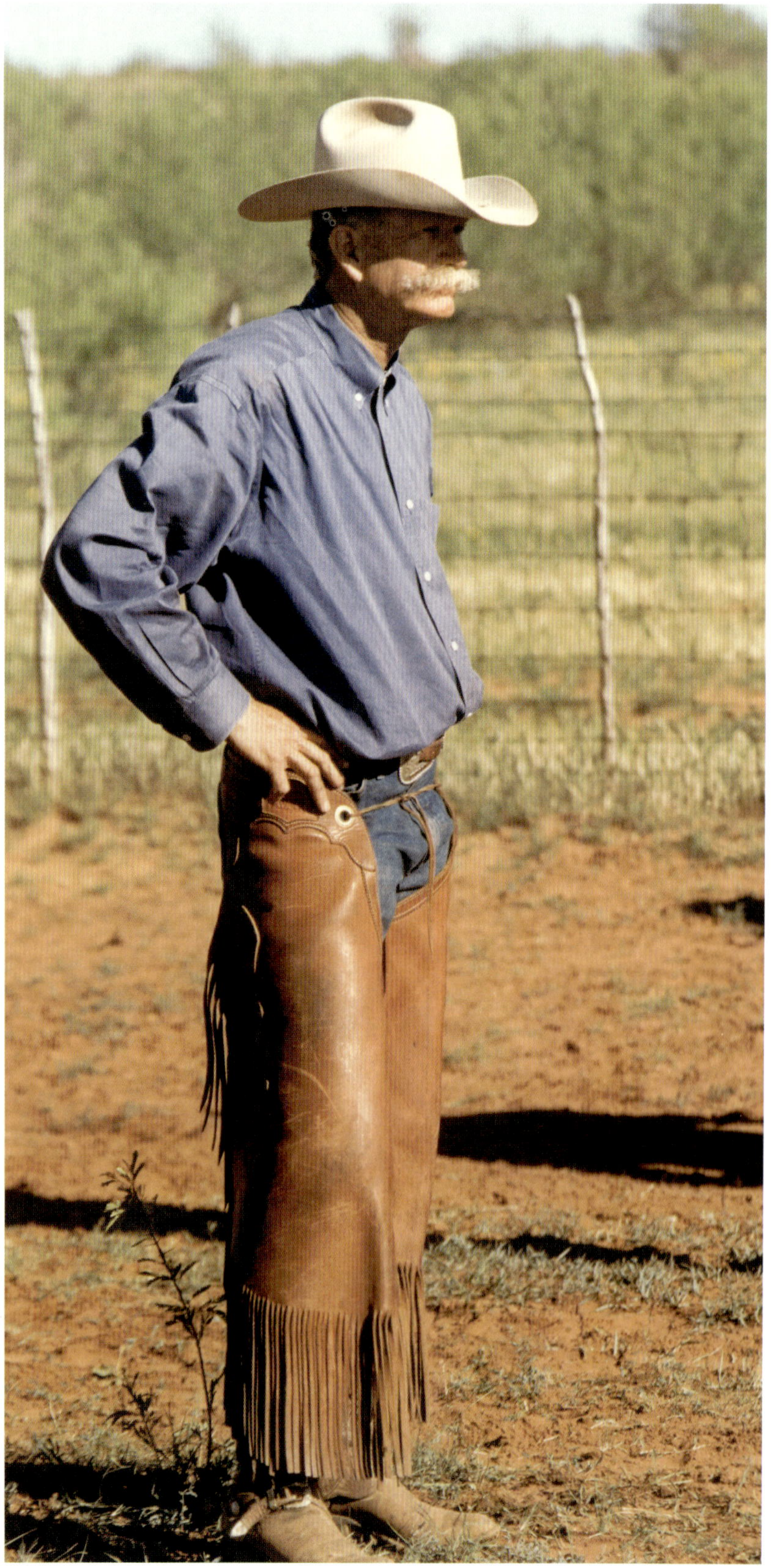

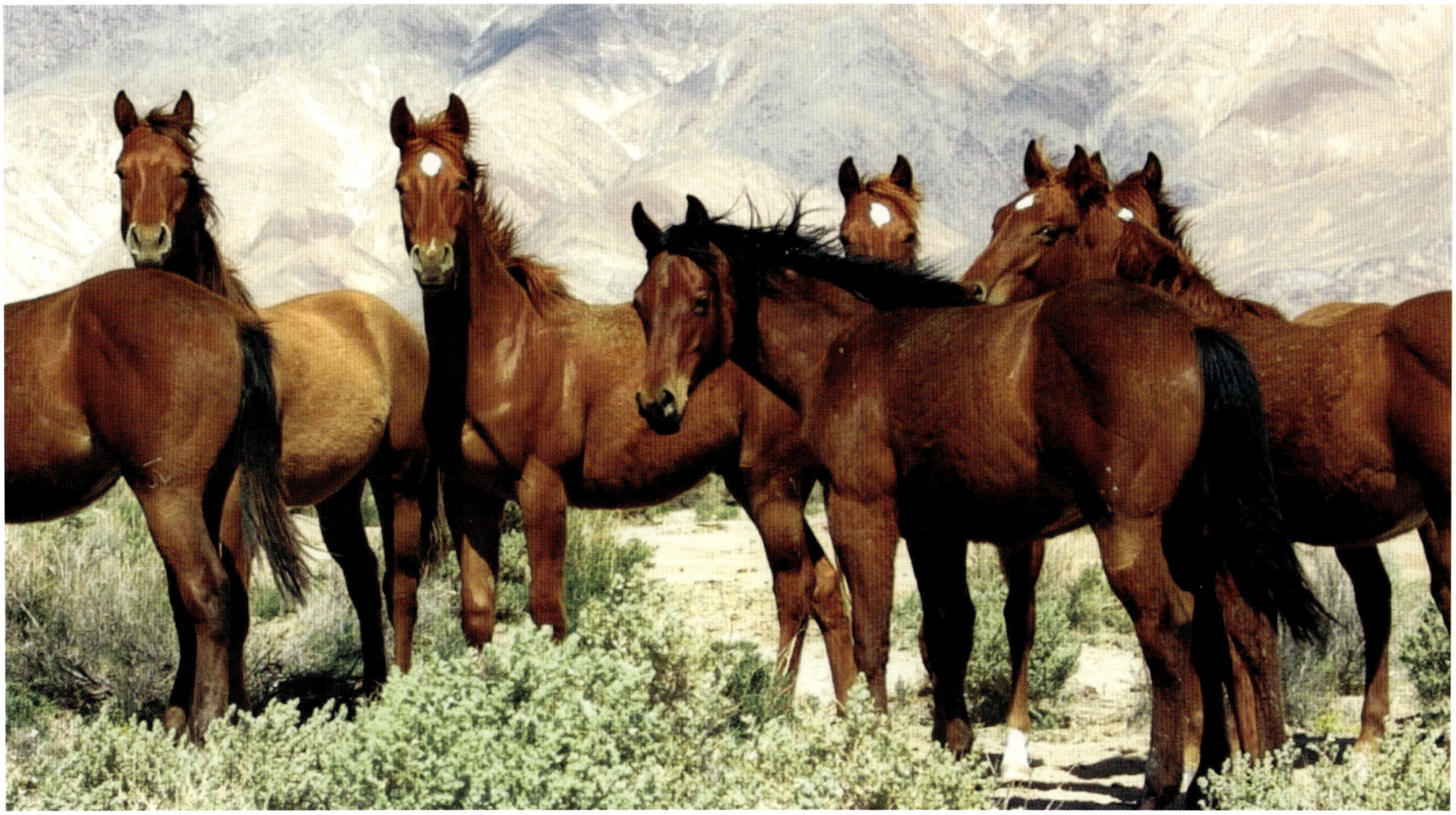

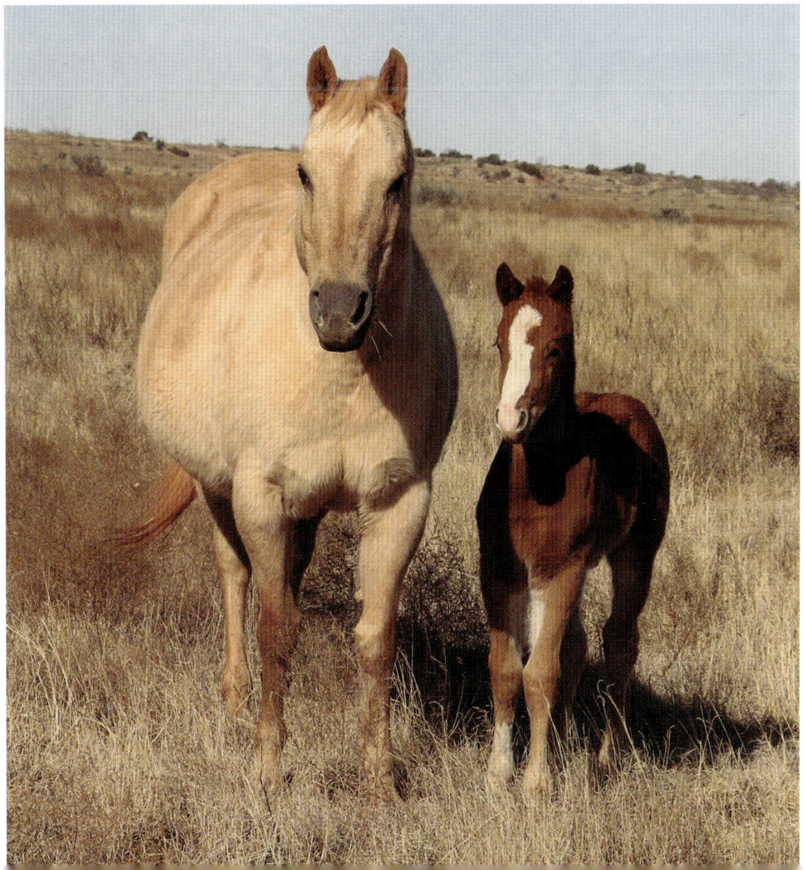

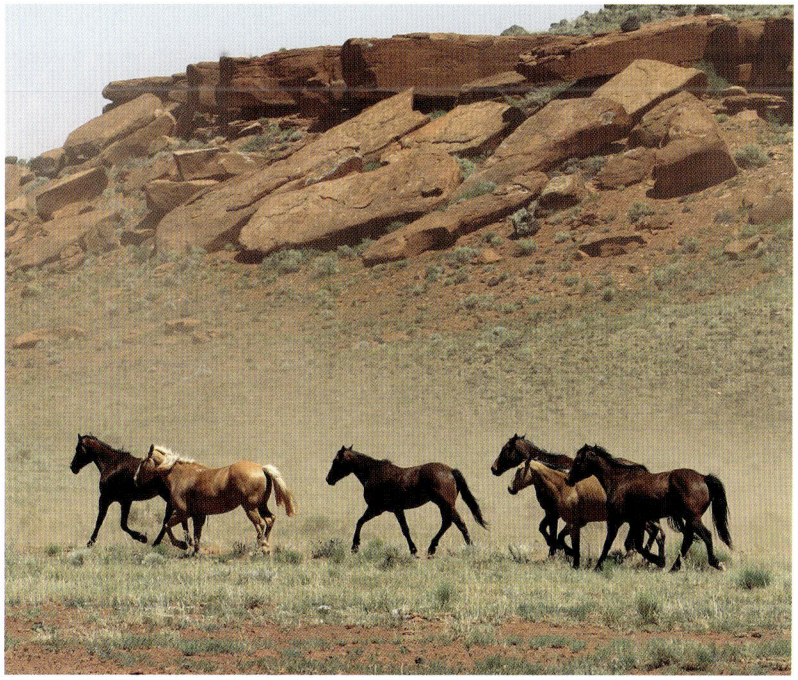

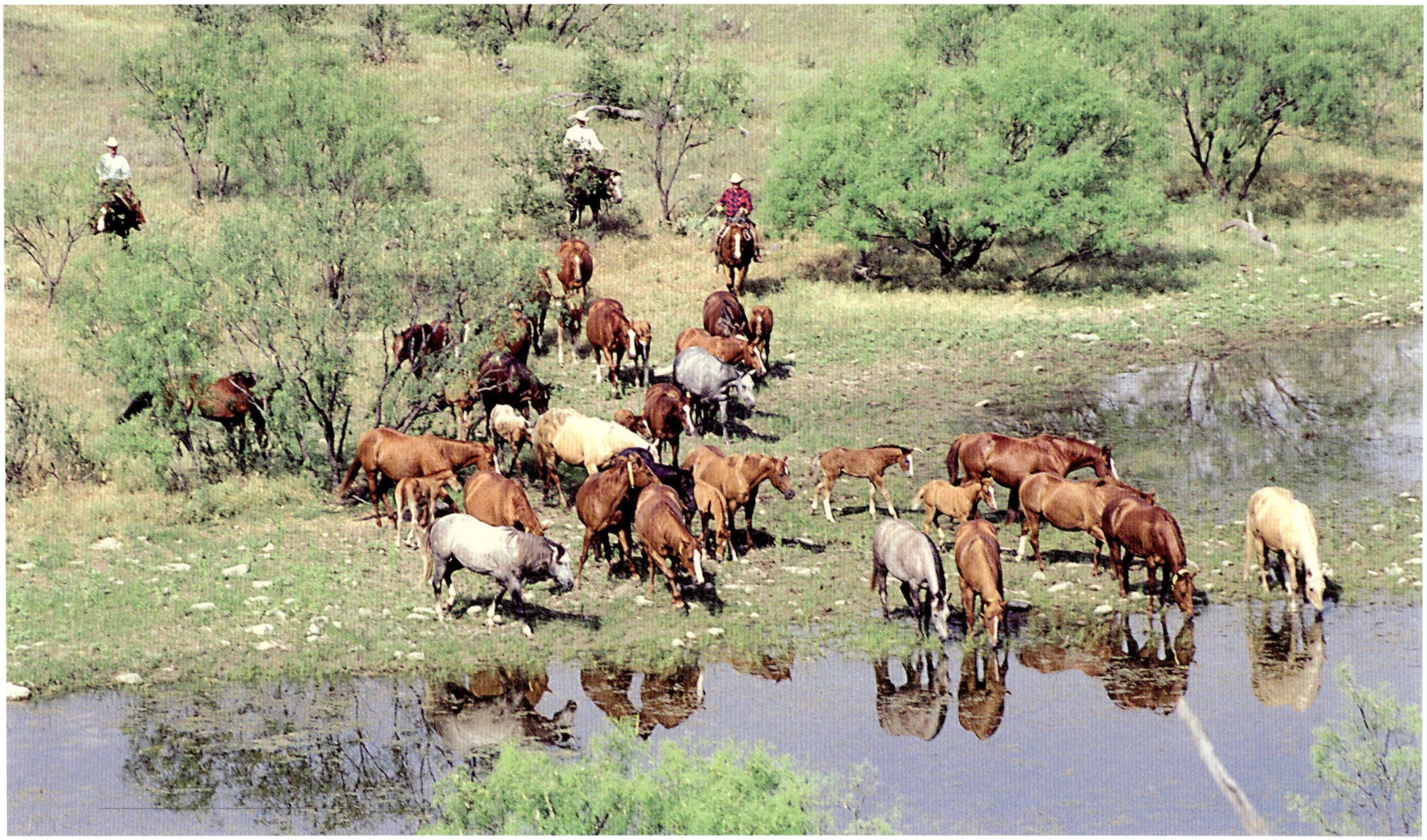

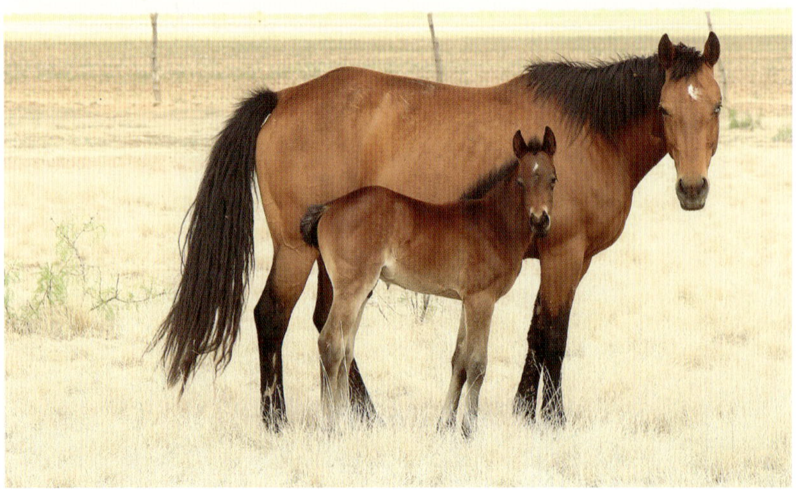

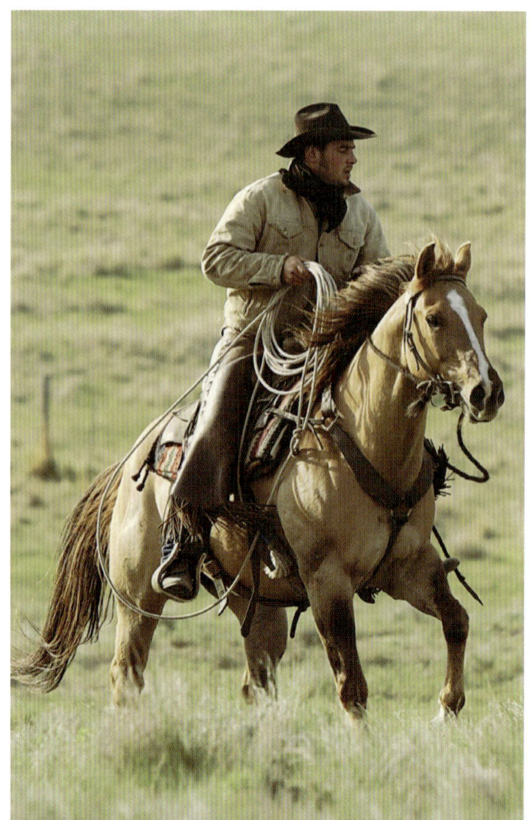

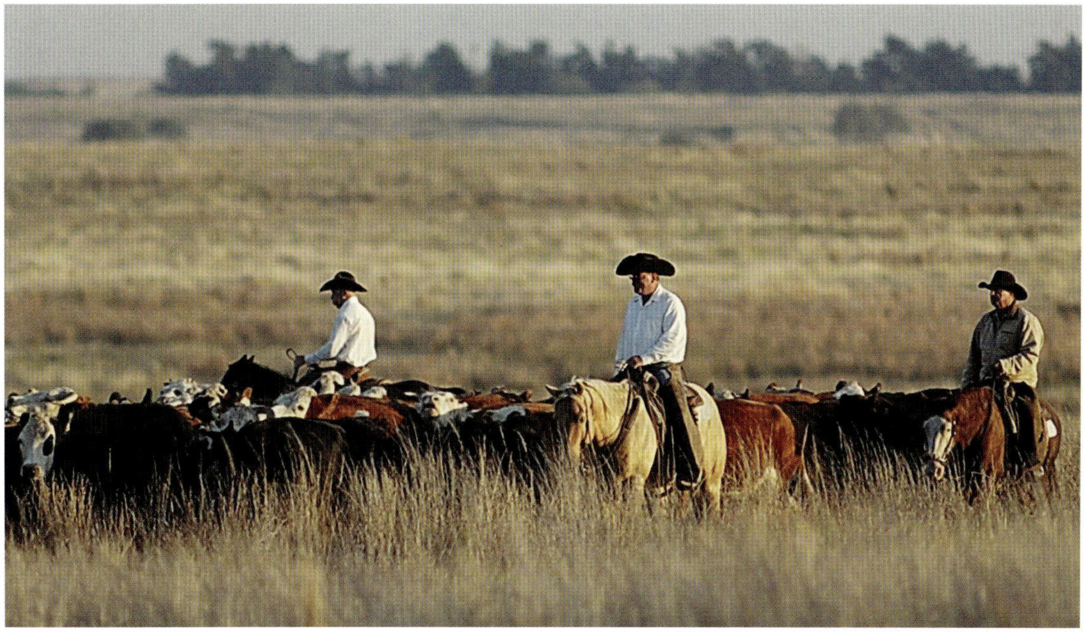

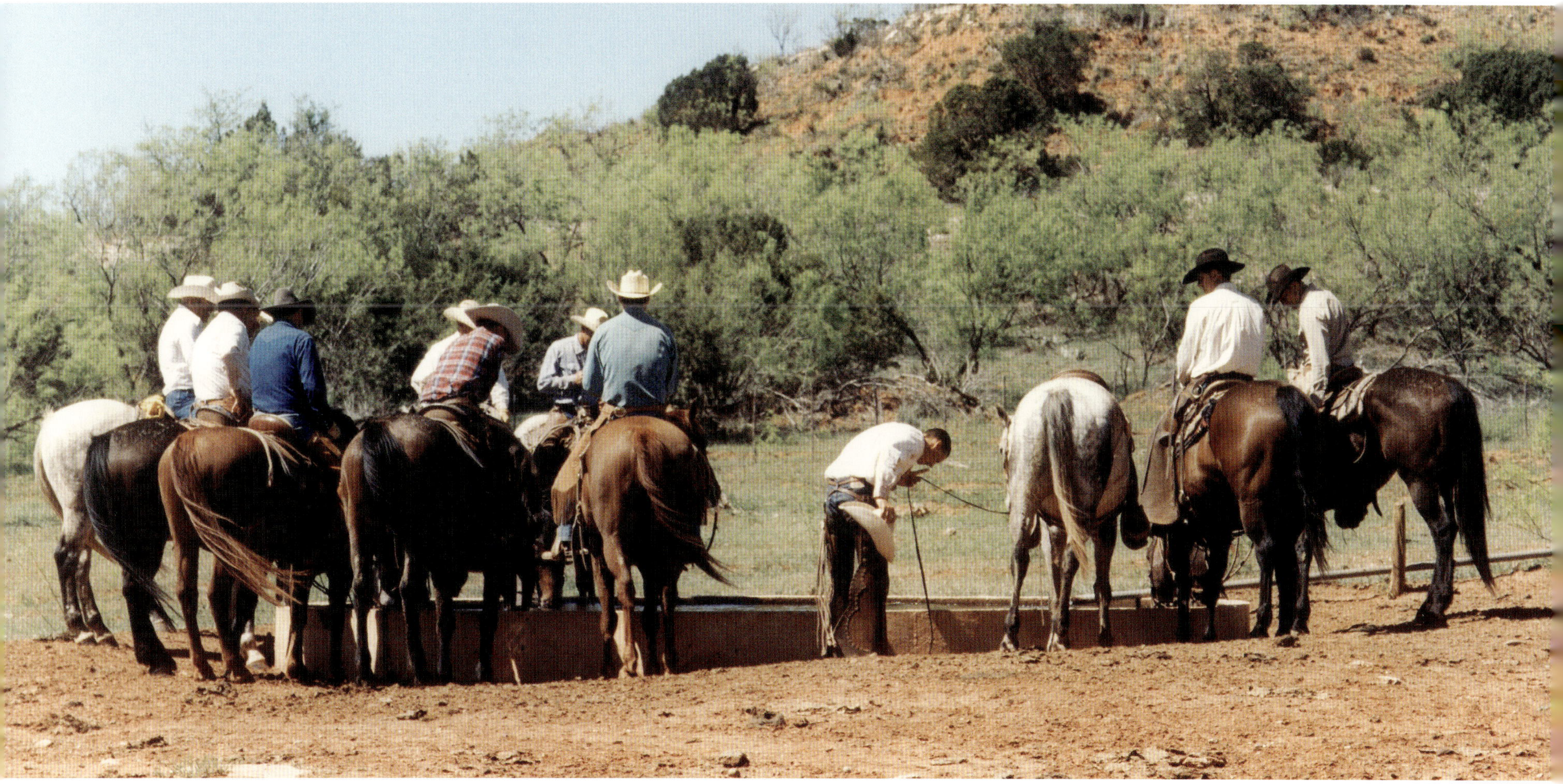

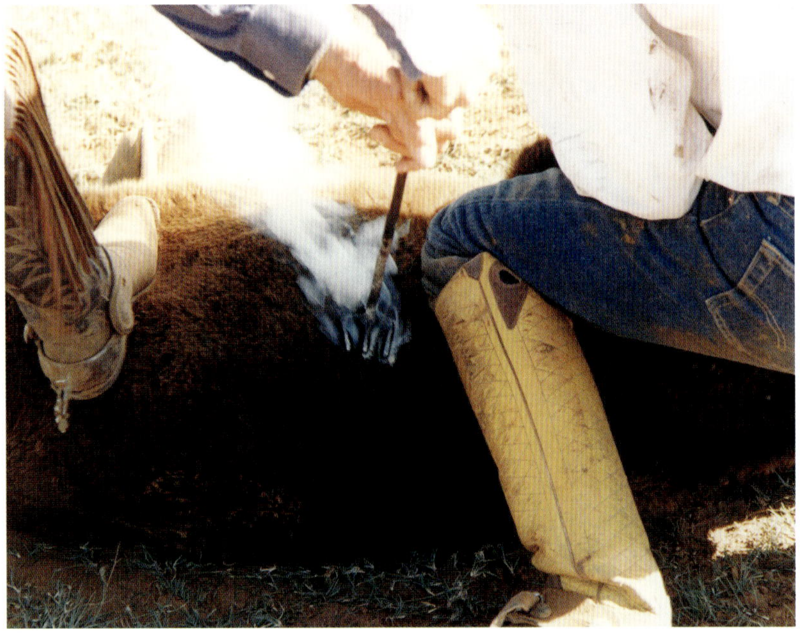

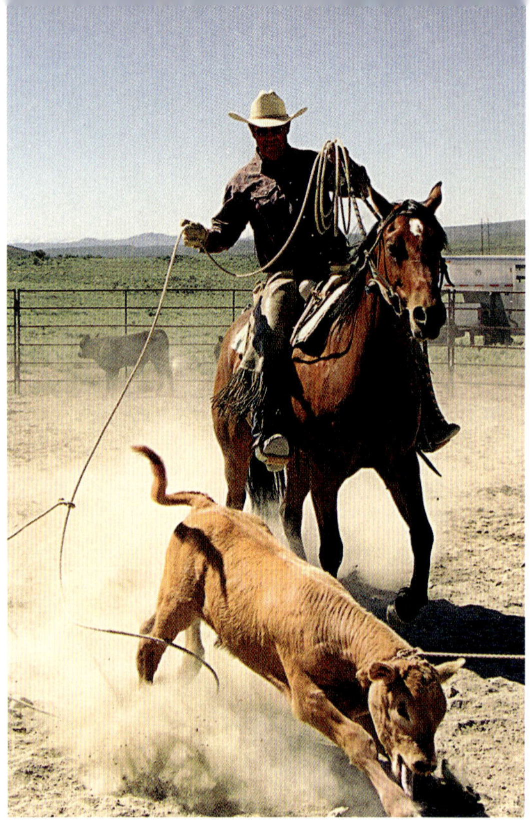

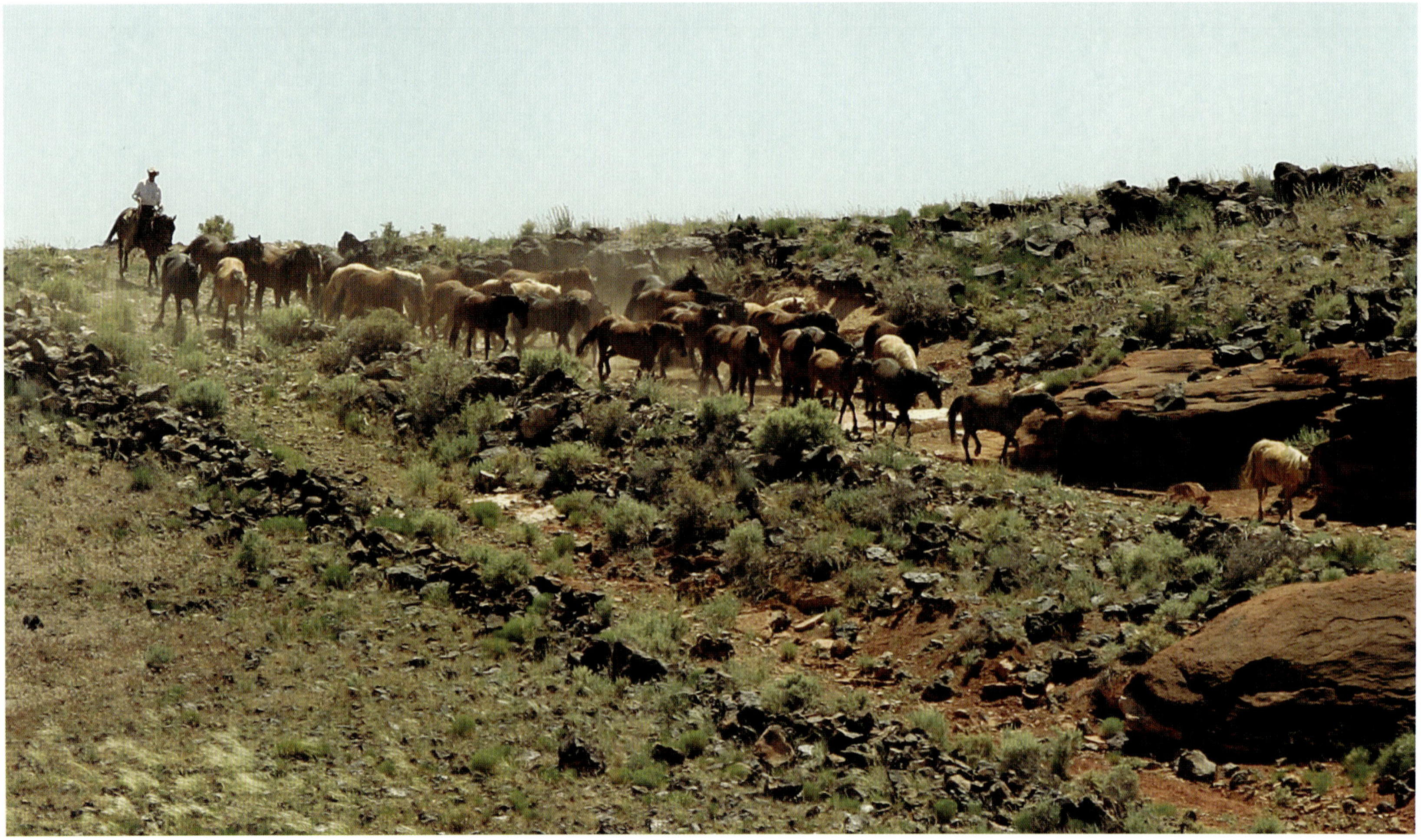

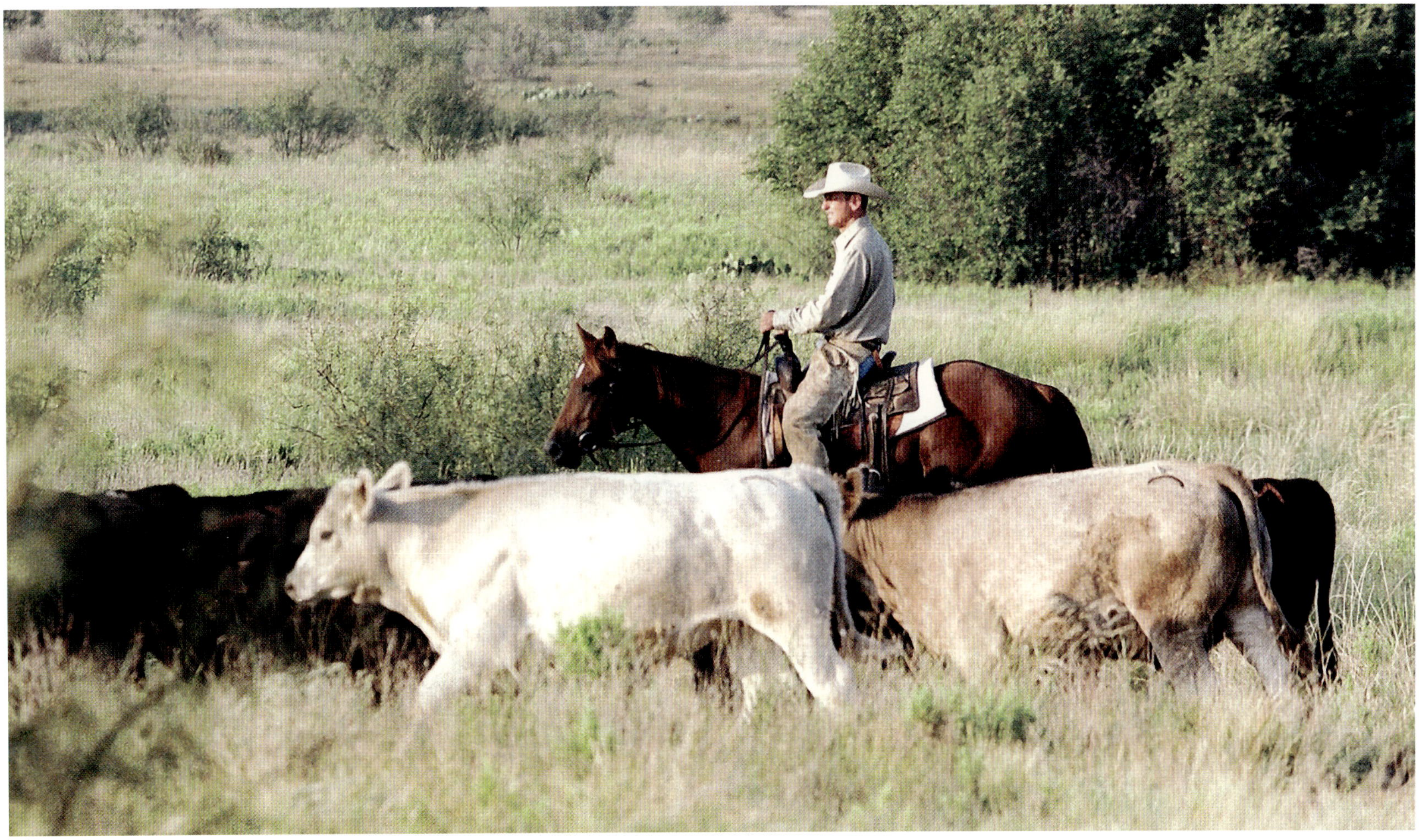
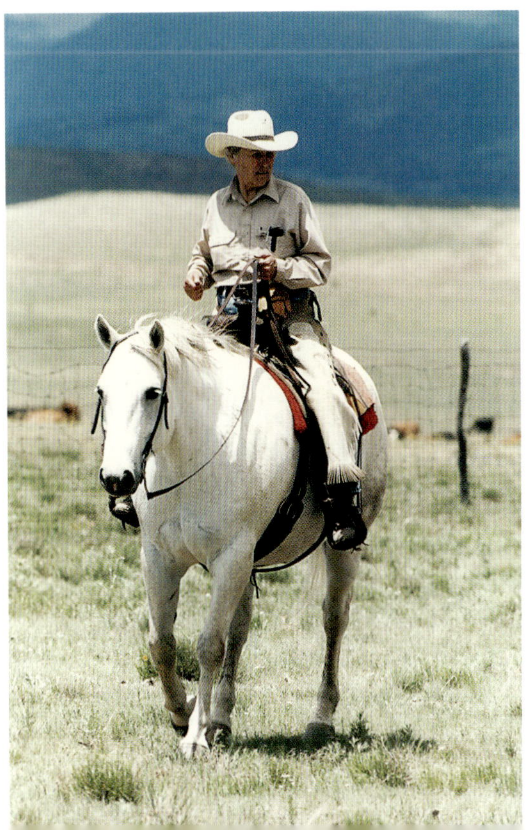
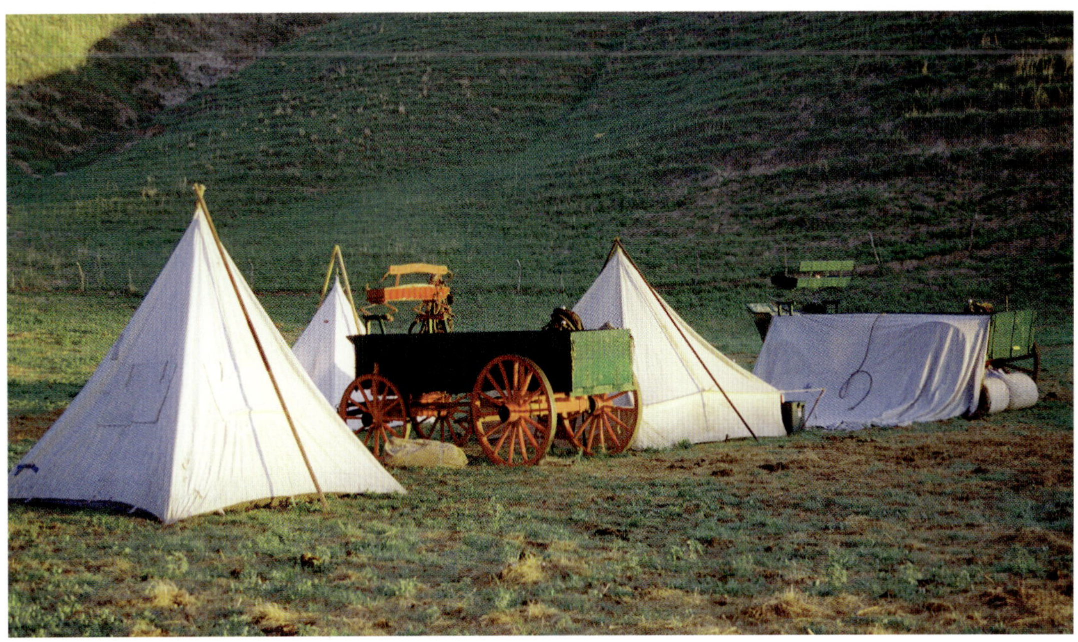

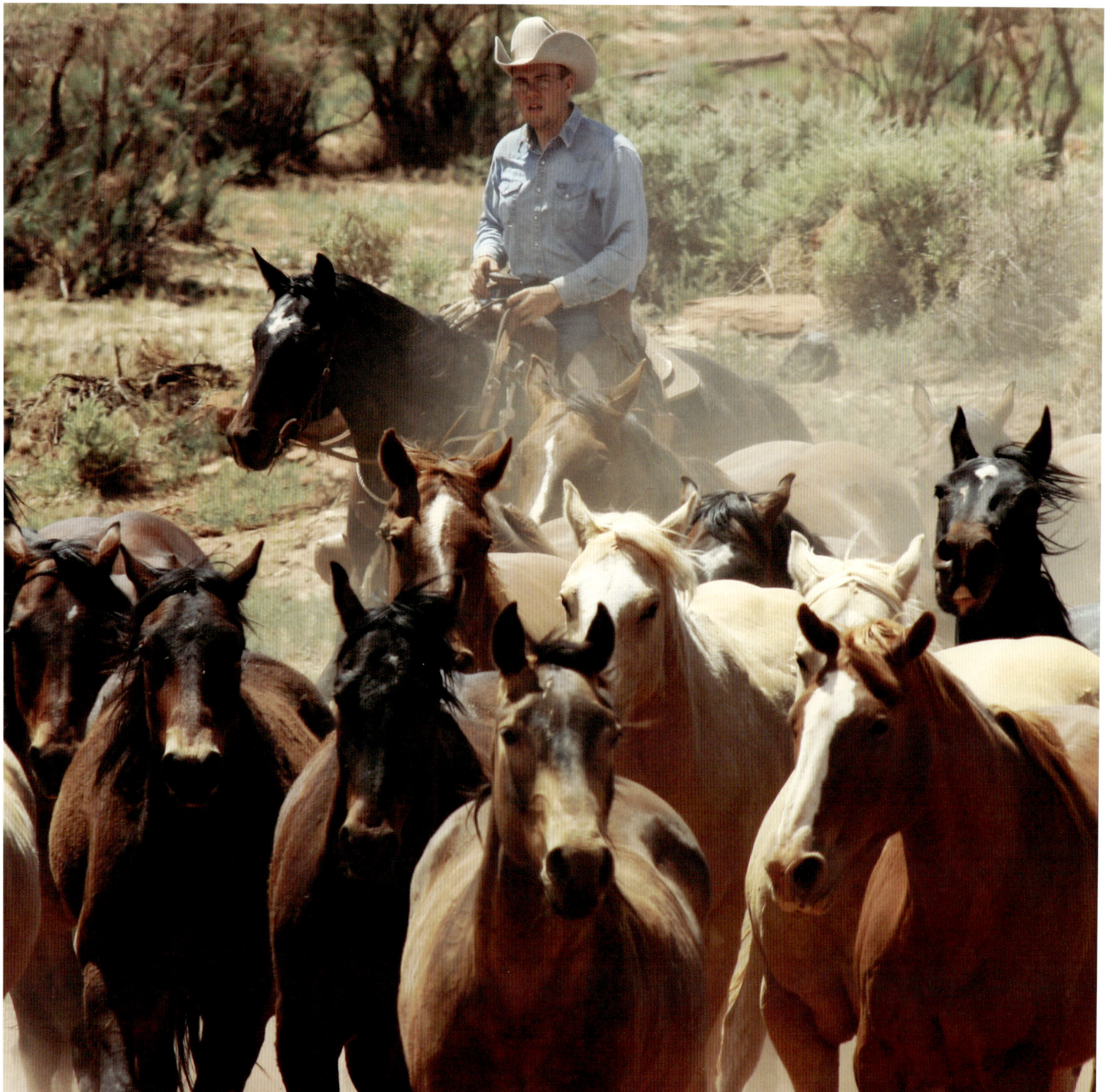

American Quarter Horse Association

BEST Remudas

Jim Jennings

Bright Sky Press

BRIGHT SKY PRESS

Box 416
Albany, Texas 76430

10 9 8 7 6 5 4 3 2 1

Library of Congress Cataloging-in-Publication Data

Jennings, James N.
Best remudas : quality Quarter horses / written and photographed by Jim Jennings.
p. cm.
Includes bibliographical references (p.).
ISBN-13: 978-1-931721-82-0 (hardcover : alk. paper) 1. Quarter horse. 2. Horse Breeders—United States. I. Title.

SF293.Q3J46 2006
636.1'330973—dc22

2006011188

Book and cover design by Isabel Lasater Hernandez
Edited by Dixie Nixon

Printed in China through Asia Pacific Offset

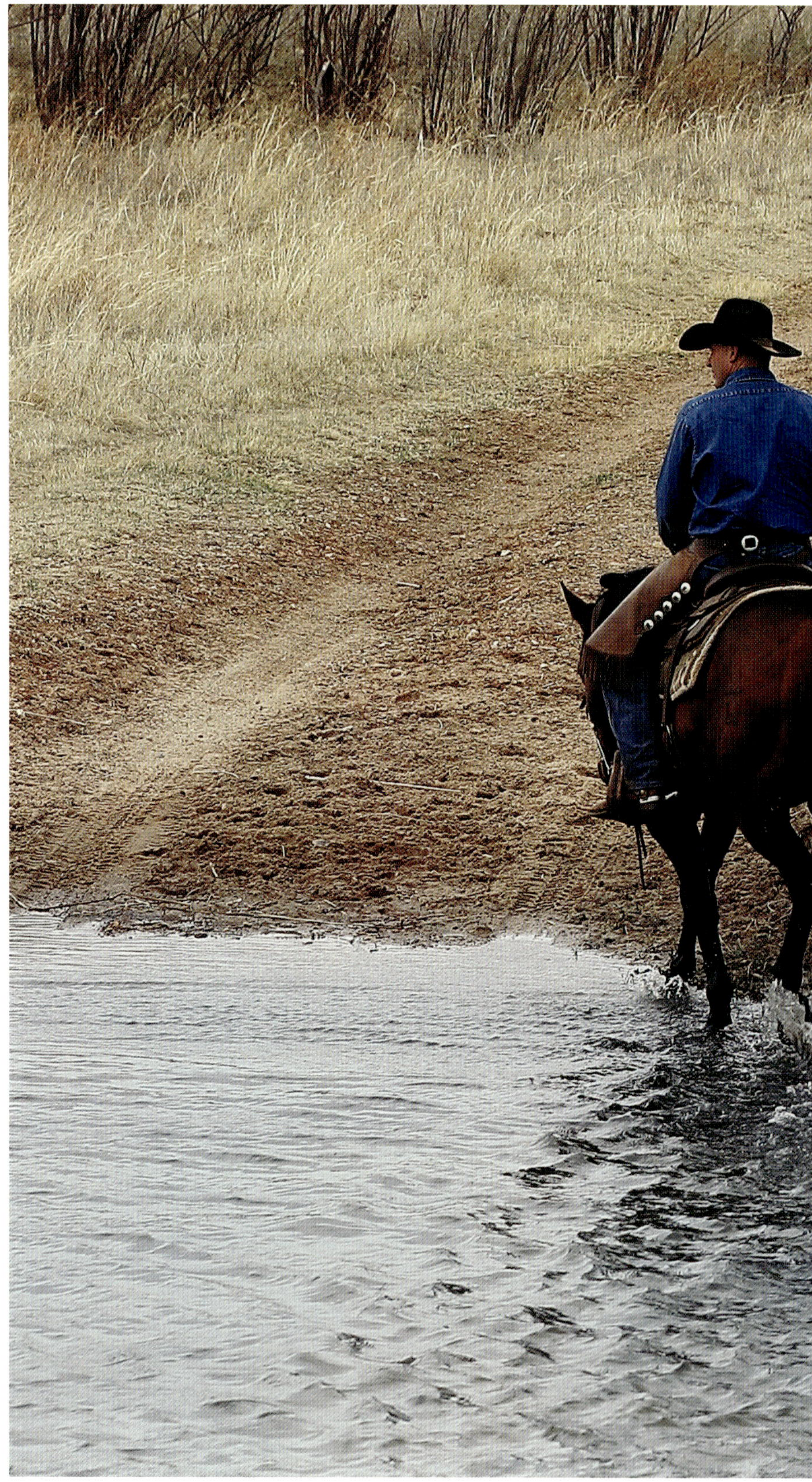

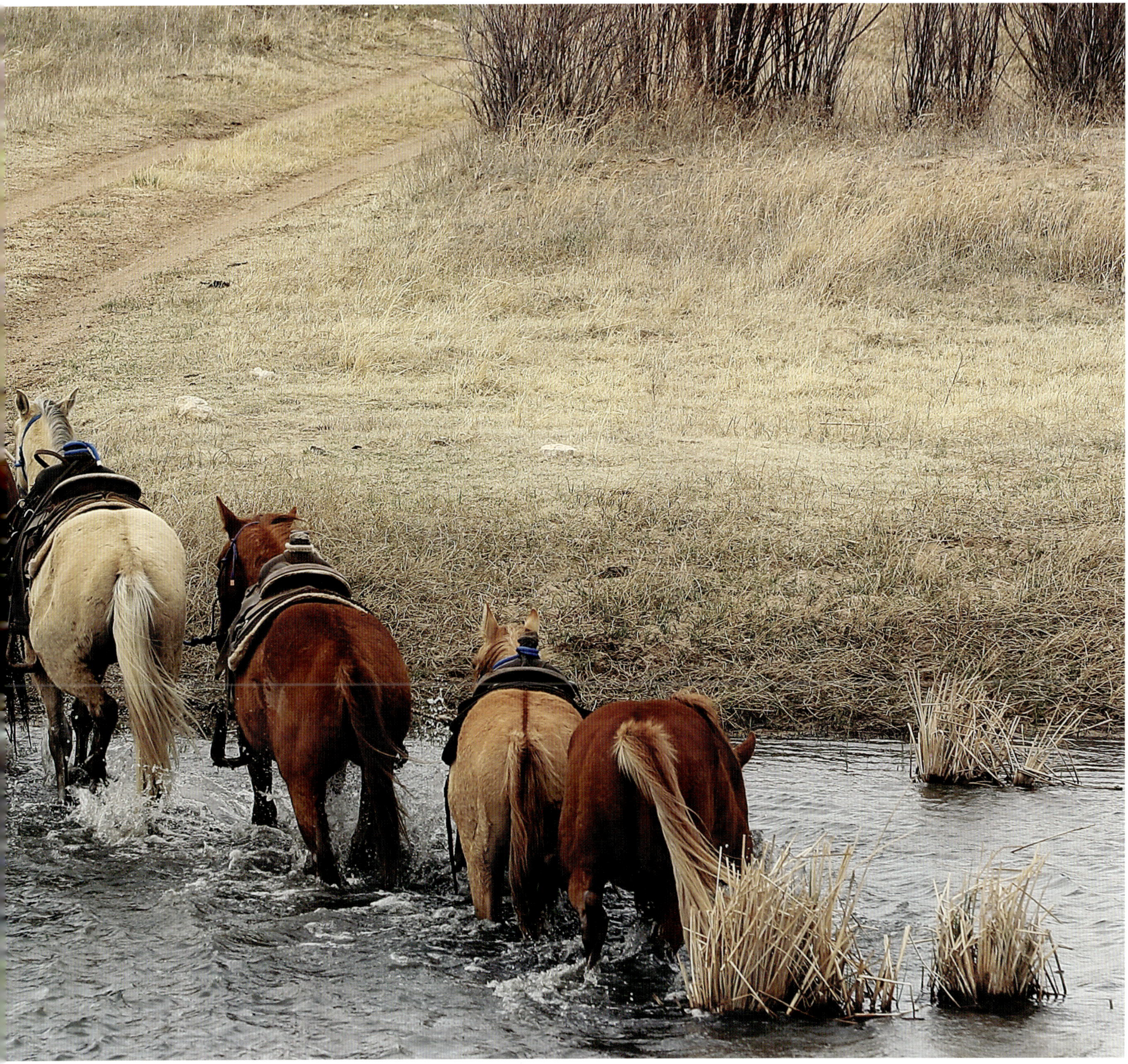

CONTENTS

Introduction

When Bob Denhardt, in the 1930s, saw the need to track the pedigrees of a group of horses that were being referred to as Quarter Horses, he went first to the ranches. That's where the good horses were. The result was the formation, in 1940, of the American Quarter Horse Association. Recognizing the importance of the ranches to the Association yet today, AQHA and corporate partner Bayer Animal Health came up with the Best Remuda Award. They wanted to honor the ranches for their horses. This book is the story of the first 15 of the ranches so honored.

A ranch remuda is defined as "the herd of horses from which those to be used for the day are chosen." The ranches in this book received the Best Remuda Award because of the quality of their remudas. Besides being a story of the ranches, this book is a story of the tool the ranches use on a daily basis—the horse. Without exception, all the ranches agree that they could not do their work without their horses, and a lot of planning goes into the breeding and training of the mounts in their remudas. These ranches are well known for the cow horses they have produced and are still producing. Some of them say that they are able to keep good cowboys because of the quality of their horses.

Each of the ranches was featured, the year it won the award, in either *The American Quarter Horse Journal* or *America's Horse,* both official publications of AQHA. Those articles form the basis of the chapters in this book. The first ranch to receive the award was the Haythorn Land & Cattle Company at Arthur, Nebraska. That was in 1992. Fourteen years later, there were obvious changes in that ranch, as well as in most of the others, so the stories were all updated.

I have had the honor of either writing or rewriting the stories of all 15 ranches and photographing them—often from horseback. I have visited every one of the ranches—some of them many times—and I count many of the owners and managers of those ranches as friends.

The stories tell of ranch life in many parts of the North American continent, and how each ranch has its own unique way of handling its cattle and horses. Some of the methods are strictly traditions that have been handed down for generations. Others are born of necessity, due to the terrain on which the ranch lies. Some of the men and women who work these ranches are called cowboys; others are called buckaroos. Regardless, they all say they do it because of the horses.

So take a look and ride with me now across the sand hills of Nebraska, the sagebrush sea of Nevada and the plains of West Texas. Let the wagon boss rope you a horse out of the remuda, one of the Best Remudas.

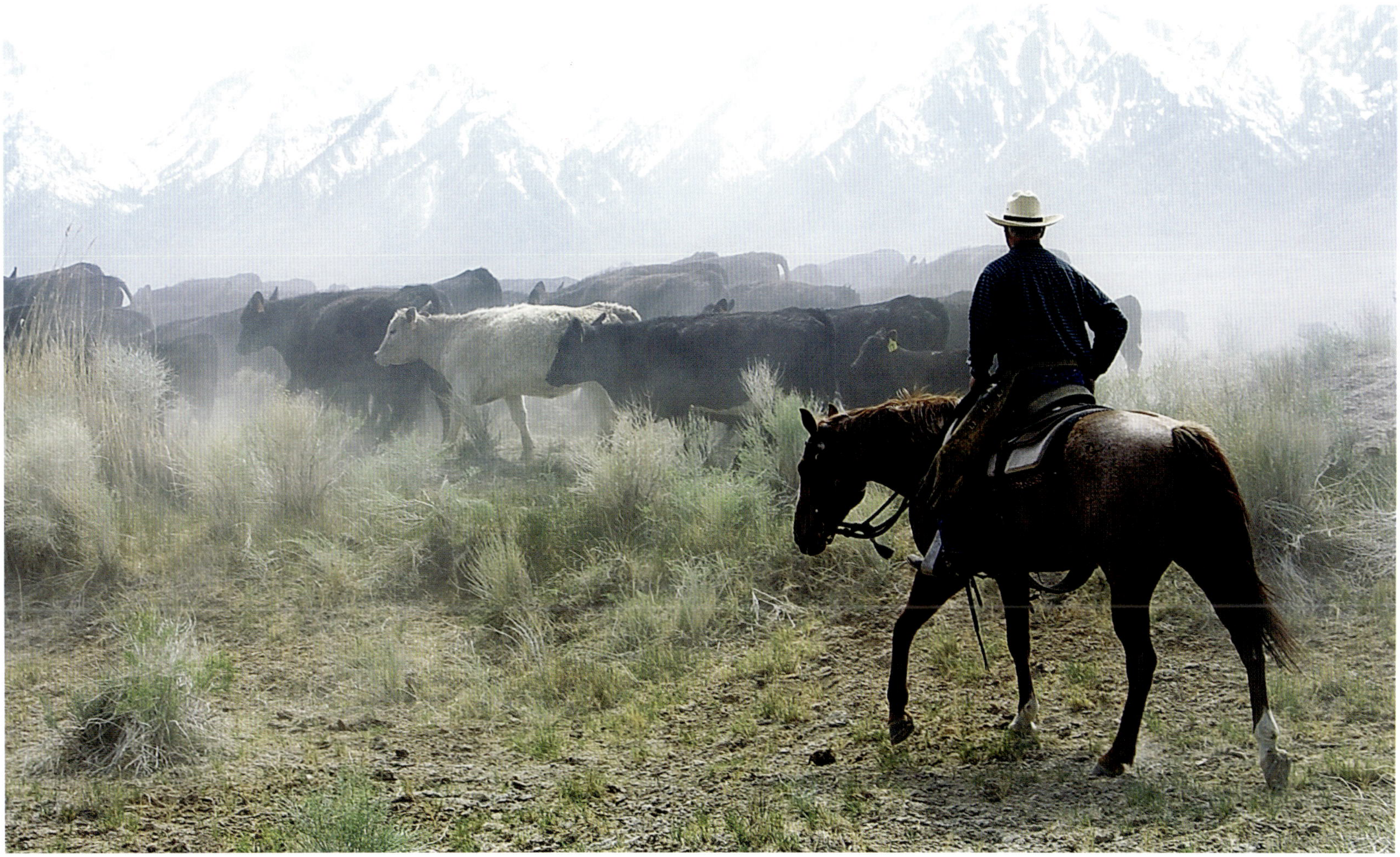

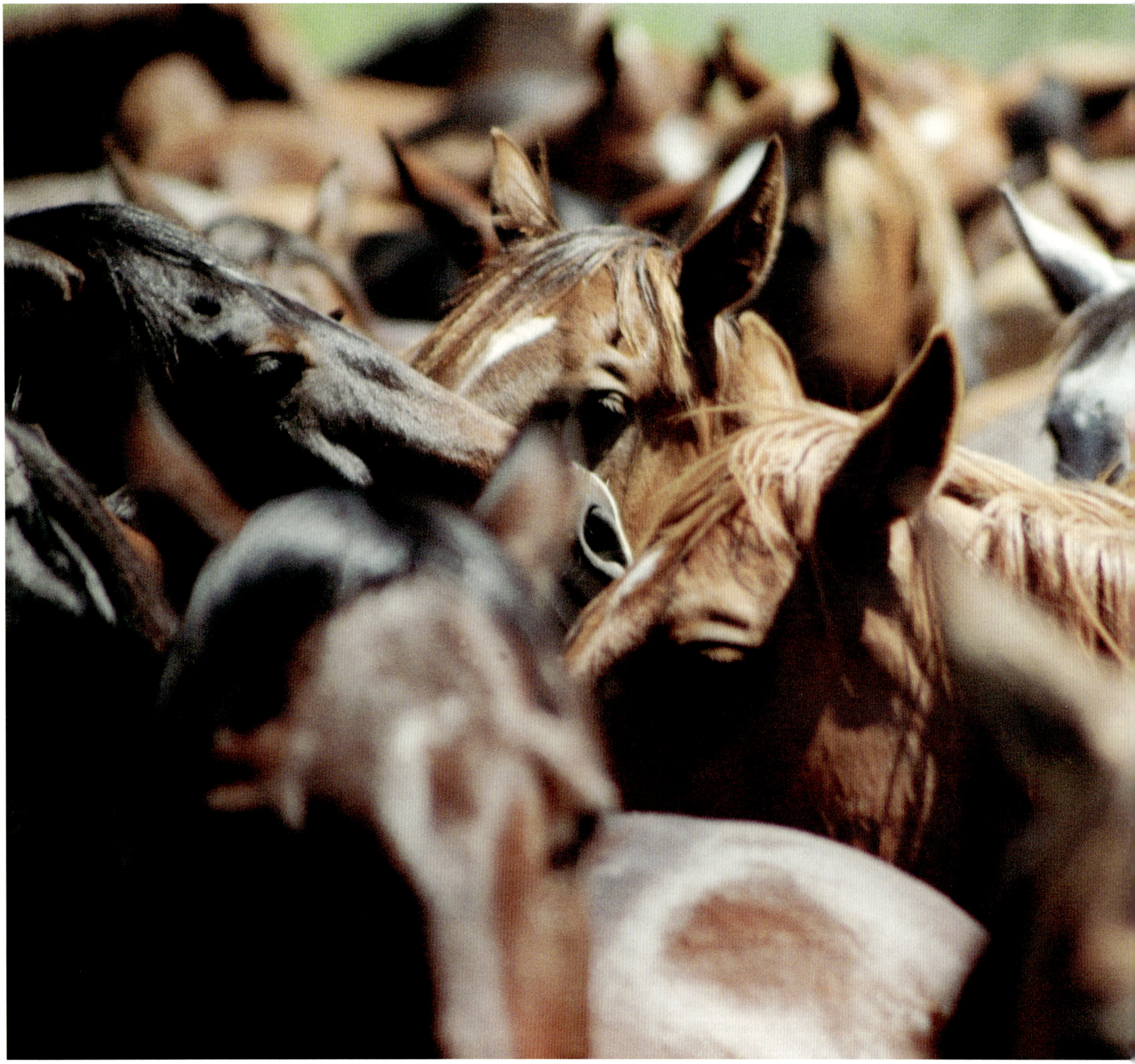

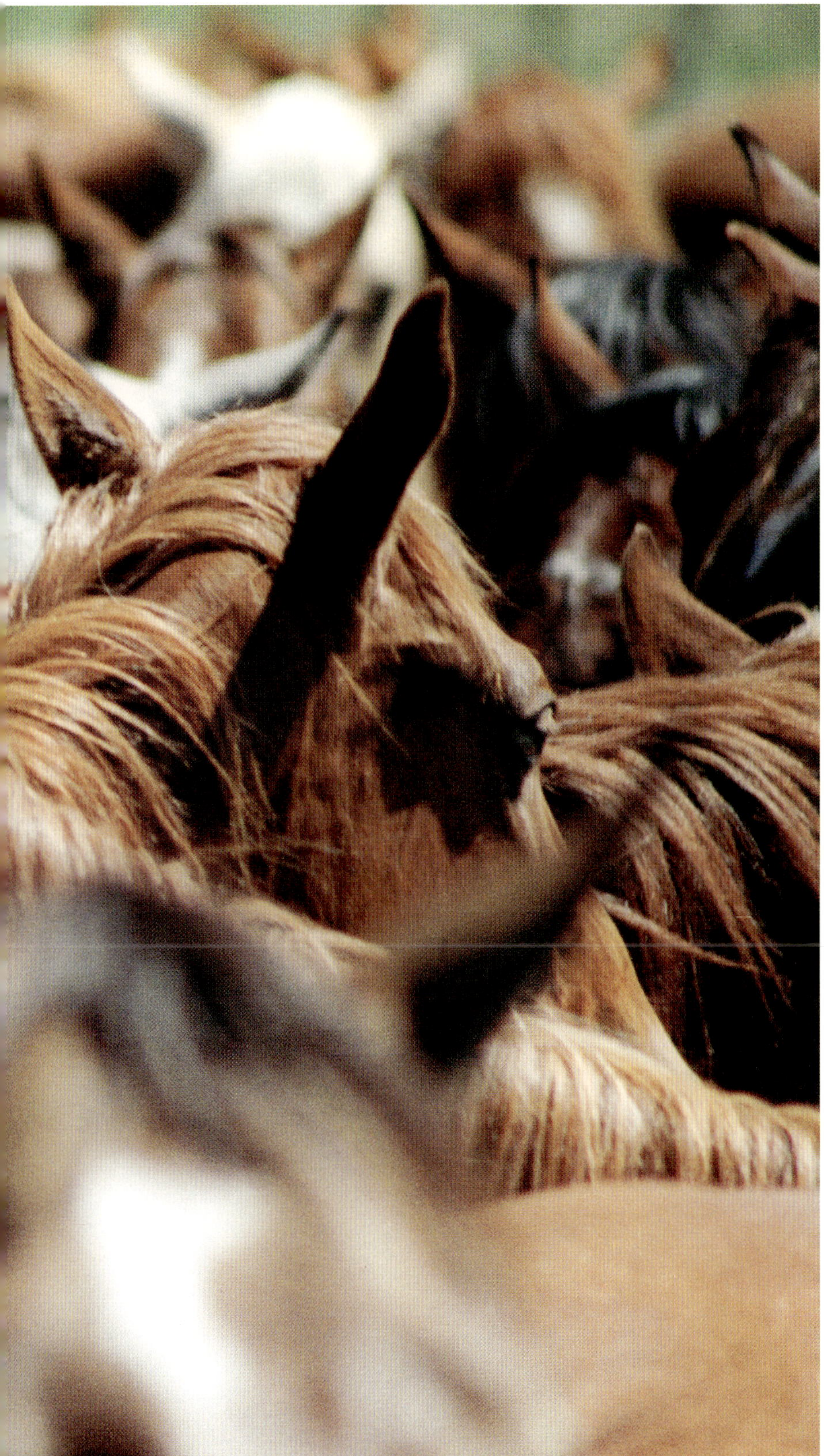

BEST Remudas

1992-2006

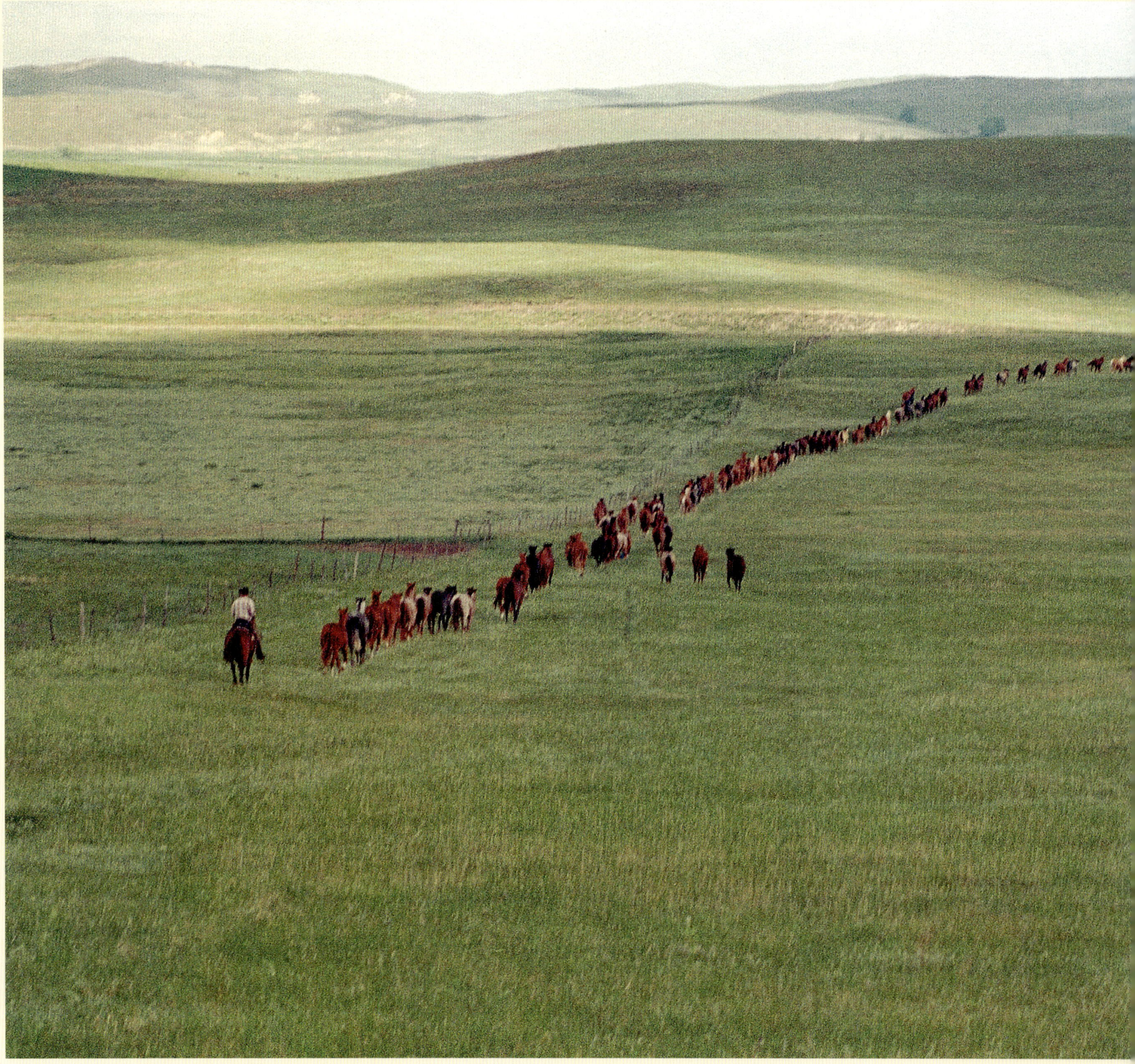

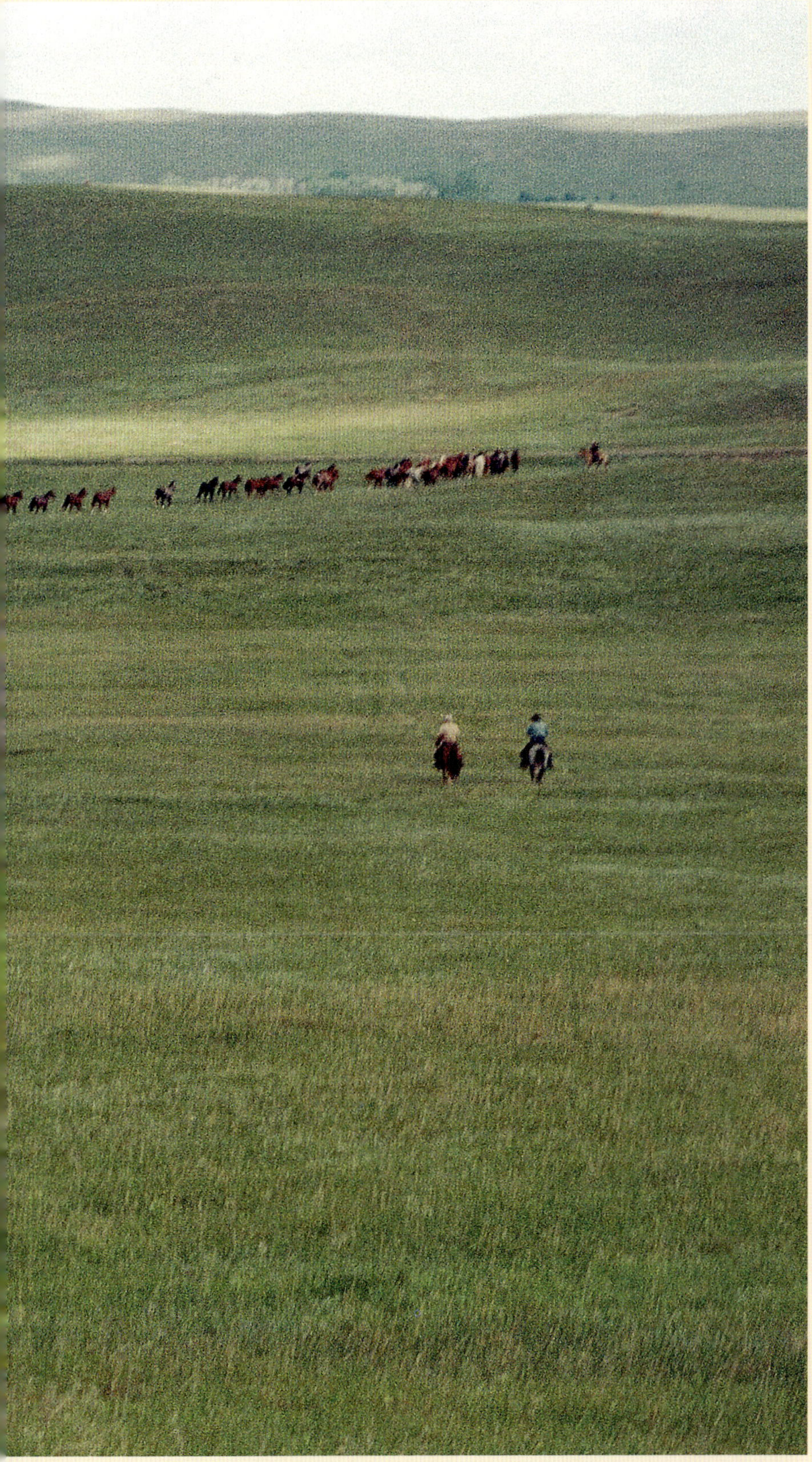

Haythorn Land & Cattle Company

Arthur, Nebraska

1992

A Sea of Grass

It was no accident that Haythorn Land & Cattle Company was selected as the inaugural winner of the Best Remuda Award.

The sand hills of Nebraska would be called sand dunes if it were not for the continuous covering of grass that paints the landscape a bright kelly green. Low hills, covered with grama, bluestem and buffalo grass, stretch as far as you can see. Only an occasional planted tree or windmill tower breaks up the horizon.

This sea of grass is what young Harry Haythornwaite couldn't get out of his mind after his first visit to Ogallala, Nebraska. He had come up the trail from Texas as a cowboy on a cattle drive. On his second cattle drive to Ogallala, he just stayed, spending his wages on the purchase of a livery stable and going into business.

The sand hills were totally alien to anything Harry had ever seen before, but for that matter, so was almost everything that he had encountered during the last few years. Harry had been born in Lancaster, England, and in 1876, at the age of 16, had stowed away on a ship bound for America. After he was discovered hiding below decks, the captain put him to work caring for some Hereford bulls bound for Texas. Harry ended up going to work for the Texas rancher who owned the bulls, and subsequently made four cattle drives north—two to Kansas railheads and two to Ogallala. It was at the conclusion of that second drive to Ogallala that Harry quit the trail driving business.

Shortly after settling in Ogallala, Harry shortened his name to Haythorn and married Emma Gilpin, the daughter of a local veterinarian. Next, Harry sold the livery stable, and he and Emma moved to the Yeast Ranch, where Harry worked as wagon boss and Emma cooked for the cowboys. Emma's wages were 25 cents a meal if there were at least 10 cowboys at the table—she cooked for free if there were fewer—but almost every penny was stashed away. In the meantime, Harry took his salary in cattle. As soon as they could afford it, Harry and Emma filed on a homestead four miles west of Arthur, Nebraska, right in the middle of that sea of grass, and in 1884, the Haythorn Land & Cattle Company was born.

Horses

The rope corral, which consists of 300-400 feet of rope suspended about 3 feet off the ground by iron stakes, was set up last night. So when some 160 odd head of saddle horses that make up the remuda came trotting through the opening in the rope this morning, everything was ready. The remuda had been turned out in a small pasture after everyone unsaddled yesterday evening, but two of the cowboys kept their horses up. They, then, saddled before breakfast this morning and gathered the remuda, driving all the horses into the rope corral.

When the horses trotted in, there was the usual jockeying and maneuvering for position before they settled down, but it didn't take long. Then, Harry Haythorn's great grandson, Craig, with an already-built loop in his left hand, the coils in his right, steps over the rope and into the corral. Horses' names are called, as Craig surveys the heads and faces of the captured remuda.

"Chick," comes a call from behind the rope.

"Cracker," voices another cowboy.

"Slick," still another cowboy calls.

Each of the cowboys who stand around the perimeter of the rope corral calls out the nickname of the horse he wants to ride that day, and Craig skillfully tosses a hoolihan loop over the head

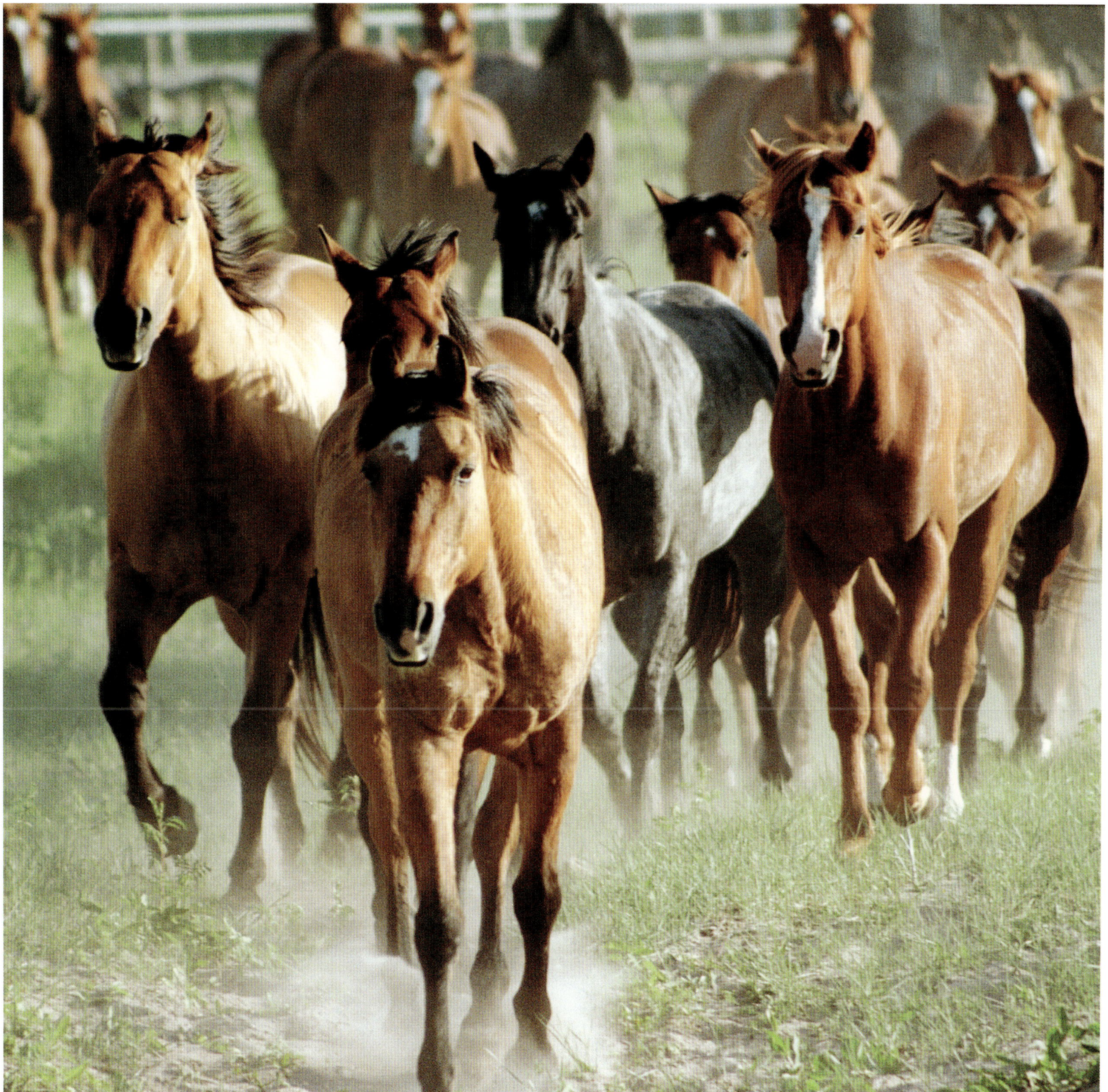

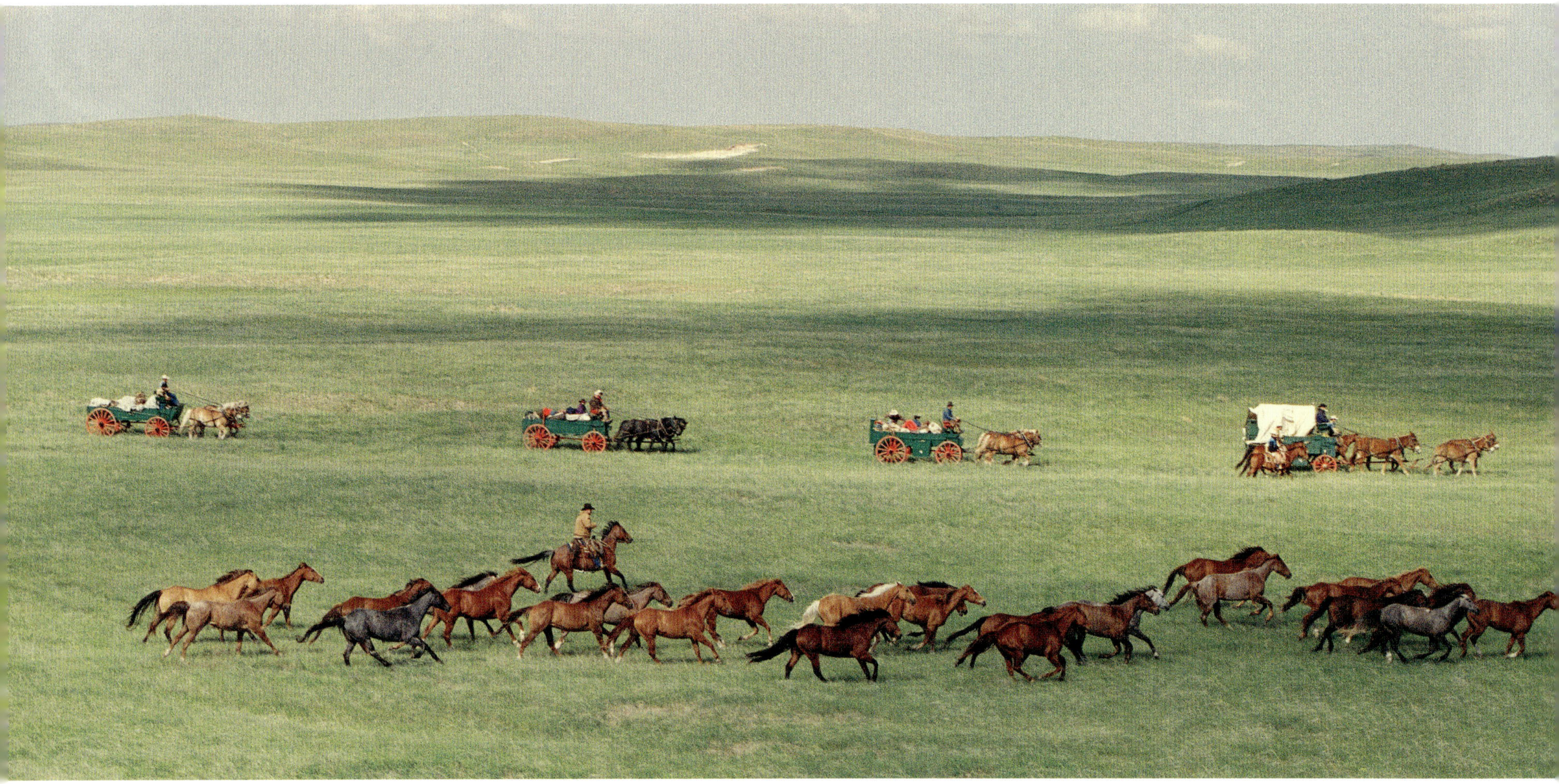

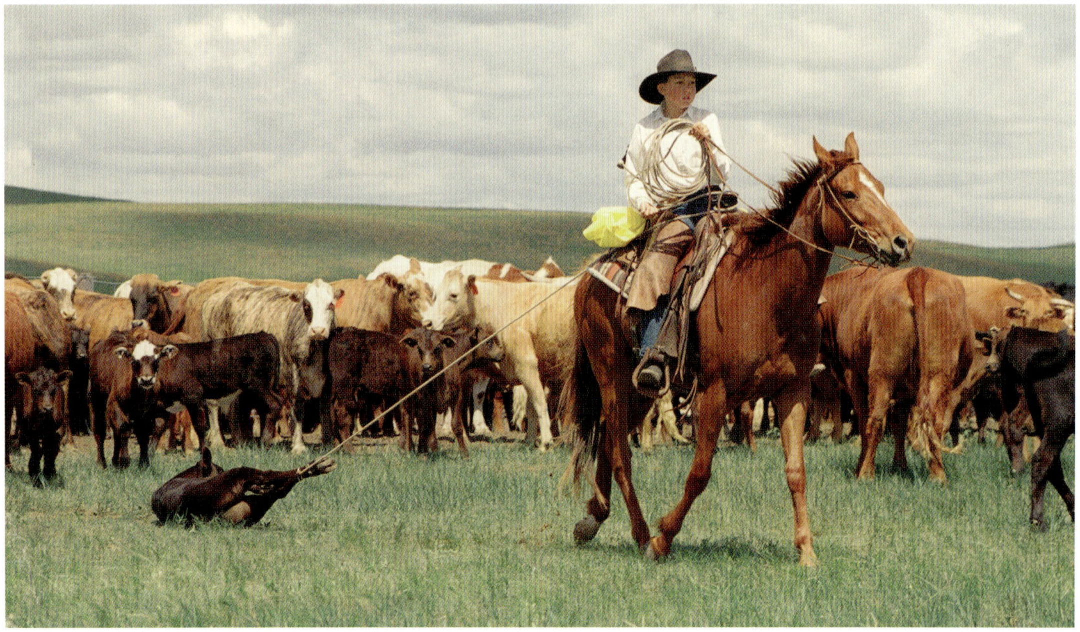

of each one. He then leads the horse to the cowboy waiting at the rope, who bridles him before Craig removes his rope from around the horse's neck.

When everyone's horse is caught, the rest of the remuda is turned back out in the pasture. The horses selected are then saddled, and a few minutes later, Craig leads the cowboys out of camp at a long lope. The pasture to be gathered that morning is five miles away. It won't take long to get there.

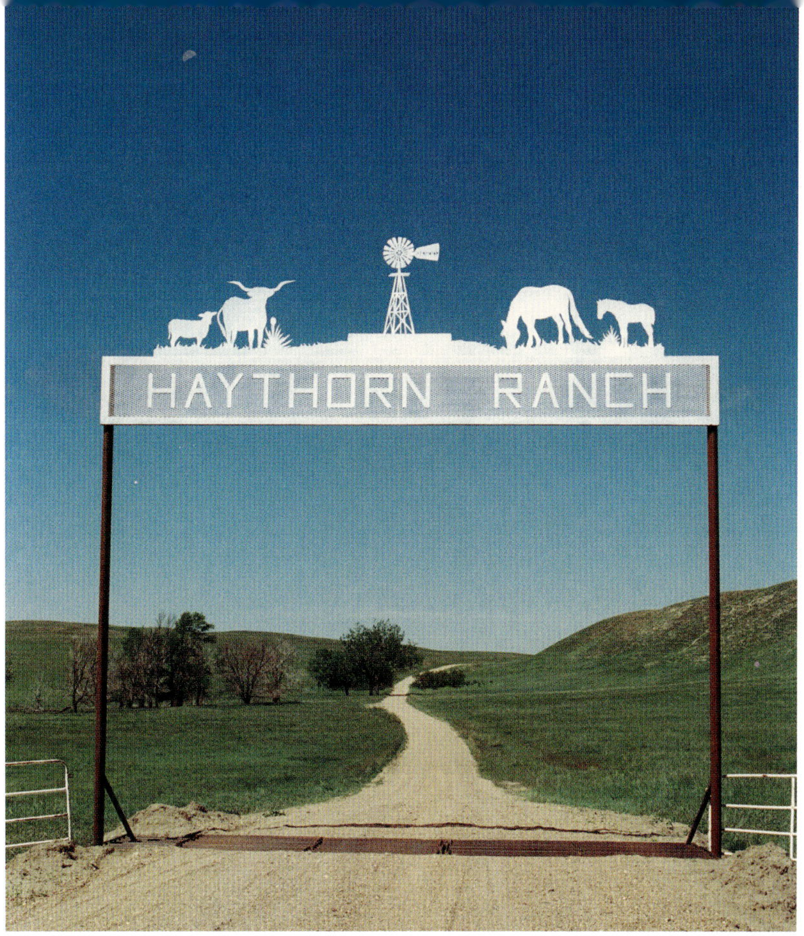

Camp is a chuckwagon surrounded by a scattering of tents and range tepees set up in the shelter of a few trees, probably remnants of those planted years ago at the site of an old homestead. A chuckwagon cook provides three meals a day, prepared over a wood fire in Dutch ovens. The menu varies, but almost always includes beef, sourdough biscuits and coffee. The ranch headquarters, and permanent home for most of the cowboys, is several miles away, but at this time of the year, the wagon is their home for six nights of the week.

Craig manages the Haythorn ranches today, having learned the ways of cattle and horses from his father, Waldo, and his grandfather, Walter. Like those before him, he puts the wagon out each spring for about three weeks, or as long as it takes to work all the calves produced by 4,000 cows on 90,000 acres. Cattle in each pasture are gathered and pushed into a holding area in the corner of the pasture, and the calves are then vaccinated and branded, and the bull calves castrated. Each one is roped and dragged to the branding fire, just like Harry did it a hundred years ago.

As each pasture is finished, camp is moved so that the cowboys are never far from the pastures in which they are working. A team of four draft mules or horses pulls the heavy chuckwagon, while single teams of draft horses pull the three bed wagons, carrying each cowboy's bedroll, tepee and extra tack. The remuda of more than 160 geldings is driven with the wagons to each new campsite.

Having plenty of horses has always been important to the Haythorns, beginning back around the turn of the 20th century when Harry took the train to Oregon and came back to Nebraska driving 500 head of horses. Haythorn Land & Cattle Company had the first registered horses in Nebraska, and today registers more horses with AQHA than anyone else in the United States.

The band of more than 200 broodmares traces mostly to stallions the ranch has owned or owns now, including even Beaver Creek and Eddie, horses that Haythorns had in the 1940s and '50s. More recent stallions the ranch has used or owned, but which are gone now, also have an influence on the broodmares. These include such horses as Snickelfritz Chex, Eddie Eighty, Eddie 40, Happy Hancock and Continental Fly. These same horses also figure in the pedigree of some of the 19 stallions that the ranch is breeding today. Other stallions are out of ranch mares bred to outside stallions Tanquery Gin and Drifts Chip, and still another stallion is by Playgun. There was yet another Playgun stallion, but he died in 2005. However, the ranch got three foal crops from him, and Craig thinks that he has some nice foals on the ground from him as well as from some of the other stallions.

All the mares are pasture bred, and the stallions are turned out about May 20. They are then picked up about the first of August.

The broodmare band produces all the horses needed on the ranch and enough for a major production sale every four years, in which a number of finished horses are sold, along with some broodmares and younger horses. On each of the three years between the major sales, the ranch sells weanlings and yearlings, all of which are eligible for a futurity that Craig started in honor of his father, Waldo, who died in 2002.

"The futurity is for 3-year-olds, and any horse that has gone through our sale is eligible," Craig says. "The pattern is a pretty simple working ranch horse-type pattern. We put $10,000 in it for a purse, and it's getting really popular."

Any horse 2 years of age or younger when it is sold is eligible for the futurity, including foals that were in mares when the mares were sold.

Cattle on the ranch range from a few Hereford cows to a lot of Angus cows, and a lot of black baldies. And, of course, they are all handled horseback, even to the extent that the draft horses that pull the wagons during the spring works are used to stack more than 6,000 tons of hay in the summer, all of which will be fed to the cattle during the winter. And those same draft horses pull the feed wagons.

There is not a lot of difference in the way Haythorn Land & Cattle Company is operated today from when it was founded in 1884. Handling cattle horseback worked then, and it works yet today for Craig and his wife, Jody, and their two sons, Sage and Cord. It's a Haythorn-run operation, and it's likely to be so for a long time.

Harry would be proud.

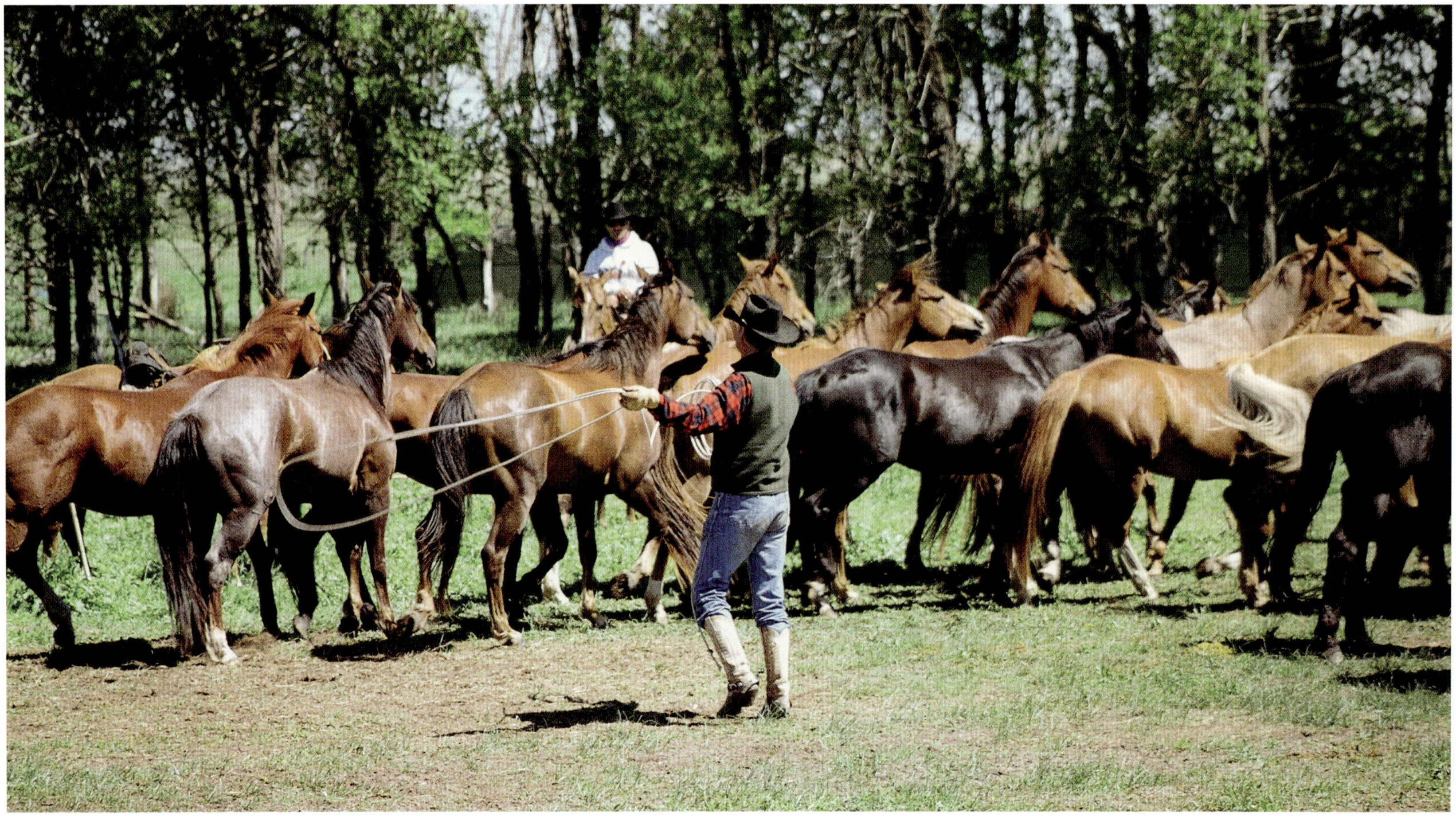

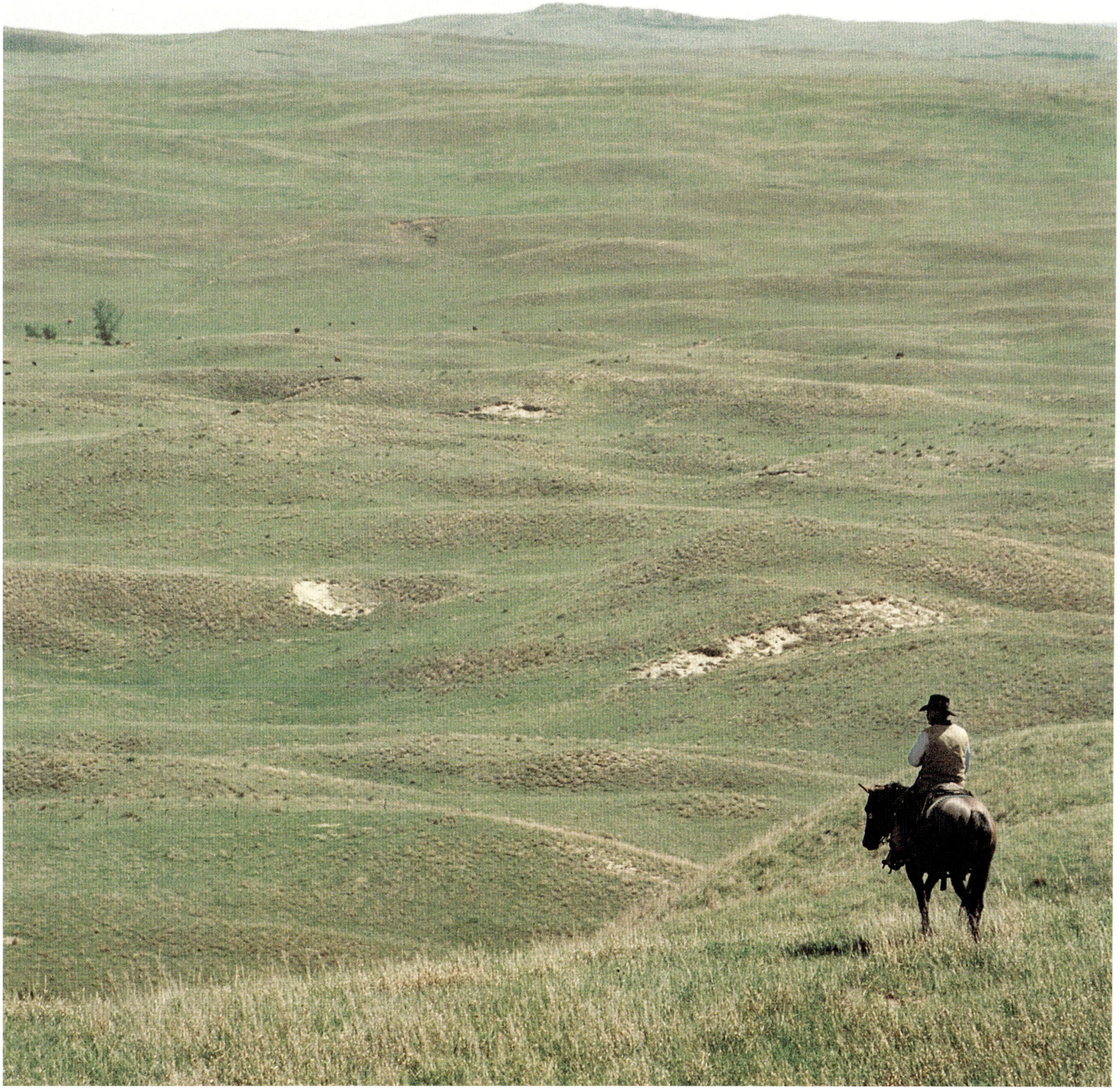

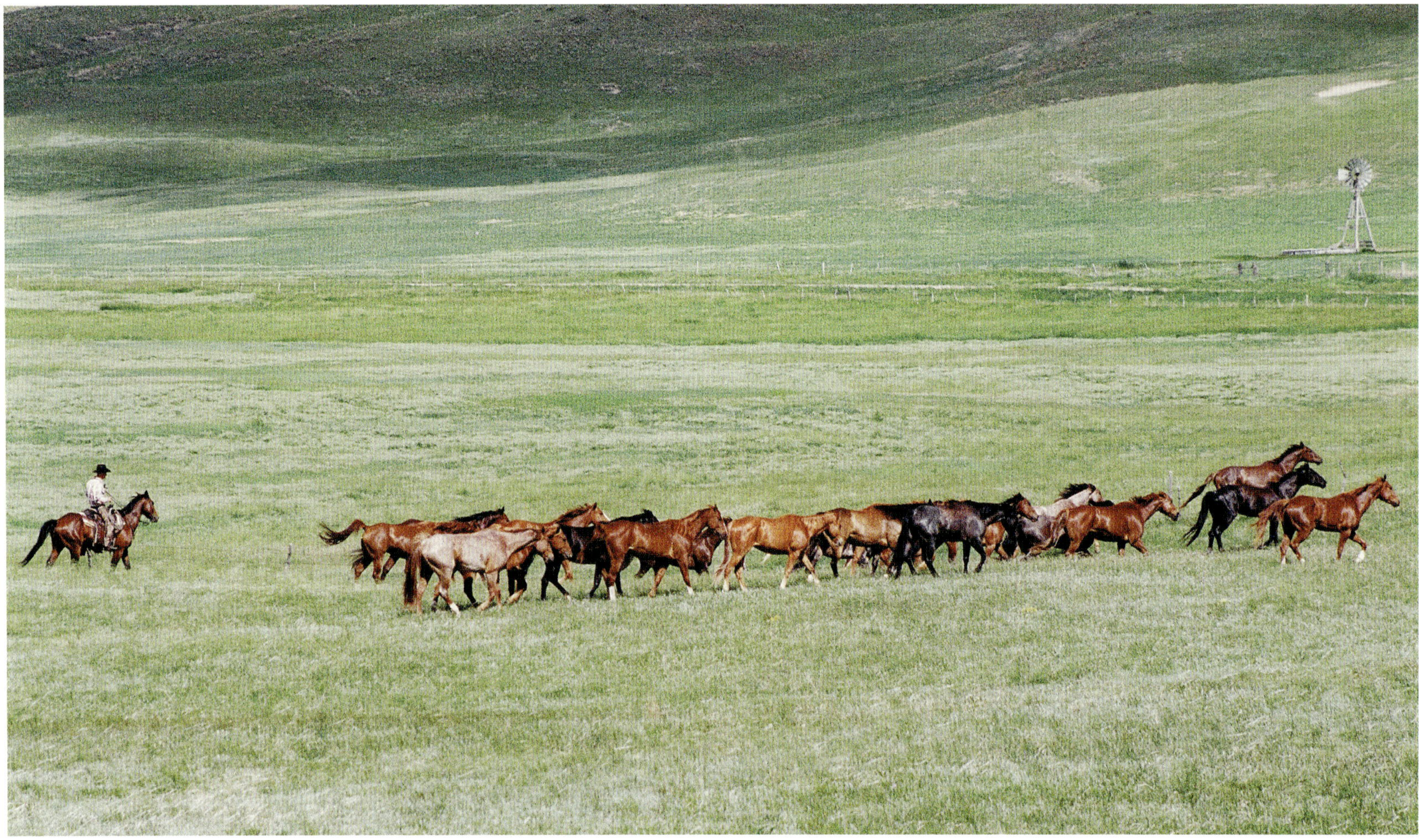

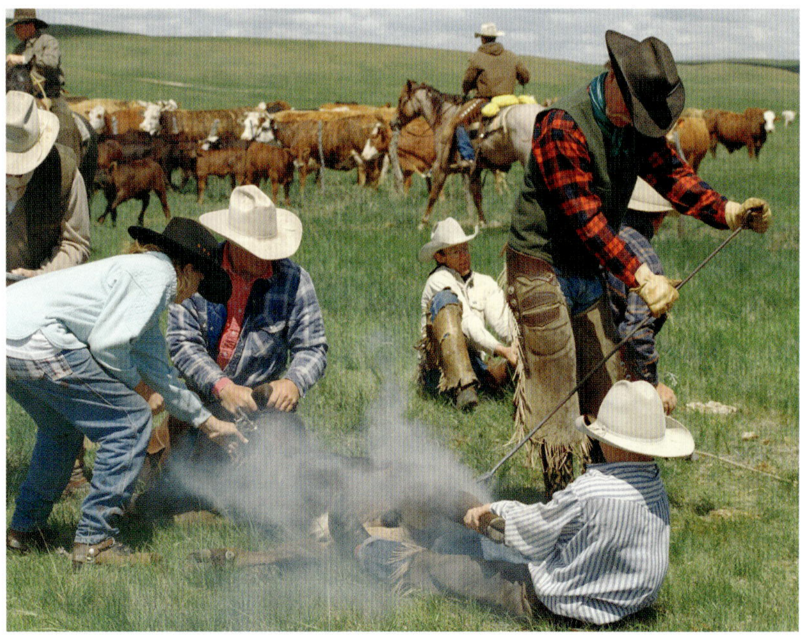

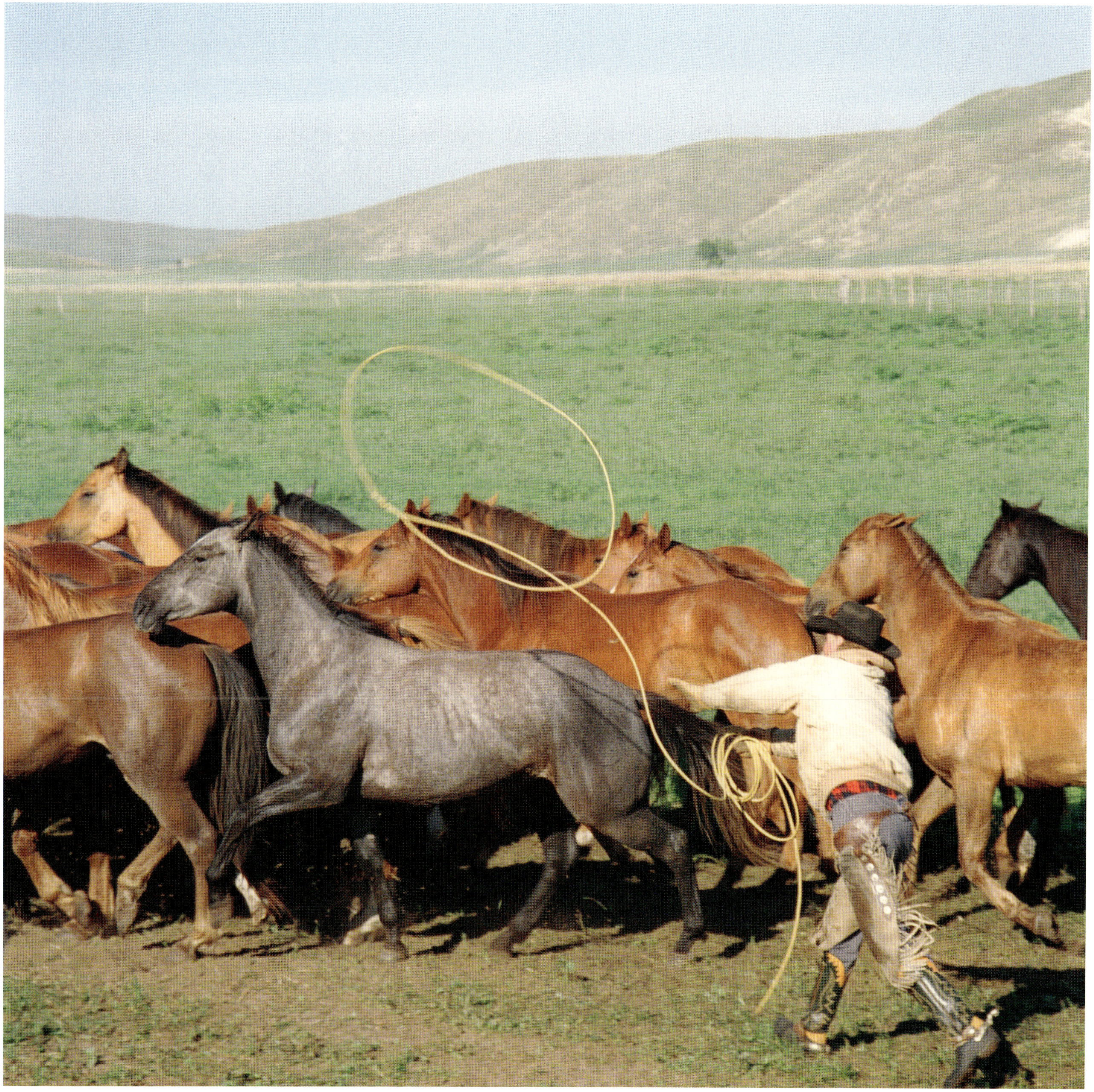

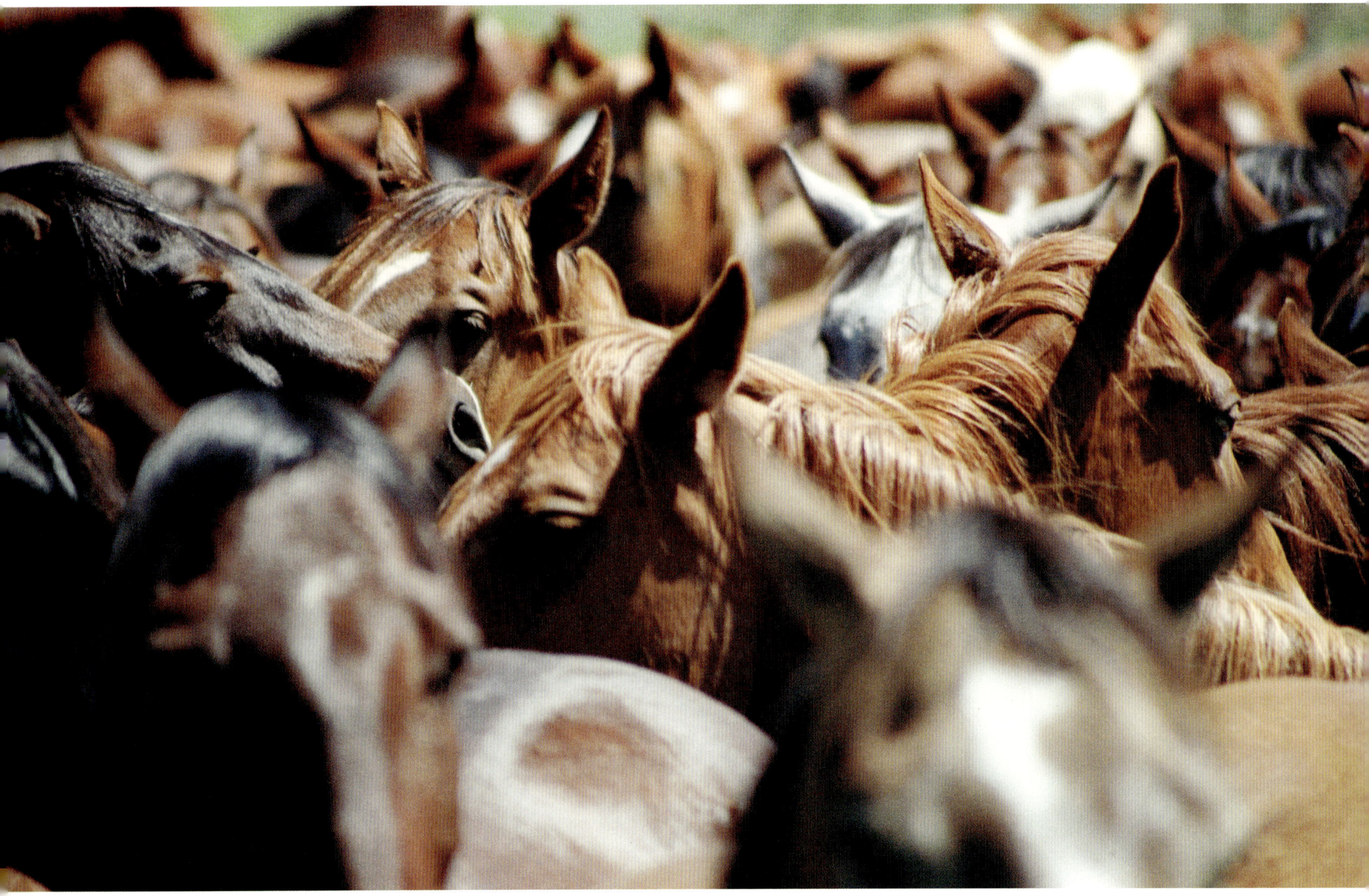

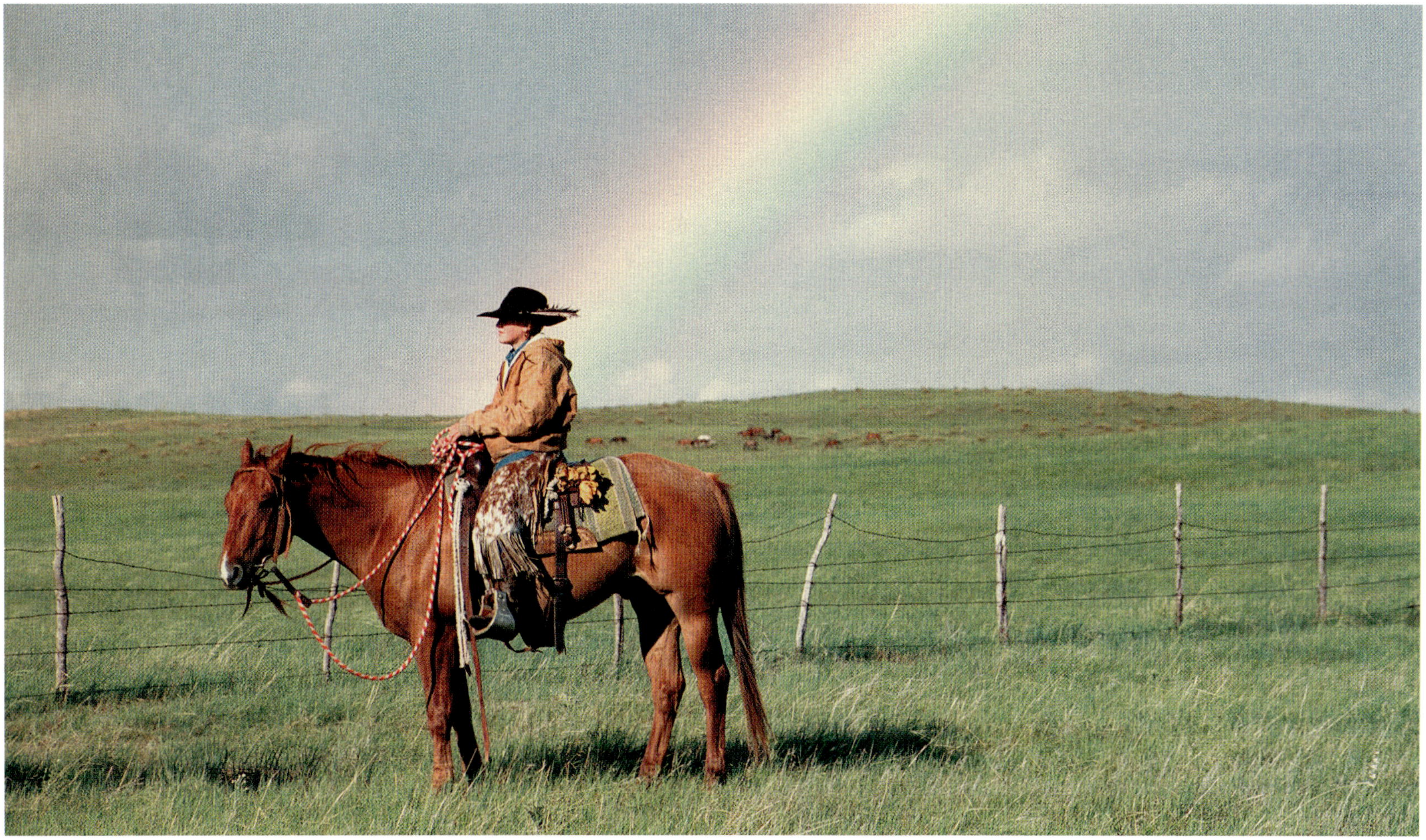

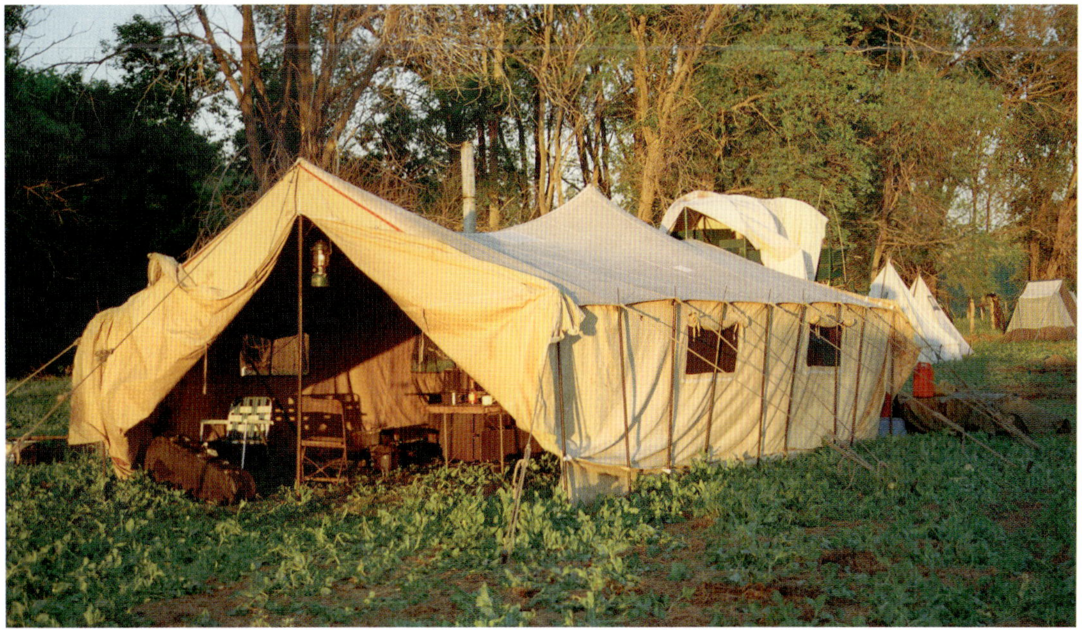

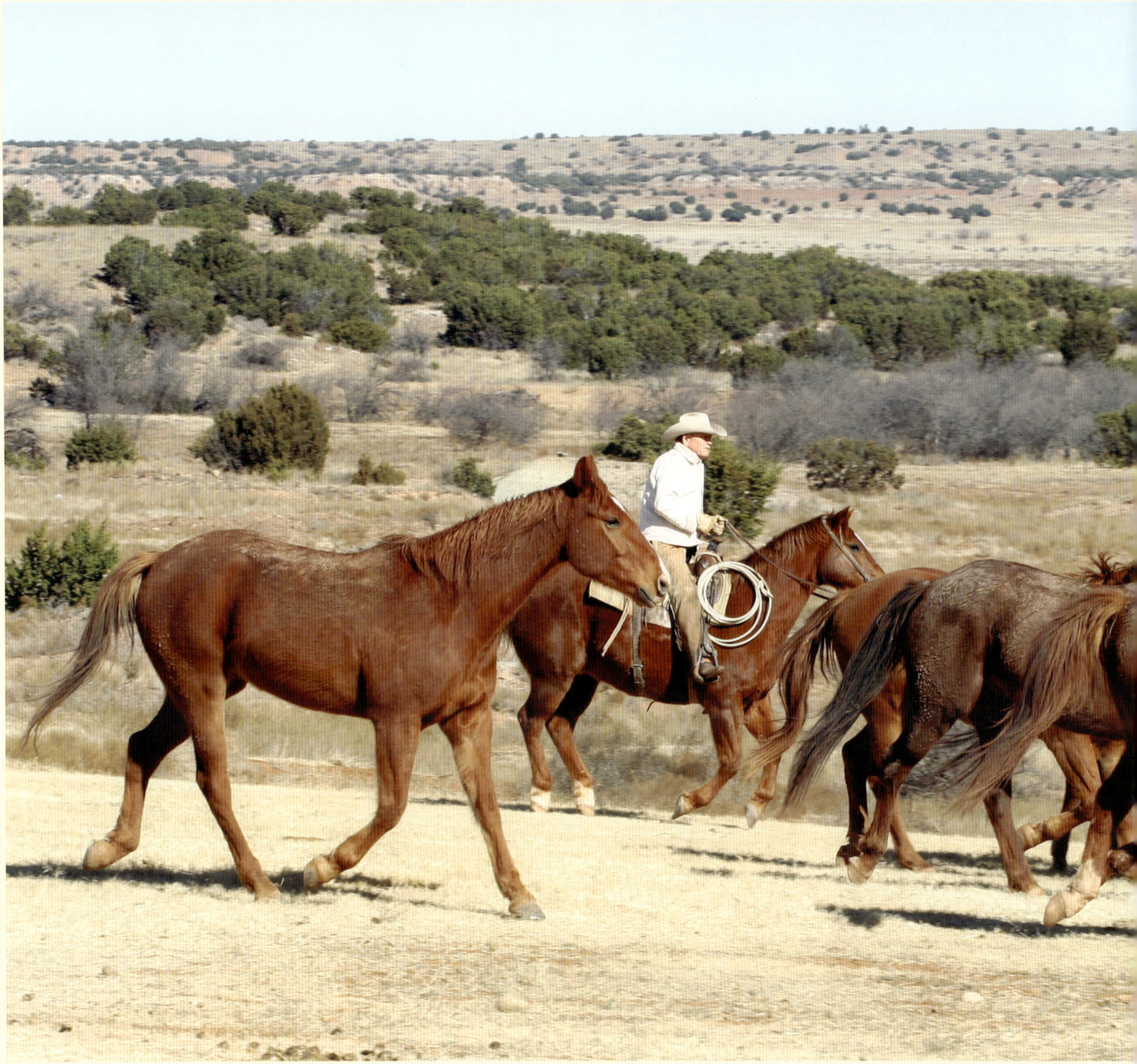

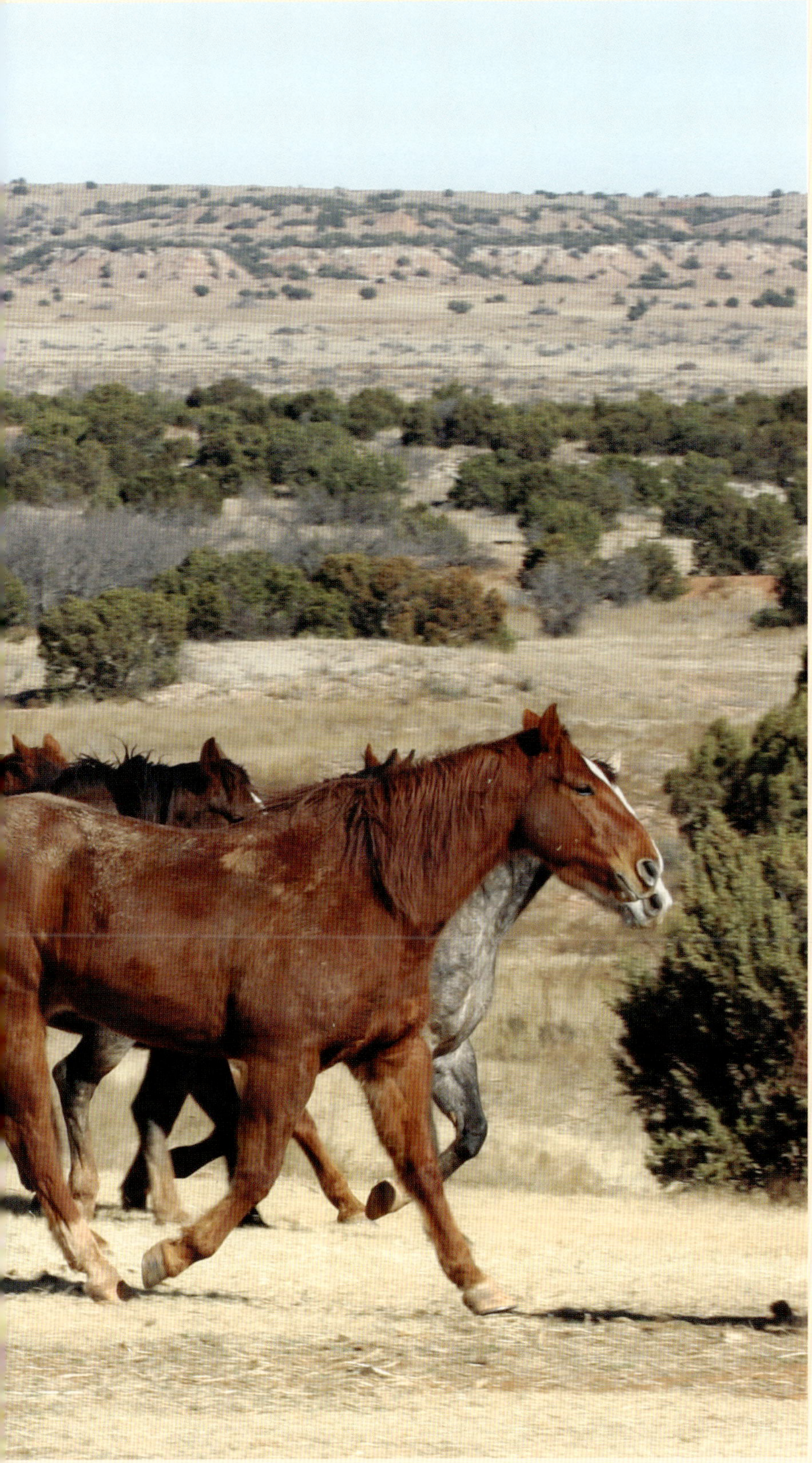

Four Sixes Ranch

Guthrie, Texas

1993

Raising Ranch Horses on the Sixes

The Four Sixes Ranch is an old one, established more than 130 years ago. It was the home of good cow horses then, and it still is.

In barns on three sides of where Dr. Glenn Blodgett sits at his desk stand 17 of the nicest stallions in the Quarter Horse industry. They include some of the top racing-bred stallions, as well as some bred for cutting, working cow horse and other events that have evolved from everyday ranch work. To a lot of people, breeding those horses would be the opportunity of a lifetime. Blodgett agrees. It is an opportunity of a lifetime. But it was not that collection of stallions that brought him to the Four Sixes Ranch. He was there long before they were. "I like to be associated with really good livestock, whether it be horses or even cattle," he says. "There's not a better feeling to me than being horseback on an early morning in one of these big pastures."

That philosophy has been around the Four Sixes Ranch for more than a hundred years, and it continues with a man responsible for breeding some of the best-known stallions in the entire Quarter Horse industry.

Blodgett, who is the horse manager at Four Sixes Ranch, says they were very proud to have earned the Best Remuda Award because, "Our goal with our ranch horse program is to produce the world's best ranch horse. And I think we're doing that."

Raising good ranch horses was always a goal with Samuel "Burk" Burnett, legendary founder of the Four Sixes Ranch. He found his calling early, and joined his father in the cattle business in Denton County, Texas, when Burk was only 10 years old.

Stories about young Burk Burnett are as many as there are cattle on the ranch today. According to one of the stories, in the 1860s, his father sent him, a lad of 17, on his first trail drive to the markets in Kansas. Two years later, at 19, Burk headed north with another herd of cattle, this time as the trail boss.

According to the story, the maturity he exhibited on the trip was probably a sign of things to come. Somewhere in the Oklahoma Territory, Indians hit the herd one night and drove off all the horses except the night horses the cowboys were riding. Burk moved the herd out the following morning, with half the cowboys walking, and continued north until he was able to purchase more mounts.

Once he reached Kansas, he found the market had bottomed out. Selling his cattle would mean a tremendous loss. He, of course, had no way to communicate with his father, so he decided to over-winter the cattle in Kansas. Arranging for temporary grasslands, he grazed the animals all winter and sold them the following year for a profit.

Another often-told story says that Burk won a herd of cattle in a game of poker by laying down a hand with four sixes, thus the 6666 brand. It's a good story, but not true. Around 1867, Burk purchased a cattle herd that carried the 6666 brand, a mark he adopted and made famous world-wide. The brand came with the cattle. How the original owners came up with the 6666 brand remains unknown.

Burk bought his first land near where Wichita Falls, Texas, would later stand, 125 miles northwest of Fort Worth. As the days of the open range ended, he increased his holdings, buying the old 8 Ranch, near Guthrie in north central Texas. That ranch became the nucleus of the Four Sixes Ranch headquartered at Guthrie today, and a later purchase added the Dixon Creek division, located in the Texas Panhandle.

Burk's son, Tom, joined his father in running the ranch, and later founded his own ranches—the Triangles, located near

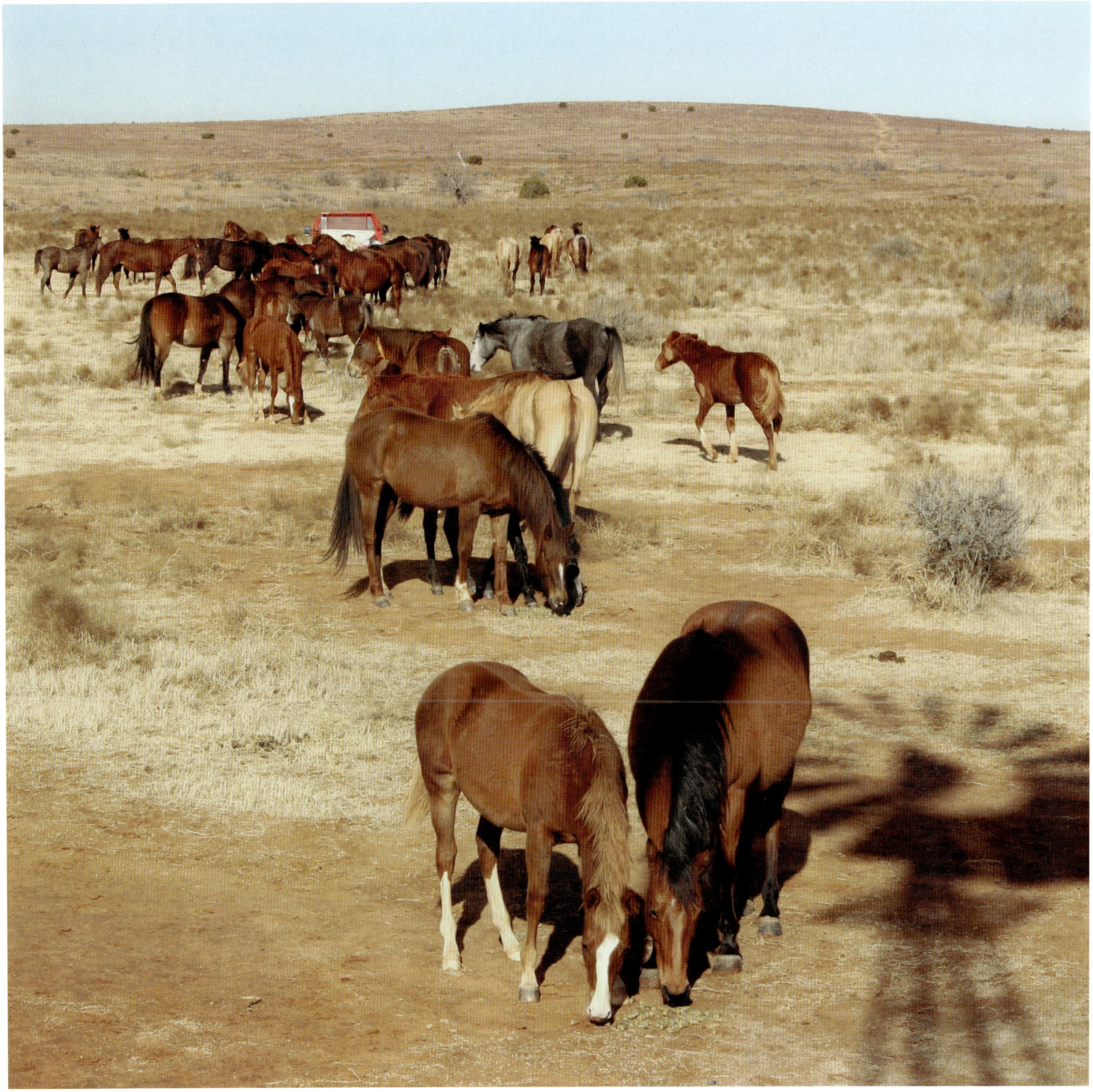

Paducah and Iowa Park. Tom had one daughter, Anne Burnett Tandy, who, upon the death of her father and grandfather, took control of both the Four Sixes and Triangle ranches. Known throughout the cattle and horse industry as "Miss Anne," she was very instrumental in the early days of AQHA. As a matter of fact, in 1940, the AQHA founders were her dinner guests the night before the meeting that originated the Association. Miss Anne was posthumously inducted into the American Quarter Horse Hall of Fame in 1990, during the Association's 50th anniversary celebration.

Miss Anne, too, had one daughter, Anne Windfohr Marion, the current owner of the Four Sixes ranches. Today the Four Sixes encompasses nearly 300,000 acres, on which run 7,500 to 8,000 mother cows and about 500 bulls. The ranch has definitely stayed up with the 21st century, with its own airstrip and hangar, pipe corrals and steel barns. But things haven't changed much in the last hundred years as far as the way the livestock is handled. It's all done by cowboys on cow horses. Some of the cowboys live at the ranch headquarters, at the edge of Guthrie. Others live with their families at the various "camps," houses placed in remote areas of the ranch. Camp cowboys usually look after the part of the ranch on which they live, while those who live at headquarters are assigned other areas. The cowboys who live at headquarters keep their horses there, usually turned out in a small pasture or "trap" near the barns. The cowboys who live at the camps keep their horses in pastures near their houses.

"Both the cowboys at the camps and at headquarters keep a couple of horses up in a lot as they are being ridden, and then rotate them with the others," Blodgett says. "They wrangle or 'jingle' horses in each day as they are needed, depending upon what needs to be done. They usually cut out two, or sometimes three, so they will have one for a spare. Then that evening they'll turn those out and get a new group in.

"We find that the horses ride better if they are turned out the majority of the time. We almost never have any sickness like colic or tying up," he adds.

When the ranch is working cattle—like in the spring, when all the calves are branded, a job that takes several weeks—the cowboys at the camps bring most of their horses into the headquarters and leave them there until the work is finished. Everyone works out of headquarters during that period, although the camp cowboys do go home to their families each night.

The ranch cowboys ride only geldings. "It's because of the way our horses are handled," Blodgett says. "Geldings get along better with each other. Any time we go somewhere on the ranch to work, we load six head in a trailer, and they've all got to get along.

"When we get cattle in a pen, some of the men have to get off and work from the ground," he continued, "and they all turn their horses loose in a corral together. If there were some mares in there, they would be squealing and kicking."

Blodgett explained that each cowboy has in his "string" six geldings older than 3 years old. Also each will have a 2- and a 3-year-old. "Some of them may have more than one 2-year-old, and maybe more than one 3-year-old, if they're riding a horse for one of the older guys who doesn't ride young horses," he says.

To get enough horses, they breed approximately 100 broodmares each year. Many go back to stallions for which the ranch is famous, such as Joe Hancock and Hollywood Gold, and some even go back to Army remount horses that were used in the '30s and '40s. However, they have all proven their worth. Quality in horses is important to the ranch, and Blodgett says that there are some specific things they look for.

"Our horses have to have a lot of stamina," he says. "Some of these pastures are several thousand acres in size, and the cowboys need a horse they can ride a long way and still have something left. Smooth riding is important. When you think of the time these men spend horseback, most of it in a trot, you've got to have a smooth trotting horse. Also, he's got to be agile, and be able to move well in a pasture."

He continues, "Here at Guthrie, we calve in the fall, and wean in the hot summer, in July—lots of 100-degree days. We've got to have a horse that's heat-tolerant, and since we still rope and drag all our calves to the fire when we brand, the horse has to have the strength to drag those calves. He's also got to have the strength to handle a big bull if we need to rope one.

"Our horses have to have cow sense; they've got to *want* to do something with a cow, and they have to be fast enough to catch these cattle, either when we're trying to turn them, or rope them.

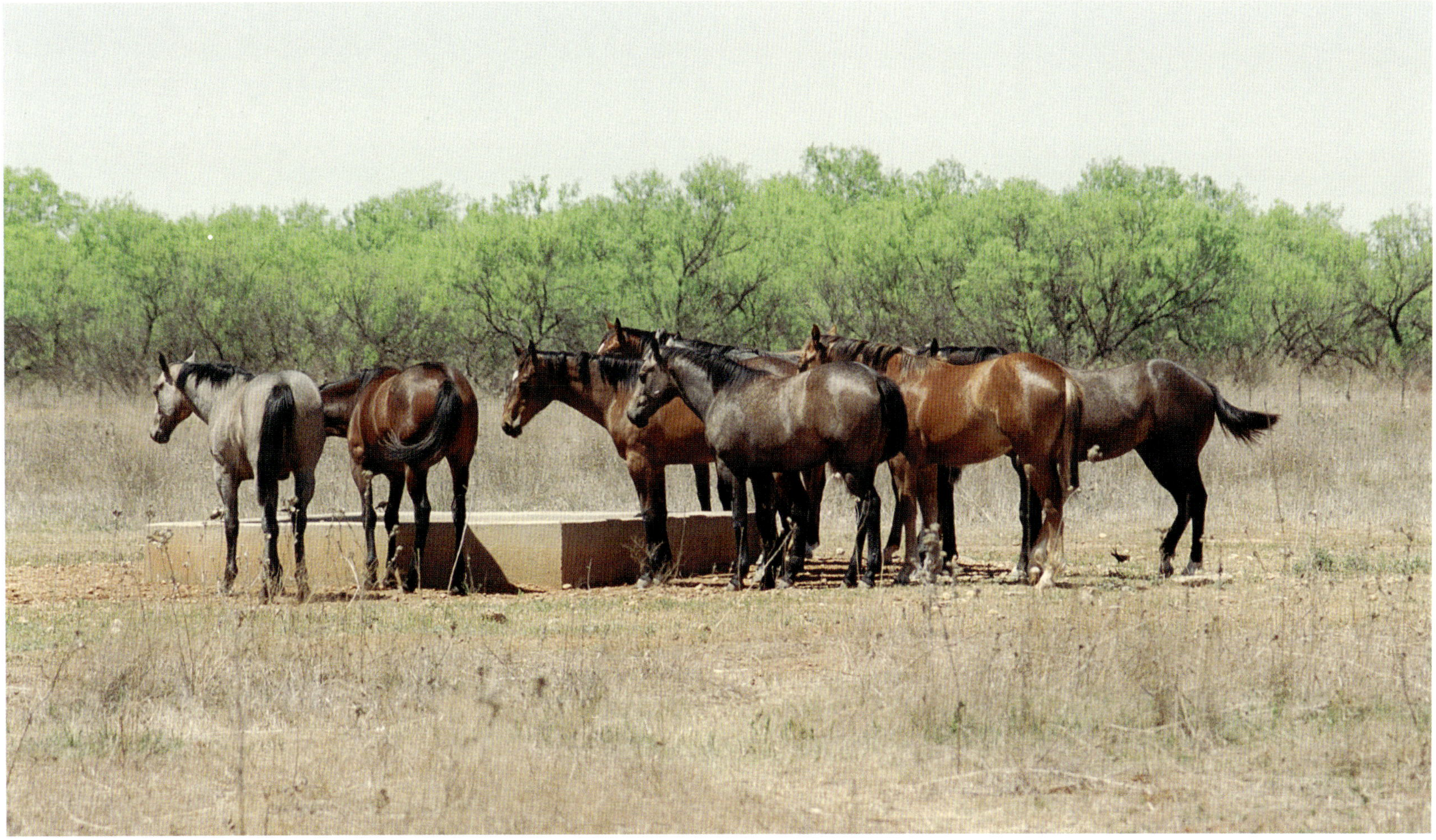

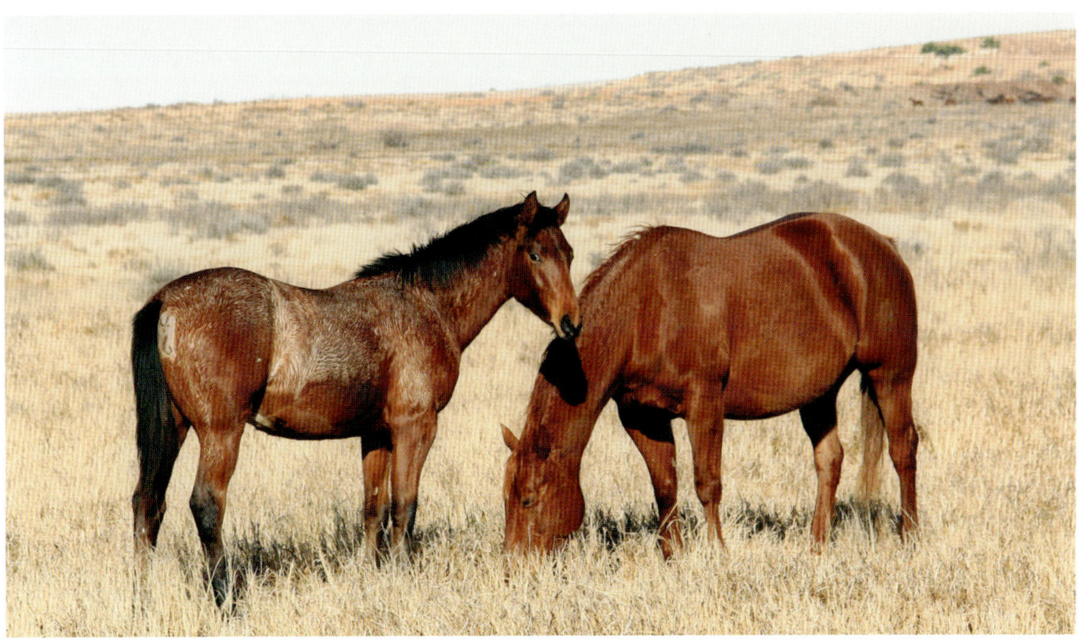

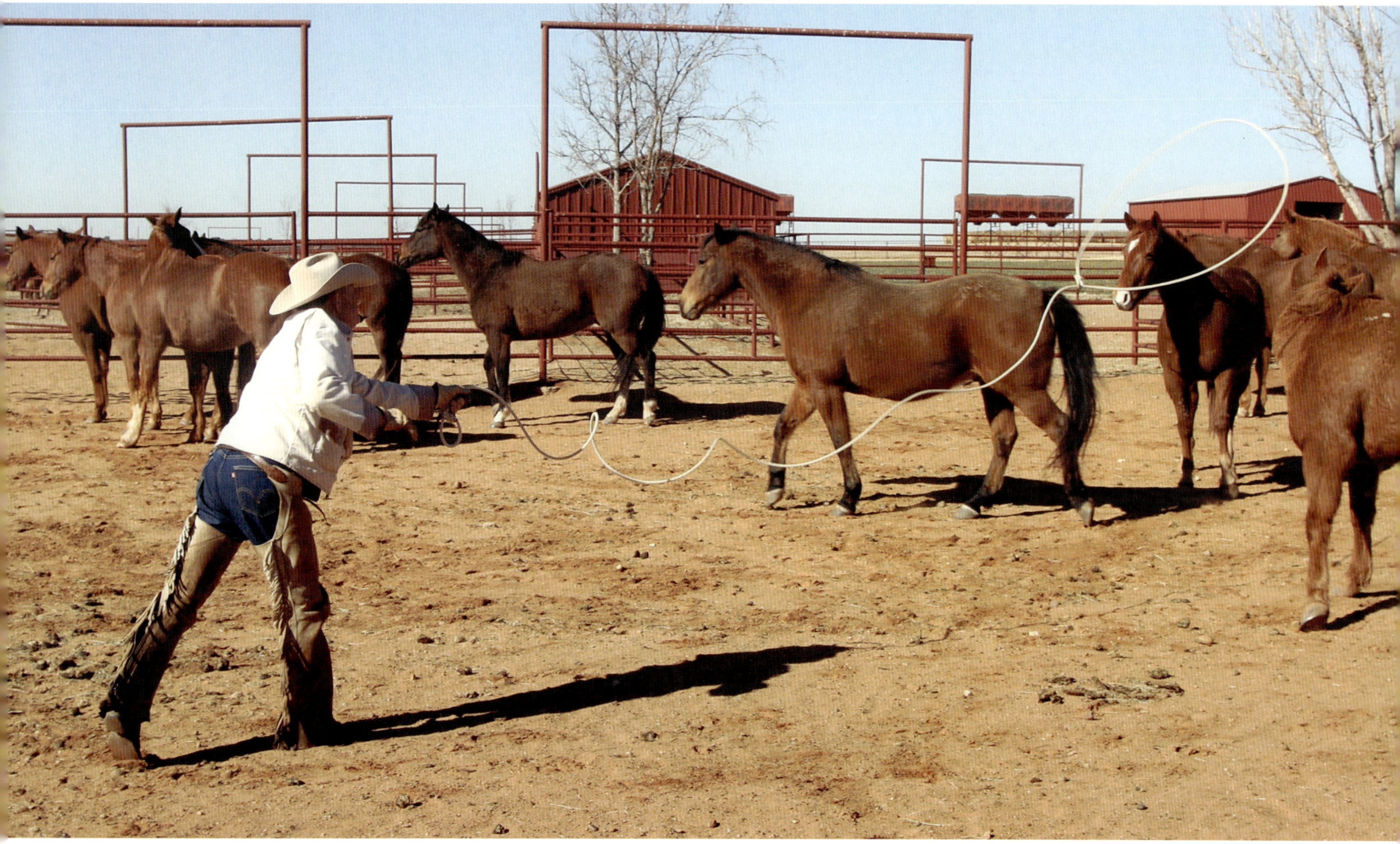

"We also want a horse that has real good hoof quality," he says. "We would prefer a horse that didn't have to be shod, but we put shoes on all of them for added protection. About half of our pastures require a horse to be shod."

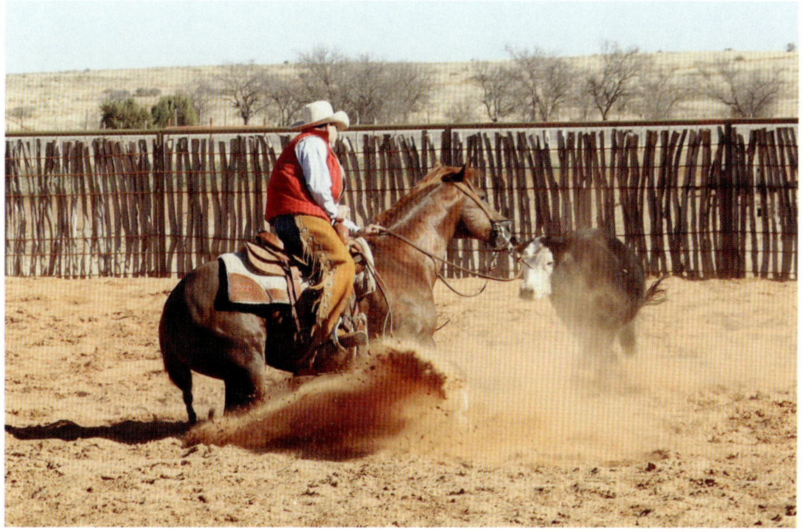

With a few exceptions, each man shoes his own horses—the ranch furnishes the shoes, the nails and the tools. "A few of the older guys might get one of the young ones to shoe their horses, but we've had guys come to work here who had to learn to shoe a horse."

Of all the foals that are born each year, most of the geldings go into the remuda, while some of the fillies are returned to the broodmare band. The Four Sixes has an agreement to ship some yearling fillies to Colorado State University each August to be used in the college's colt breaking class. The fillies are used in a new program at CSU now, where they are broken to ride, and then some of them are fitted for a sale, which is held as a part of class work. Those not sold will be returned to the ranch.

The Four Sixes partners with ranch neighbors Pitchfork, Tongue River and Beggs, and holds an annual Return to the Remuda sale each year in October, and they put some horses in the annual Best of the Remudas sale, which is limited to those ranches that have won the award.

Seven stallions make up the Four Sixes stallion battery, as far as the ranch horses go. However, three of them—Sixes Pick by Tanquery Gin; Paddys Irish Whiskey by Peppy San Badger; and Seven From Heaven, a world champion cutting horse by Playgun—are used the heaviest. Other stallions used on a limited basis include Playin Stylish by Docs Stylish Oak; Four Roan Fly by Continental Fly; Royal Fletch, an NCHA Futurity champion by Jae Bar Fletch; and Ikes Double Drift by White Lightning Ike.

Blodgett says, "Those first three more closely depict what we are looking for here—pedigree, size, conformation, good feet and legs, overall balance—and they are projected to have the best impact. But I am impressed with what we are getting from Playin Stylish and Royal Fletch. They are both working out better than I thought they would. We are riding some of Playin Stylish's foals, and we are starting to ride some by Royal Fletch. The cowboys really like the Playin Stylish colts."

Blodgett says that Sixes Pick's sire, Tanquery Gin, had a far greater impact on the ranch horses than any other stallion, including those for which the ranch is famous, like Hollywood Gold, Joe Hancock and Grey Badger II. But, he acknowledges that Tanquery Gin had a long life so they used him for a longer period of time, and with artificial insemination, they were able to breed more mares.

"He was a good horse, and sired so many more foals than the older horses," Blodgett says. "He had a lot more opportunity.

"But," he continues, "at this stage, it appears that Sixes Pick may have even more potential than Tanquery Gin. He's a better looking horse than his daddy was, he's a good moving horse and when we use him in the pasture, he has plenty of speed to go head some cattle. And I really like how his foals look and how they are riding."

All horses on the ranch have been branded since Burk Burnett's day. Through the brands, each horse can instantly be identified as to its breeding and age. On the left jaw is a one- or two-digit number that indicates the animal's mare family, and on the left shoulder is an "L"—in honor of Burnett's father-in-law, Captain M.B. Lloyd. On the left buttock is a single digit indicating the horse's year of birth, and the right buttock contains a two-digit number indicating his sire.

When the Four Sixes cowboys pick the new horses for their string each year, it works much like the NFL draft. The picking order is based on seniority at the ranch, but a cowboy will occasionally trade one of the horses in his string to another cowboy in order to get a higher pick.

"To a large extent, these guys will get to liking horses out of a certain mare," Blodgett says. "They seem to get more aligned to a mare than they do to a stud, but, of course, they get on and off these studs, too.

"It's been real interesting to me to sit back and watch things play out," he continues. "Sometimes the men really do better not knowing what a colt's breeding is, if they pick the horses on how they move, how they react. But if you take everything into consideration, look at the breeding and then go out and study these horses and watch them move out in the pasture, watch them react to you, and really study them over a period of time, you can do a pretty good job of picking a good mount for the future."

The Dixon Creek division, located in the Texas Panhandle, is, like the Guthrie ranch, a cow/calf operation. Also, replacement heifers for both ranches are grown out on this division. Cows on that ranch calve in the spring, while cows at the Guthrie ranch calve in the fall. After the fall calves are weaned, they are shipped to the Dixon Creek ranch, where they are grown out until the following fall. Then, the steers and those heifers not selected for replacement are sent to the feedyard.

There are no broodmares at the Dixon Creek ranch, and those cowboys, too, get their picks from the colts at the main ranch.

Unlike most ranches of this size, Blodgett artificially inseminates all the mares. He was hired by Mrs. Marion in 1982 because, Blodgett says, "she wanted to really develop the horse program, and intensify and modernize the horse production here on the ranch."

Joe Leathers is foreman of the Dixon Creek division, and Mike Gibson holds those duties at the ranch in Guthrie. Mike's father, J.J. Gibson, held the position for years. Although Mike Gibson's primary responsibility is the cattle, he, too, has a soft spot for the horses. He grew up on the "Sixes," and started helping his dad work cattle as soon as he was big enough to get horseback. Today he doesn't get to ride as much as he would like, due to his job—he relies on a pickup truck to see as much of the ranch as possible every day, checking on pasture condition, water, etc.—but he's still very aware of how the horses are handled.

He explains that everyone starts their own 2-year-olds in the round pen, plus there will be some extras that the younger men will pick up. At one time, famed horseman Ray Hunt came to the ranch each year to help the men start their young horses. Today, Joe Leathers, who spent a lot of time with Ray Hunt when he was at the ranch, helps the others some.

"I've been here long enough to see a definite change in the remuda," Gibson says. "We started using a Ray Hunt way of things about 1985. The remuda has turned over now, and it's a whole different remuda. The general disposition and attitude toward people is a lot better. If something goes wrong, say, someone drops a rope, or a horse gets over a rope, or a saddle turns a little bit, it doesn't seem to bother them as much. Those old horses, where we tied a foot up, sacked them out, and got on and rode off, any little thing that went wrong, they threw a fit. It's not that way anymore."

There have been some changes on the Four Sixes Ranch in the last hundred years, but Burk Burnett would still recognize it. The home that Burnett built in 1917 for $100,000 still sits atop the hill overlooking the town of Guthrie and the headquarters of the ranch; the ranch is still in the family, owned by Burnett's great-granddaughter; the brands remain the same, both on the cattle and the horses; and the philosophy about good cow horses is still there.

Perhaps Blodgett says it best: "Something I'm always doing is rethinking how I did something, and trying to figure out how to do it better. I'm always trying to figure out a better way to do things. However, sometimes I may change and go back to the way I was doing it before. The old way might have been better."

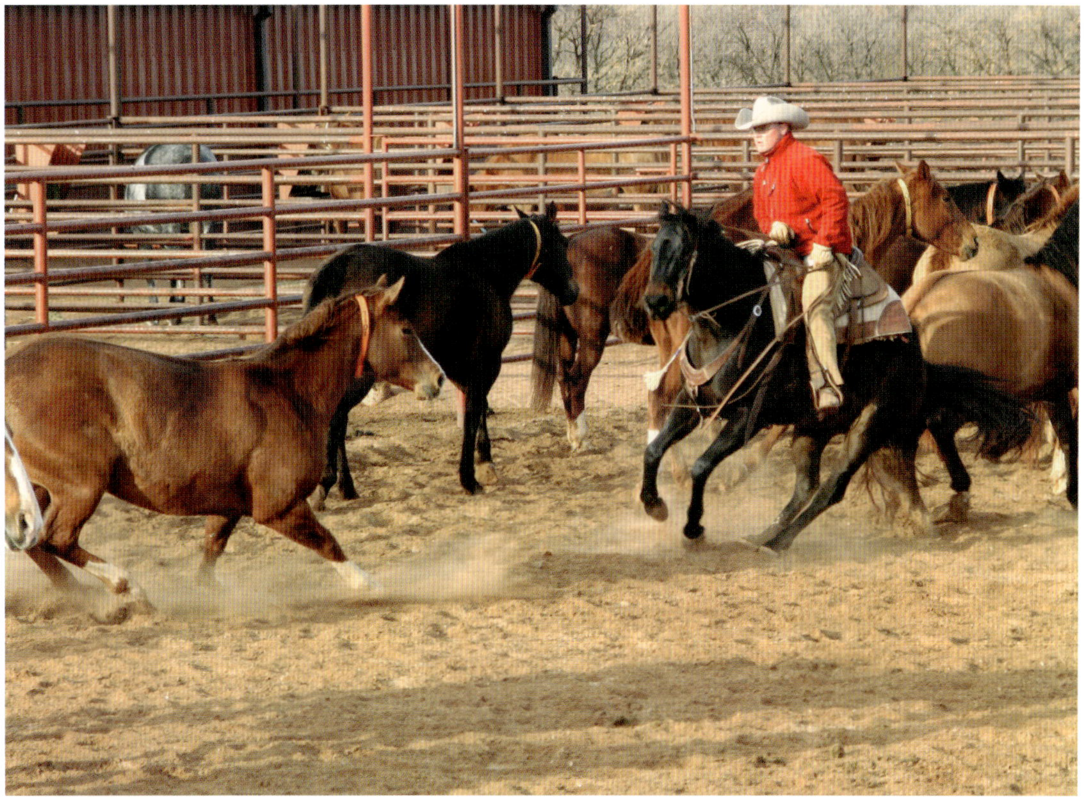

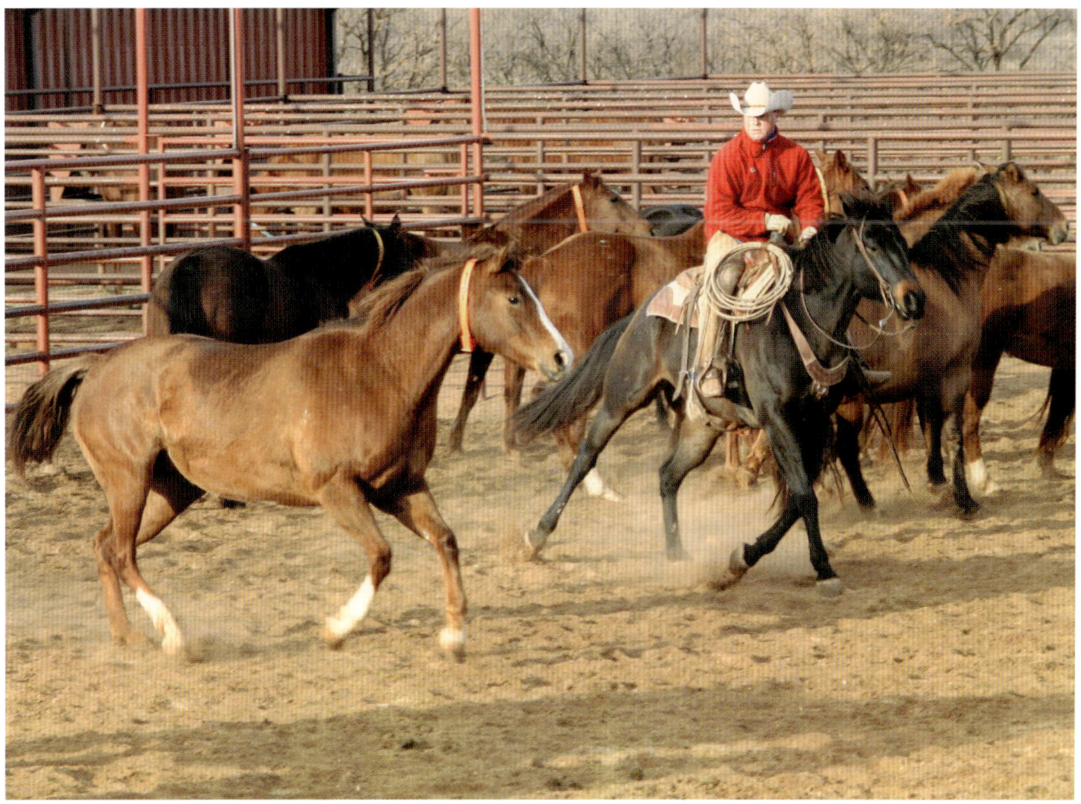

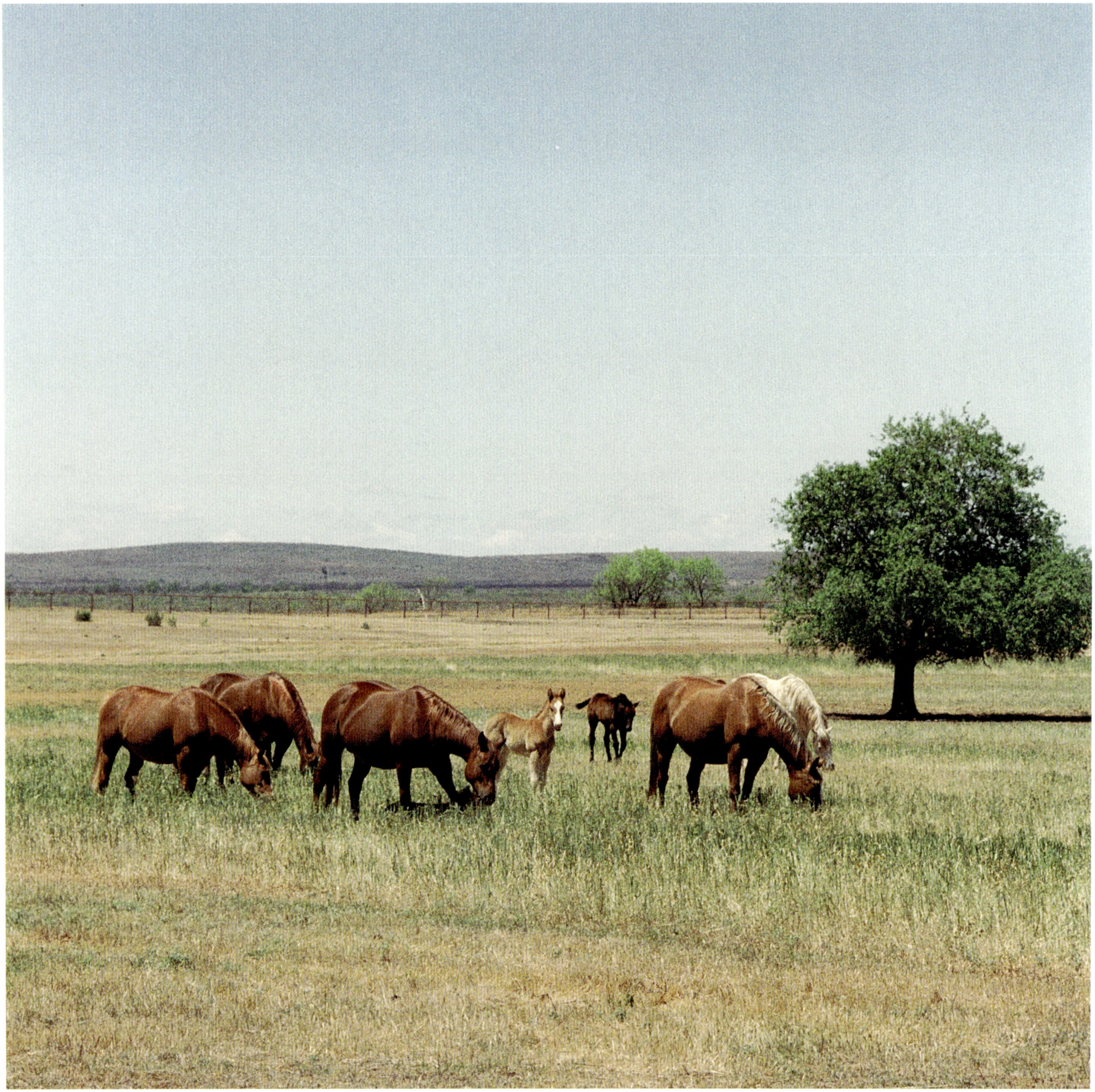

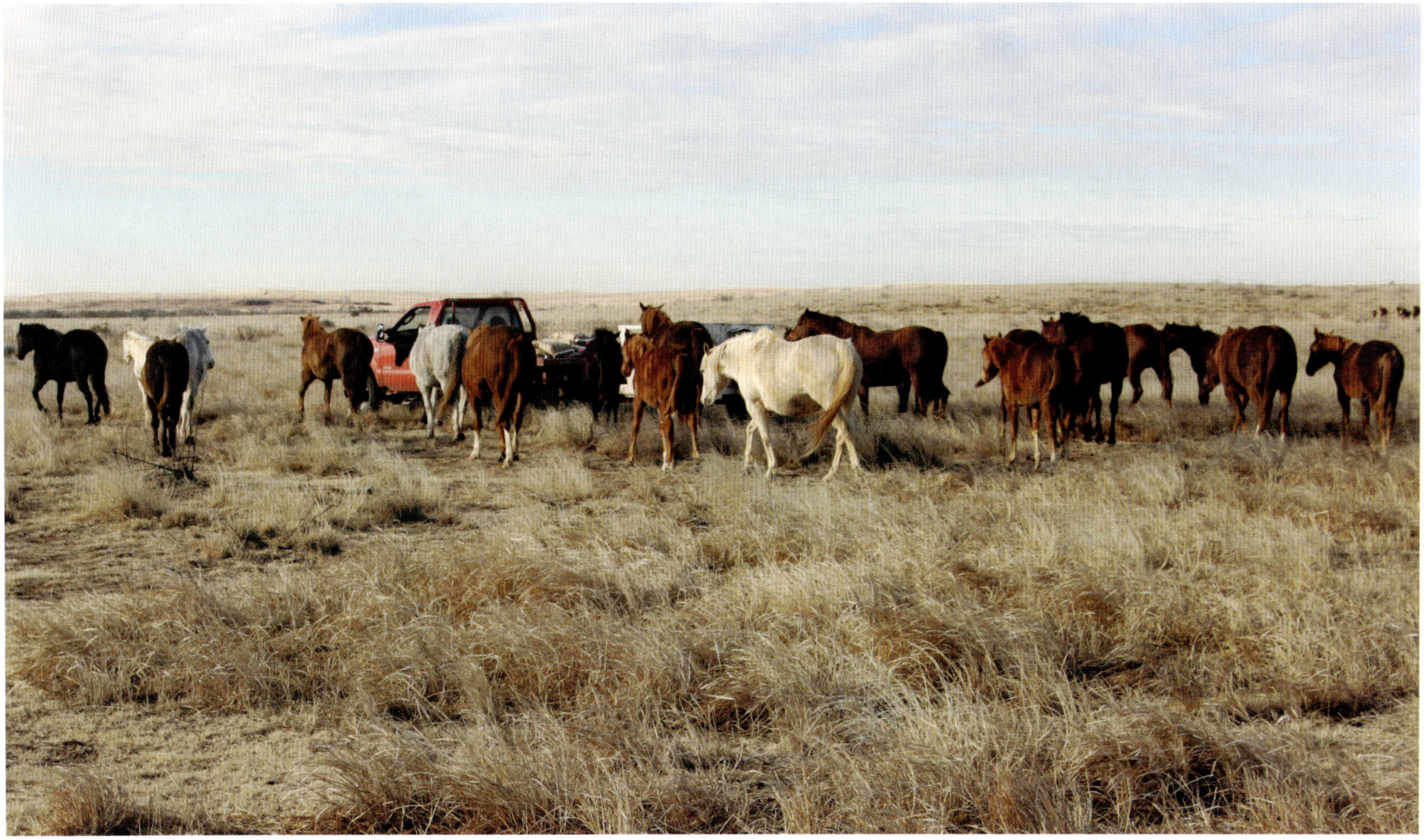

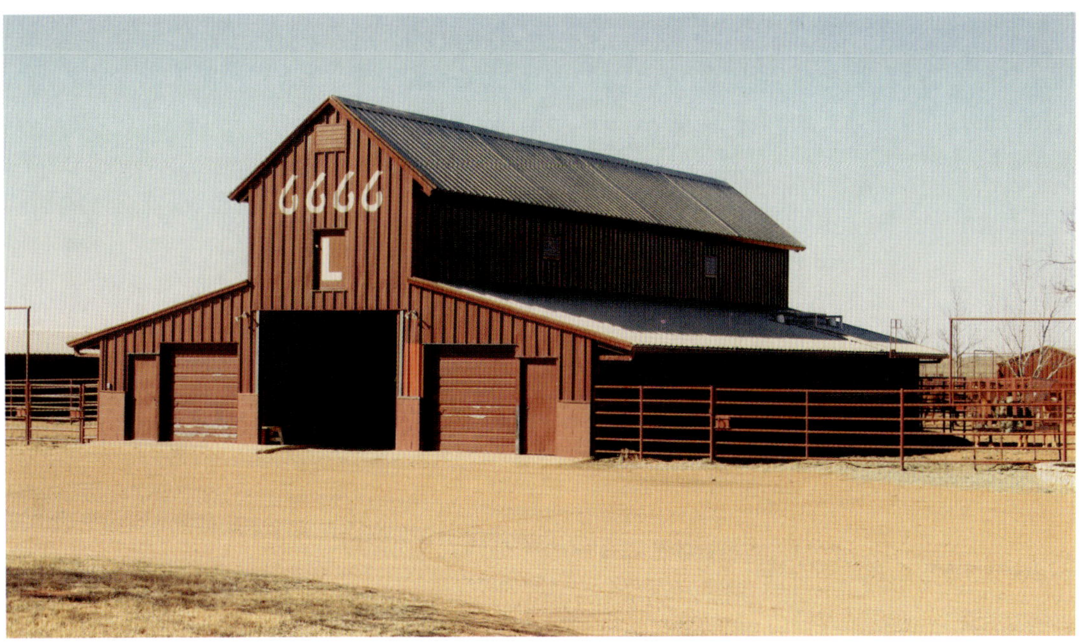
6666
L

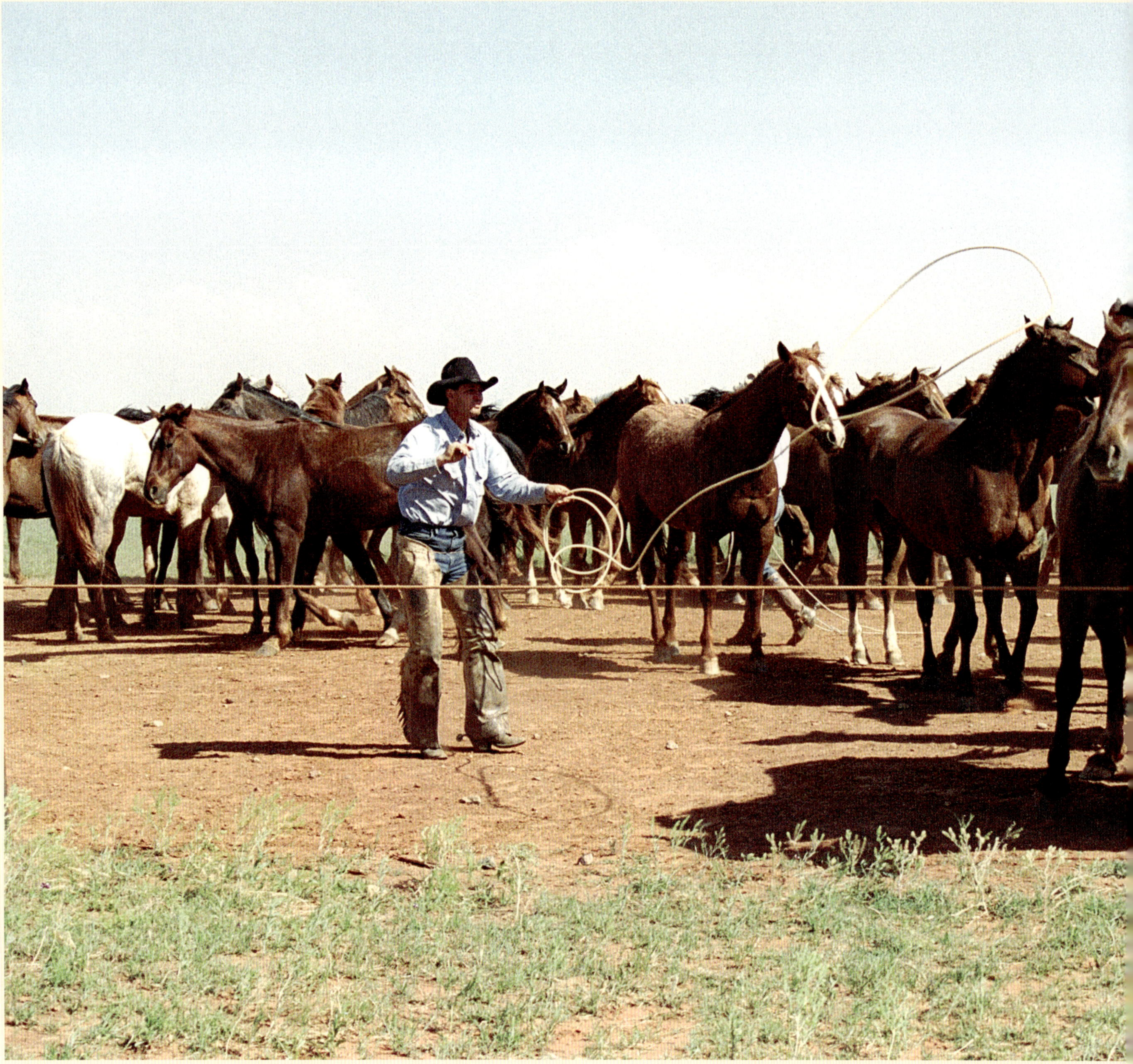

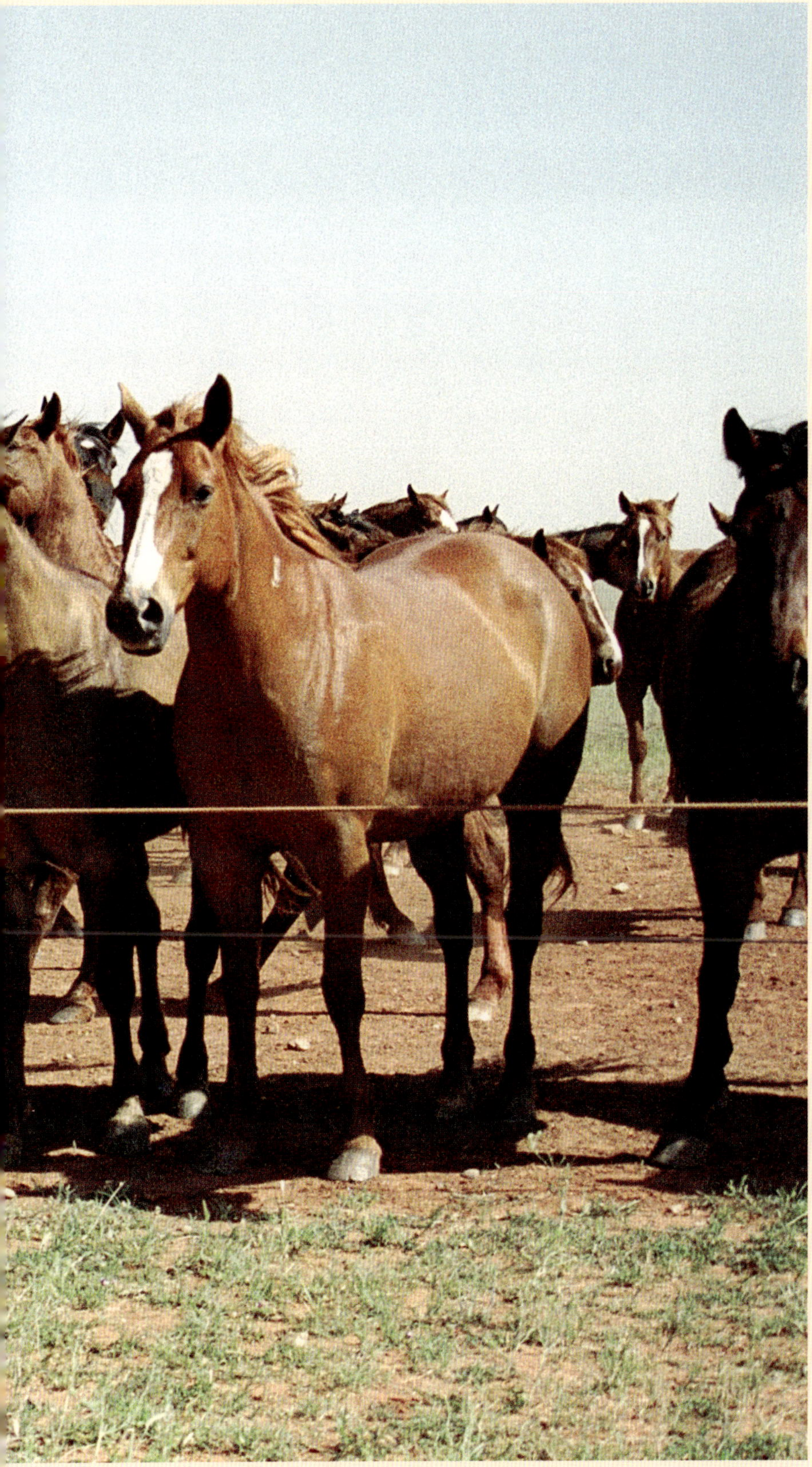

WAGGONER RANCH

Vernon, Texas

1994

Working for the Waggoners

Back in 1903, when W.T. (Tom) Waggoner made 18-year-old Lige Reed wagon boss of Waggoner's 500,000-acre ranch, Reed was understandably a little nervous. After all, most of the men he would be bossing were older than he was.

When Reed voiced his concerns, Waggoner gave him two pieces of advice: First, he told Reed, always keep the wagon ahead of the drive. That way, the boys would be moving forward all the time; they wouldn't be slowing up or turning back when they got hungry. Second, Waggoner told Reed, always have the cook, when he parked the wagon each night, face it in the direction he was going to travel the next day.

"That way," said Waggoner, "he won't have to waste time turning around in the morning."

The Waggoner family has never been one to waste time.

Tom Waggoner knew what he was doing when he gave an 18-year-old that kind of responsibility. After all, it was nothing less than what happened to him.

Tom's dad, Dan Waggoner, migrated from Tennessee to Texas in 1840, settling first in the area north of Fort Worth. In 1852, Dan started a ranch in the North Texas region around Vernon. His stock at that time included 230 head of Longhorns and a few horses. He also had one slave, a young wife and an infant son, Tom. The ranch was located just south of the Red River, which separated Texas from Indian Territory, and both the Comanche and Kiowa made frequent raids into the area.

By 1868, the Waggoner outfit had grown considerably, but help was still short. Dan sent Tom, who was only 16 at the time, to Kansas City with a herd of 5,000 steers. To make the trip, Dan gave Tom $12, a group of drovers and 50 saddle horses, all of which had sores on their backs from heavy use. In later years, W.T. Waggoner was heard to say, "There I was with more than 5,000 big, wild steers, 50 sore-backed horses, $12 and 16 years of my life."

Writer Franklin Reynolds, in a 1956 article in *The Quarter Horse Journal,* noted that it was possibly Tom Waggoner's experiences with those sorry horses that caused him to resolve to always have good ones. Even when he was a young lad of 14, W.T. had said, "I want to run the most cattle, breed the best horses and work harder than anyone." When he was 17, his dad made him a full partner. By the time he was 27, Tom had full reins of the Waggoner cattle empire.

Reynolds further said: "Turning back through the pages of Texas horse history, it may be said that in his time, he [W.T. Waggoner] owned or used most of the great Texas horses of his day." That claim can also be made by his descendants.

Nation's Largest Ranch

The W.T. Waggoner Estate is the nation's largest ranch under one fence. It comprises 520,000 acres in six counties, and is the home for 12,000 mother cows.

The ranch originally operated under the name of Dan Waggoner & Son. Dan died in 1903, leaving everything to Tom. In 1923, Tom formed what today is known as the W.T. Waggoner Estate, naming himself as trustee. Three of his children—Guy, Electra and E. Paul—were appointed to the board of directors. Tom died in 1934. Today, the estate is owned by E. Paul's granddaughters, Helen Willingham and Electra Moulder, and A.B. Wharton, grandson of Electra Waggoner Wharton. Its interests include the cow/calf operation, farming and oil operations—and horses.

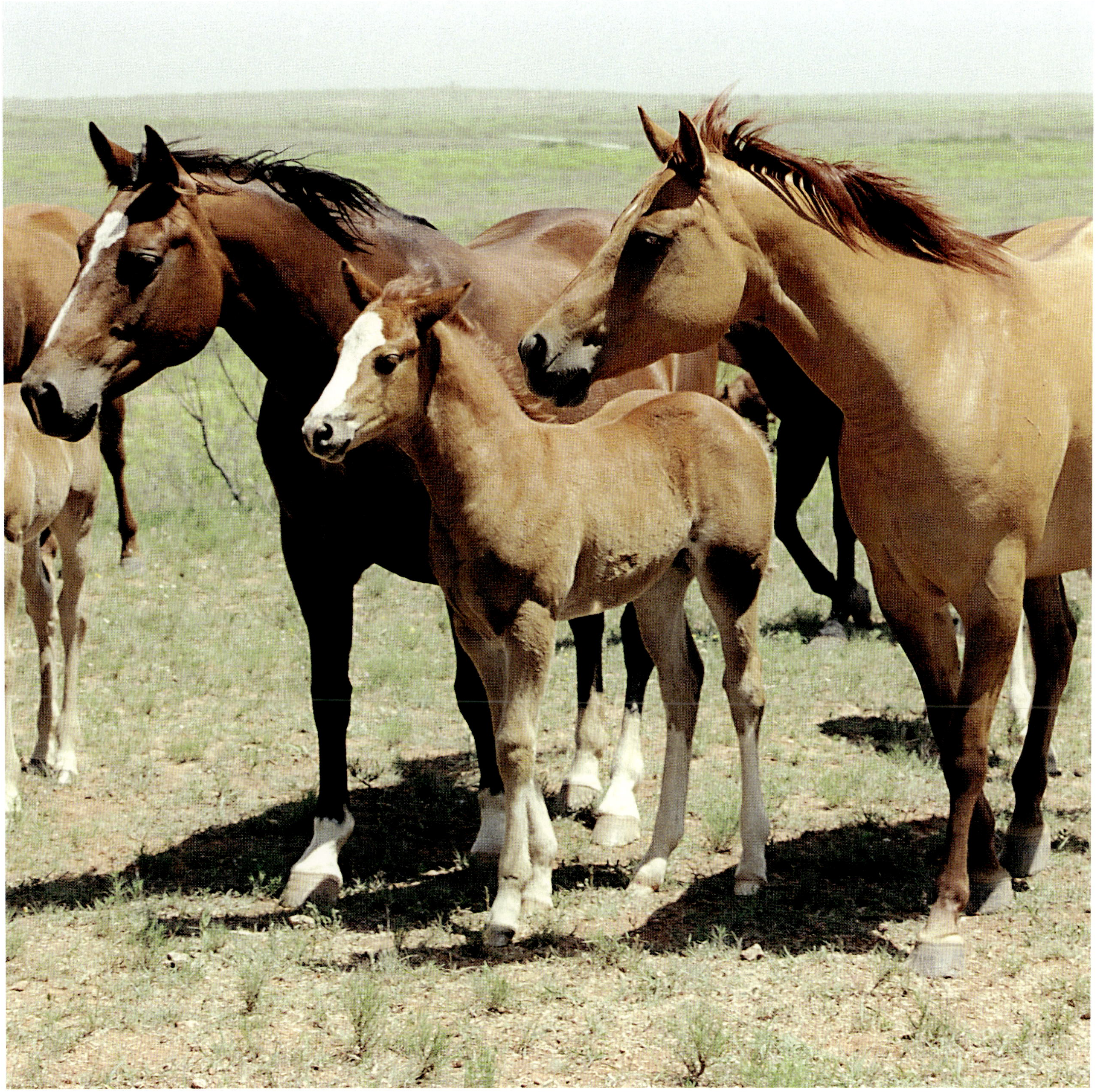

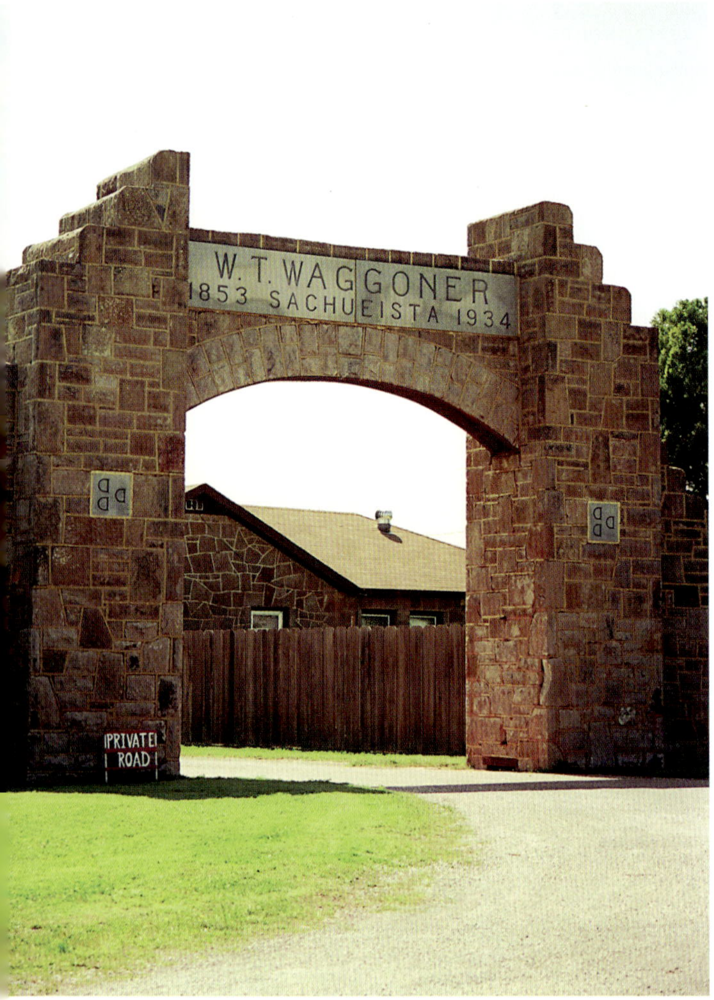
W.T. WAGGONER
1853 SACHUEISTA 1934
PRIVATE
ROAD

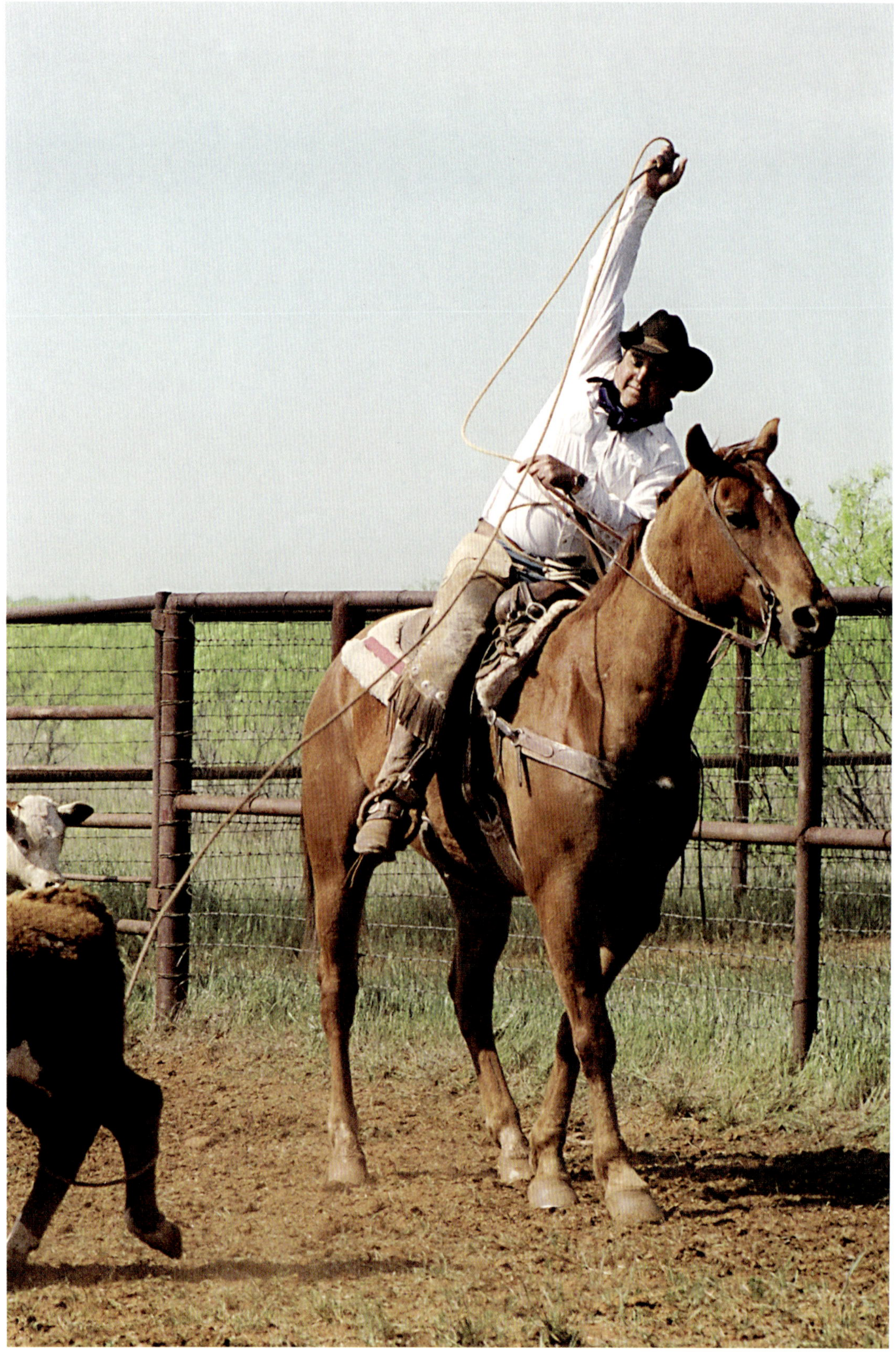

Especially horses. The Waggoner contribution to the Quarter Horse industry has long been recognized. Horses like Waggoner's Rainy Day, Midnight, (One Eyed) Waggoner, Yellow Wolf, Yellow Jacket, Pretty Boy, Pretty Buck, Snipper W, Pep Up, Jesse James and, of course, Poco Bueno still appear in the pedigrees of many great horses of today. Some of these were bred strictly for ranch work, and others for the show arena, but they all had one thing in common—they were cow horses.

Cow horses were not the only equine interests of the Waggoner family. W.T. Waggoner loved horse racing. He reportedly tried to buy the Thoroughbred Man O'War, but was unsuccessful in his efforts. However, he is credited with being the primary force in legalizing pari-mutuel wagering in Texas in 1934. The year before, so sure was he that the law would pass, Waggoner started construction on a $2-million racetrack on property he owned in Arlington, Texas. Arlington Downs, located between Dallas and Fort Worth (approximately where Six Flags Over Texas and The Ballpark at Arlington are today), opened only a few months after the passage of the bill.

W.T. Waggoner died shortly after his racetrack opened. His sons, Guy and Paul, continued to operate the facility until the law was repealed in 1937.

It was at that site that E. Paul Waggoner built his Three D Stock Farm, which headquartered the estate's horse showing operation. It was there, too, that he brought famed halter and cutting horse Poco Bueno, which he purchased from the Hankins brothers in 1945 for $5,700; it was also in Arlington that cutting horses Jesse James and Snipper W, and rope horse Pretty Buck, got their starts.

E. Paul Waggoner died in 1967, truly a legend in the Quarter Horse industry. In 1991, he was posthumously inducted into the American Quarter Horse Hall of Fame.

Poco Bueno

Perhaps the horse that is most associated with the W.T. Waggoner Estate is Poco Bueno, which was foaled in 1944 at Jess Hankins' ranch near Rocksprings, Texas. A son of King P-234, he was out of Miss Taylor by a son of Little Joe. E. Paul Waggoner bought the horse as a yearling for $5,700 in the Hankins brothers' first production sale.

Poco Bueno was shipped to Waggoner's Three D Stock Farm in Arlington, Texas, where Bob Burton broke him to ride. However, Burton left Waggoners later that year, and probably the man who put the most saddlings on the stallion was Pine Johnson.

Poco Bueno stood grand champion at halter many times, and he and Johnson won numerous cuttings throughout the Southwest, along with an AQHA Championship.

As a sire, Poco Bueno's get included 405 registered foals from 24 breeding seasons. More than half of those foals earned points in AQHA halter and performance competition. Thirty-six became AQHA Champions, and three are in the National Cutting Horse Association's Hall of Fame: Poco Mona, Poco Stampede and Poco Lena.

Poco Bueno died in 1969. He is buried near the entrance to the Waggoner Ranch, south of Vernon.

Ranch Horses

Ranching at the Waggoner Estate is big business. A half-million acres and 12,000 cows add up to approximately 12,000 calves that have to be gathered from some rough and brushy pastures, worked, and shipped to market each year. As in their other business interests, in the cattle operation the Waggoners use all the modern conveniences at their disposal—including even a helicopter. However, some things never change. No one has ever come up with a replacement for the cowboy and his horse when it comes to pushing a cow and her calf out of a mesquite thicket so dense that you can hardly see into it.

Dan and W.T. Waggoner knew this a hundred years ago, and cattle manager Weldon Hawley knows it today. That's why it's important to him for his cowboys to be well mounted.

There are 200 saddle horses in the ranch's remuda. Of these, each cowboy can have up to eight head that he calls his own. The ranch runs 85 broodmares, and is currently breeding five stallions.

Trace Cribbs is the horse manager for the ranch. He lives at what has always been Waggoners' horse headquarters, the Whiteface Camp, located near the town of Electra. It's his responsibility to care for the mares, handle all the breeding and wean the

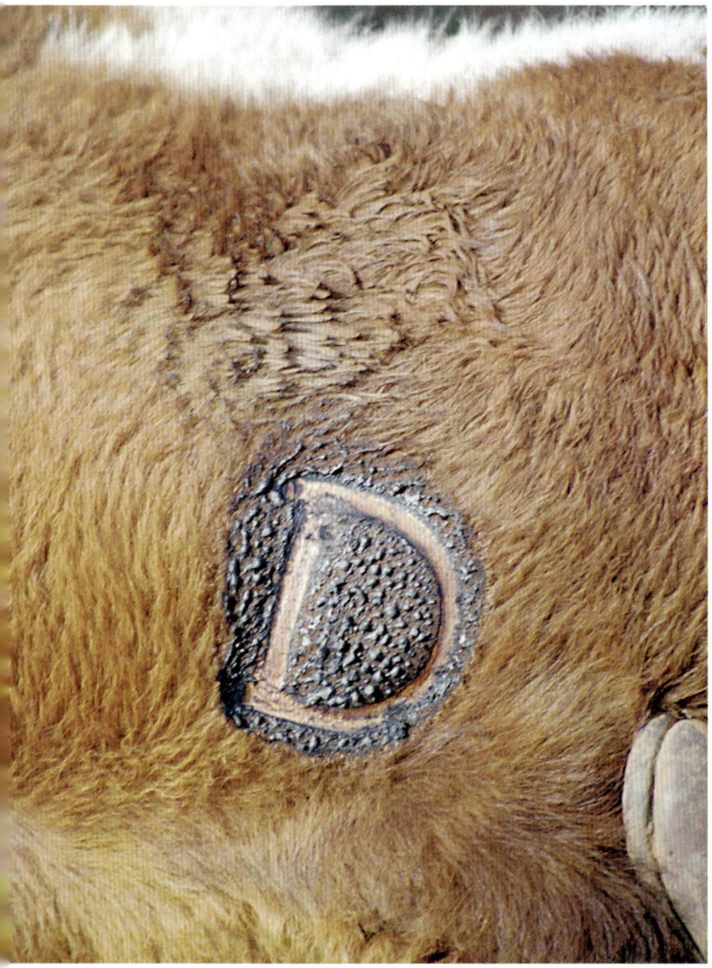

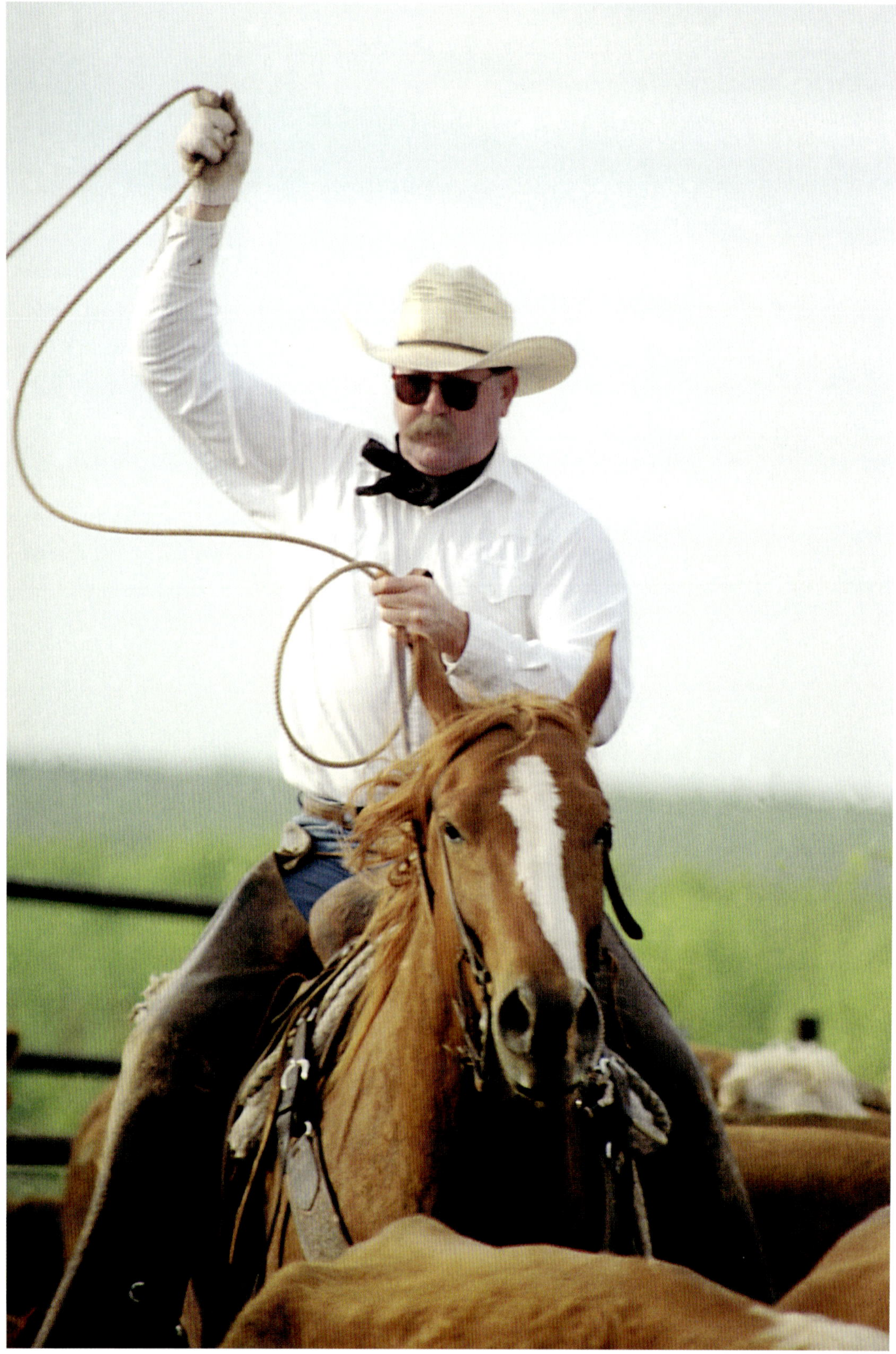

foals. He is also in charge of halter breaking and branding the foals, and seeing that they have all their immunizations.

Trace holds the job that long-time ranch cowboy Wes O'Neal had for years. Wes, today, is semi-retired, but he still comes to work every day. "I don't know how to describe Wes' position," Trace says. "I guess he's an adviser or consultant, but the main thing he does is work with the horses that we put in some of the sales. He rides about four horses a day.

"But he has a lot of value to me," Trace continues. "I just feel blessed to have him around. He is a world of knowledge."

The horse breeding operation has changed tremendously over the past few years. No longer are the stallions turned out with the mares. Instead, nearly all the mares are brought to the Whiteface Camp and are artificially inseminated.

"We try to breed on foal heat," Trace says. "As soon as a mare foals, we bring her in and doctor the foal's navel, and then put her in a pen right behind the barn. Then we check those mares regularly, and when they ovulate, we breed them. We check them again in 14 days. If they are pregnant, they go back out to pasture."

The foals are not weaned until around Thanksgiving, but they are halter broken and branded in July. Waggoners have a series of brands that go on each foal, the history of which goes back to the days of W.T. Waggoner.

A number that denotes the foal's dam is placed on the left shoulder, and the ranch's famous D brand is placed under that. Another number brand is placed on the left hip to denote the year. Finally, a sire brand is placed on each foal's left thigh, behind the stifle. From a quick glance, nearly any cowboy on the ranch can determine a horse's breeding and age.

The colts are gelded at about 12 months. When the colts turn 2, someone rides them all 30 to 40 days before they are assigned to different cowboys.

"The cowboys get to pick their colts by seniority," Trace says. "And our policy is that when a gelding turns 15 he has to be turned in, unless it happens to be one of their kids' horses.

"We also have an incentive now," he adds. "We have a production sale every two years, and if a cowboy turns in a horse prior to age 15, he gets a 10 percent commission on what that horse brings in the sale. We call those our 'top cut' geldings. After about age 10, the peak value on a horse starts going down, so we give them an incentive to turn in one earlier for the production sale."

Trace says that the broodmares are also cycled out and sold in the sale after they are about 15, an age at which they can still get in foal and be a value to someone.

Although the colts are broken as 2-year-olds, they are not ridden very much at that time. "You know, you can't do too much on a 2-year-old," Trace says, "they just can't go all day. The boys will ride them some in the evenings, maybe gather some bulls on them. If they get 30 or 40 rides during the year, they won't forget that."

When the horses turn 3, they are used a little more, but they are normally 4 or 5 years old before they are really used much in the ranch work.

Waggoner cowboys ride only geldings for the cow work. However, all the fillies are also ridden as 2s to see if they might turn out to be the types of mares the ranch wants in the broodmare band.

"I need to know, when breeding the mares, what is working and what isn't working," Trace says. "We will keep 10 to 15 fillies each year, and after we ride them, we have an idea which ones we want to breed. But handling them like that also makes them gentler and easier to handle in the breeding barn."

Many of the Waggoner broodmares go back to those stallions that have made the ranch famous, including Poco Bueno. The five stallions bred on the Waggoner ranch this year included Greyt Whiz by Top Sail Whiz out of an Easter Gentleman daughter; Poco Tuck Zero by Poco King Tuck, a son of Poco Bueno that was used on the ranch for a number of years; Mr Sorrel Bo by a son of Peppy San Badger; and Doc O Boots by Grey Starlight and out of a Doc O'Lena mare. The fifth stallion is a 3-year-old son of Justa Swinging Peppy out of a daughter of Poco King Tuck. His first foals were born in the spring of 2006.

"We have about 14 daughters of Poco King Tuck left," Trace says, "and we sure do like them. They have a lot of size and bone, and they are really good minded and pretty. Most all of our mares go back to some of the stallions we have had over the years, including Mr Solano Smoke, Poco King Tuck and Tee J Jack Steel, a son of Jackie Bee.

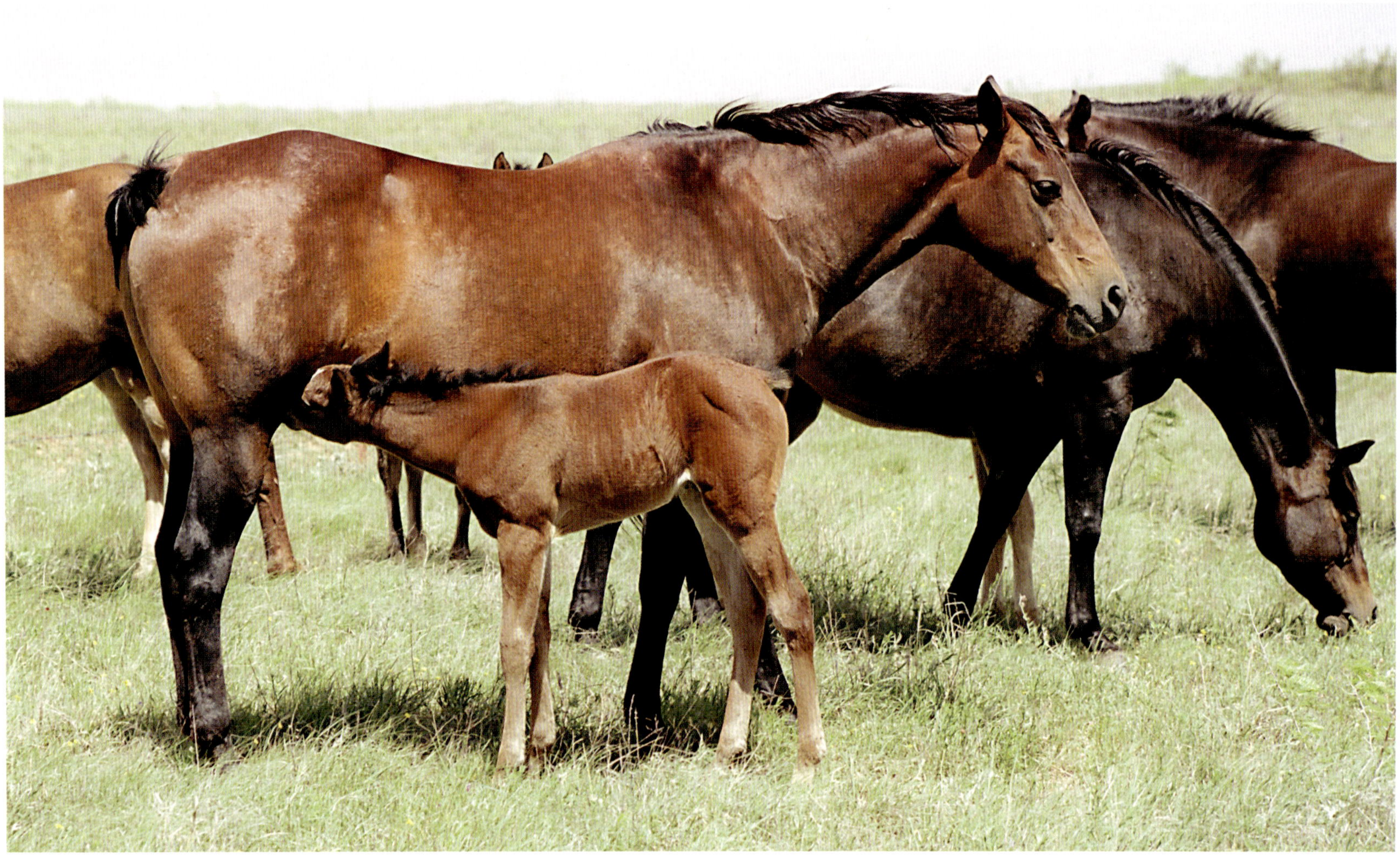

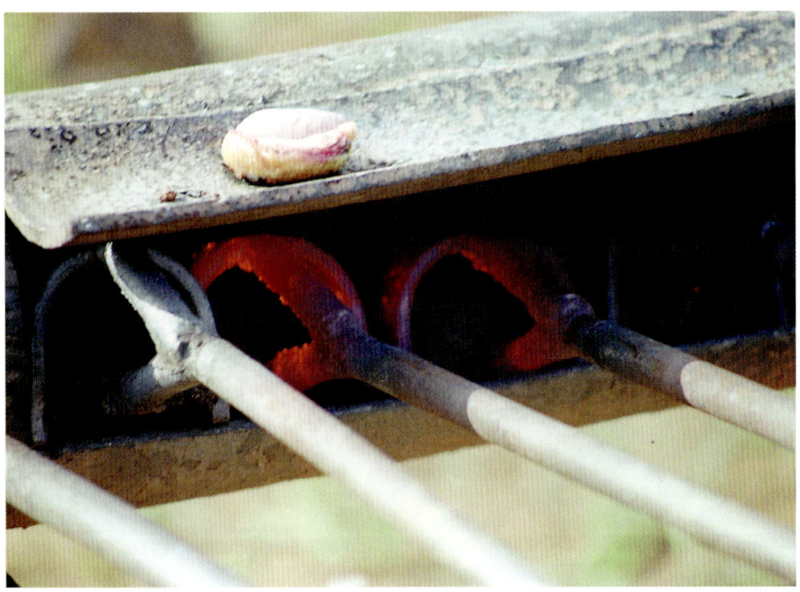

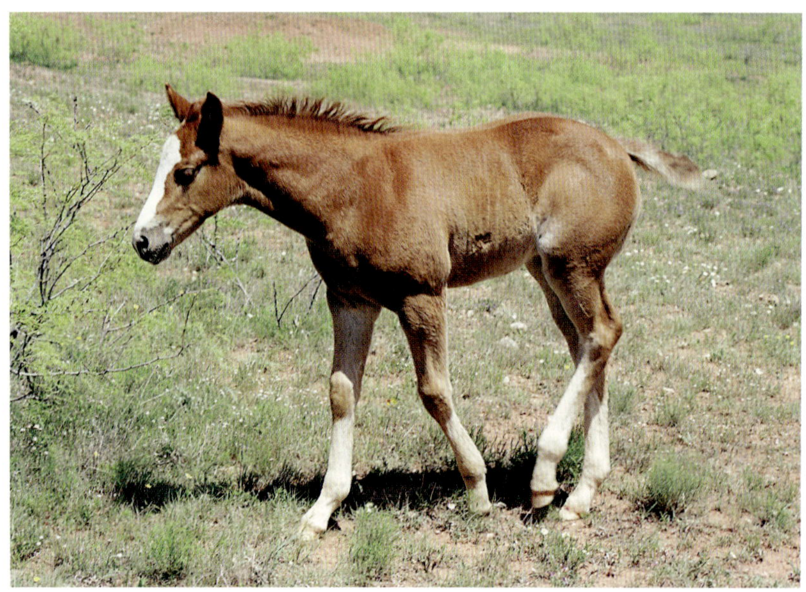

"Greyt Whiz is a really nice horse," he continues. "He's being shown in reining, and we're going to do some other things with him, too. One of the things we really like about him is that he has no 'Little Peppy' (Peppy San Badger) in him. So many of the performance lines are saturated with Little Peppy, including some of our own horses, so he's going to be a great outcross for us.

"We are also breeding some of our Poco King Tuck daughters to some outside stallions to try to raise some stallion prospects. We own a share of Paddys Irish Whiskey, and we have also bred some mares to 'Rooster' (Gallo Del Cielo) and to Wimpys Little Step."

Cowboys

In some ways, the job of a cowboy on the Waggoner Ranch hasn't changed much since 1903, when W.T. Waggoner gave Lige Reed that advice about how to manage the hands. It's still a horseback job that starts with each cowboy feeding his horses before breakfast, which is served at 5:30. Following breakfast, the horses are hauled to wherever that day's work is going to take place.

In Lige Reed's time, and until the mid '70s, the ranch sent a chuckwagon out with the cowboys each spring and fall when the major cattle work was in progress. The men slept and ate at the wagon, and it became their home for up to weeks at a time. The wagon no longer goes out. Each cowboy sleeps in his own bed at night, and either eats breakfast with his family or at the cookhouse at headquarters. Lunch, however, is normally brought to where the cowboys are working. The day usually ends when the remuda is penned in a rope corral, from which the wagon boss or straw boss ropes each cowboy's mount for the following day.

Because of the size of the ranch, more than 800 square miles, some cowboys live with their families in various "camps" around the ranch. Camps today are houses with a small barn and a couple of corrals, located in areas distant from the headquarters. Cowboys assigned to camps are usually responsible for the cattle on their areas of the ranch. They will, however, join the other cowboys at times of the year when large pastures are being worked, and others will assist them in their areas when needed.

All single cowboys live in the bunkhouse at the headquarters and take their meals in the cookhouse. Married cowboys not assigned to camps live in houses at the headquarters. They usually eat breakfast and lunch with the other men, but return to their homes for dinner with their families.

Camp men keep their strings of horses at their houses. All those living at headquarters have their horses turned out together in a horse pasture. When the next day's mounts are roped each evening, those horses are brought to the corrals at the headquarters so they will be easy to catch before daylight the next morning.

"Easy to catch" means it won't take much time. The Waggoners have never been ones to waste time.

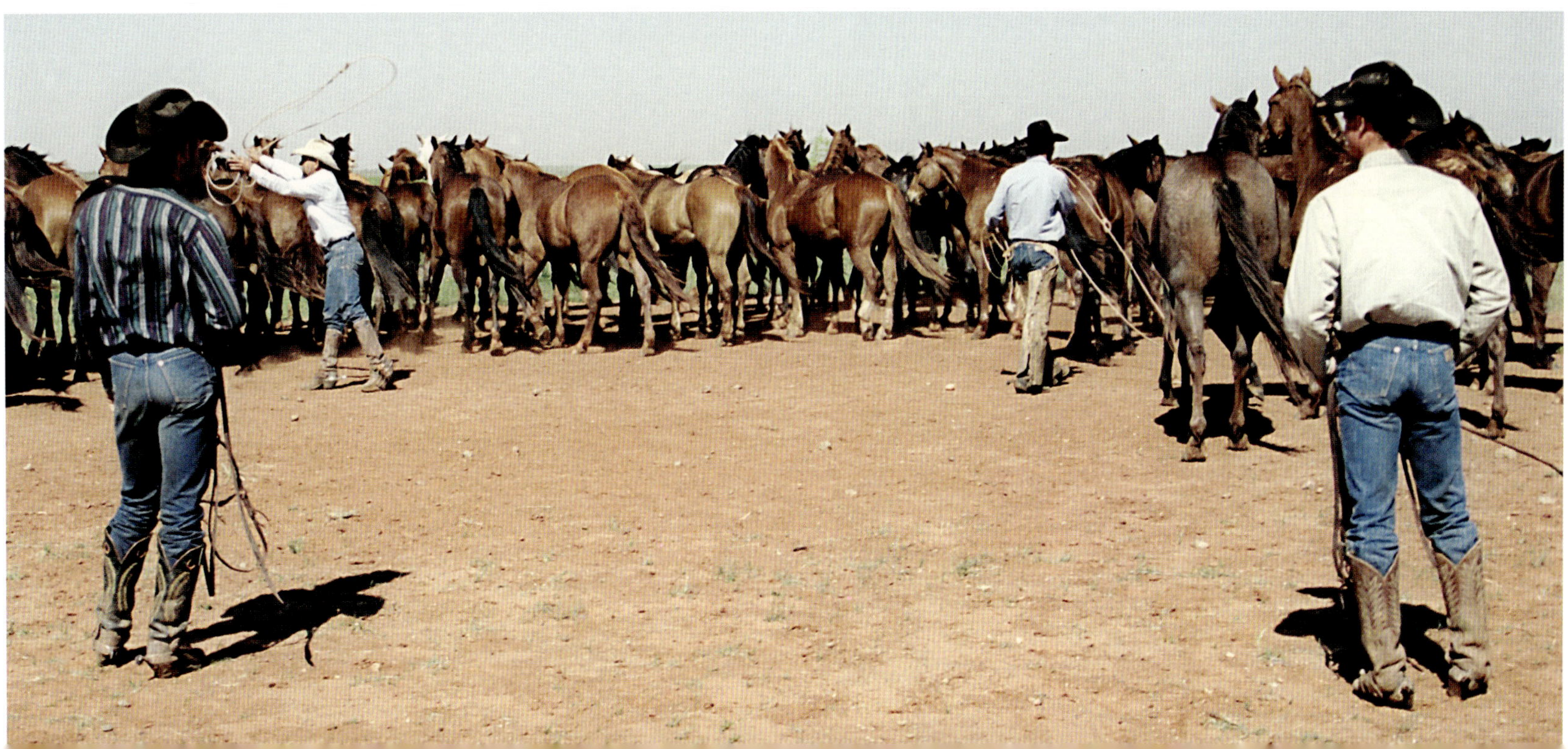

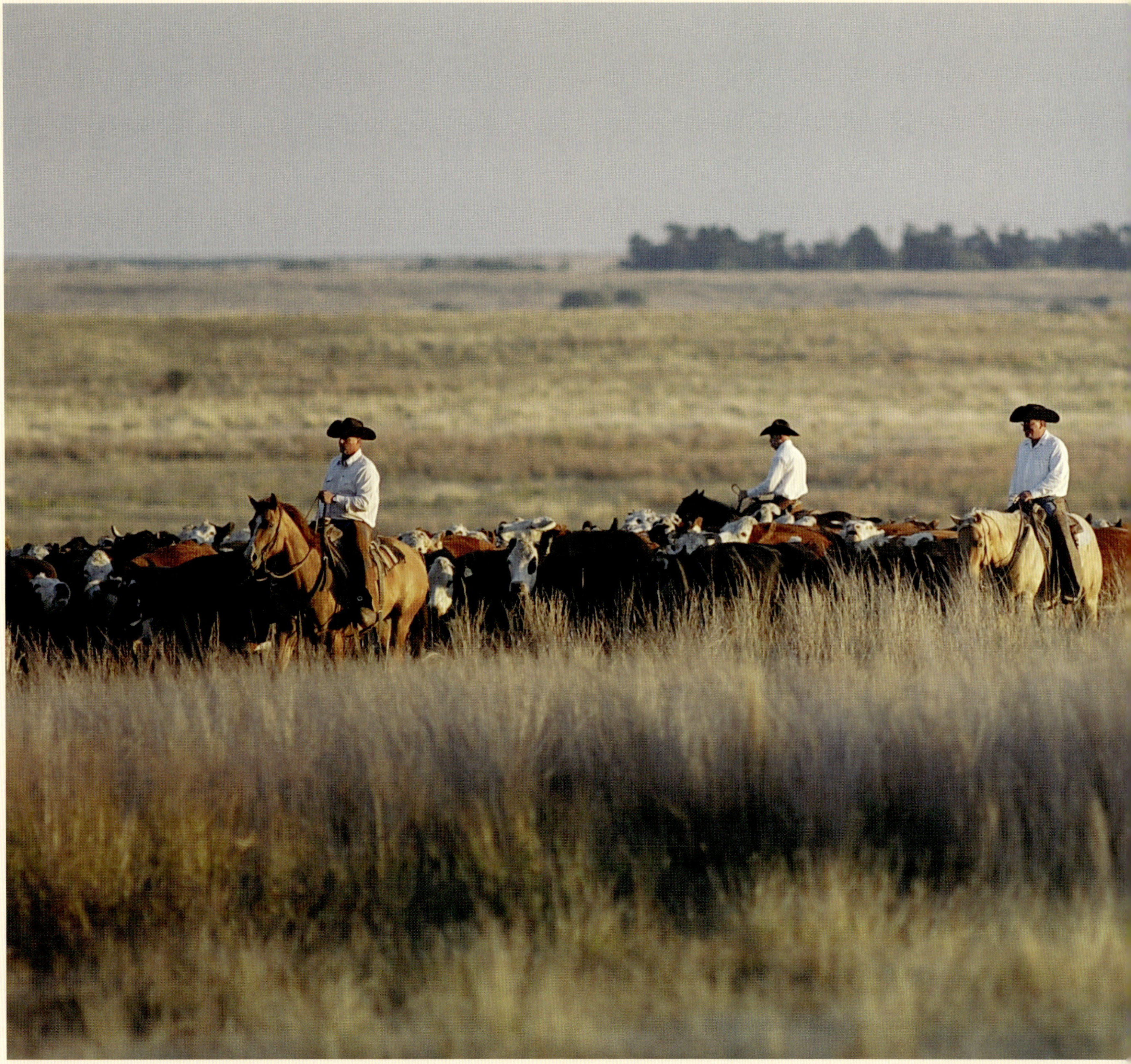

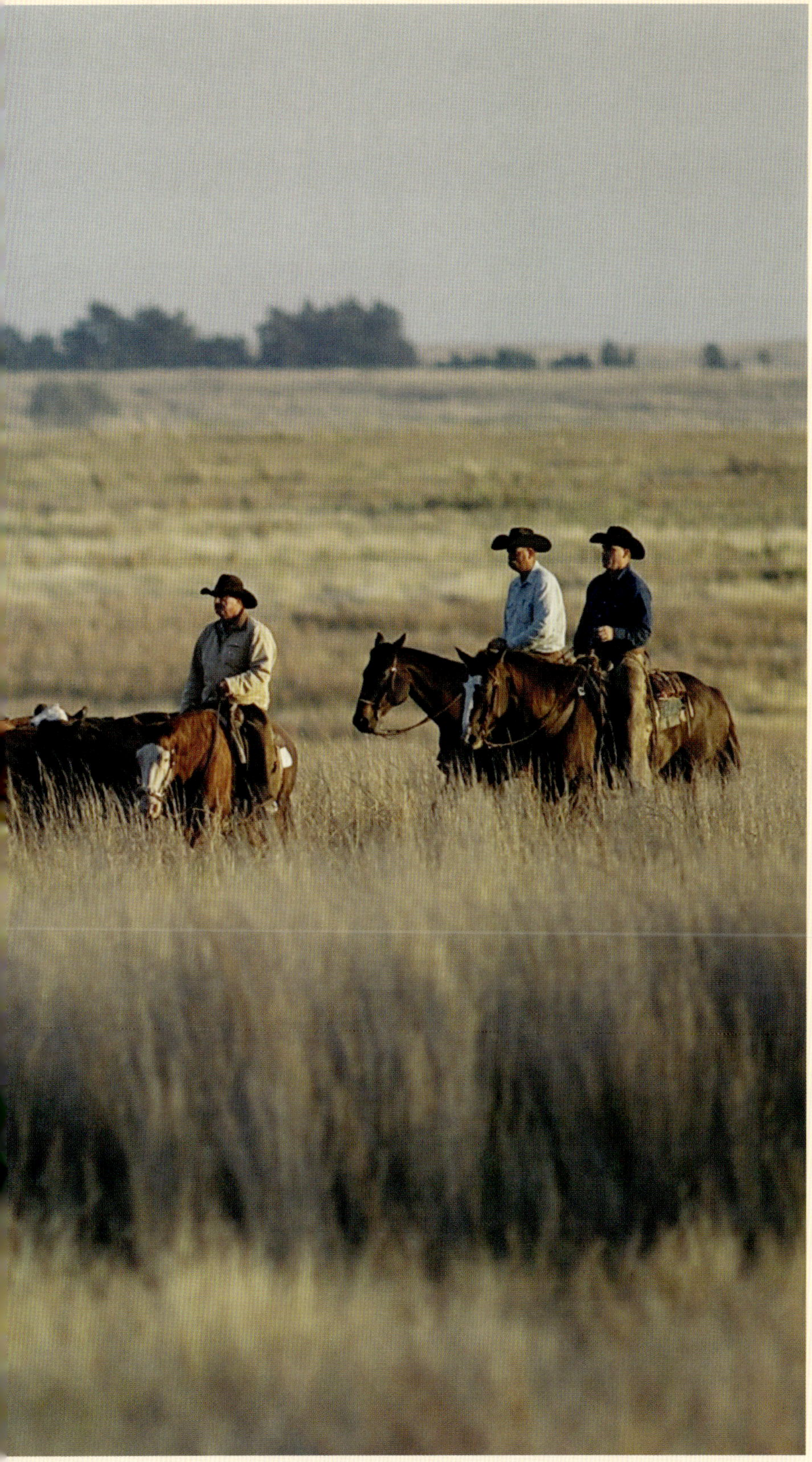

Stuart Ranch

Waurika, Oklahoma
1995

The Super Remuda

Stuart Ranch has a "super" herd of horses.

It was early morning, mid-October. One of those mornings when you needed a jacket early, but you shed it as soon as the sun had a chance to dry the dew on the grass. A mostly dark sky showed just a touch of pink—or maybe it was orange—hard to tell because the color was changing so fast. The sun was letting us know that it was heading toward the southern Oklahoma prairie, but it would be a little while yet before it arrived.

Stuart Ranch cow boss Bob Benton led the cowboys away from the barn at a trot, but the color in the sky did little to light their way. Although the full moon helped, it was still too dark to see much farther than a few feet before the blackness closed in. Bob, however, knew it would be daylight soon. It would sure be light enough to see the cattle by the time they reached the backside of the pasture.

The 8-year-old dun gelding Bob was riding trotted almost noiselessly through the tall grass. Head down, ears forward, he peered into the darkness. It was as if he was looking forward to the day as much as his rider was. He was the same way later, when he and Bob sat in the gate of the corrals, cutting the steers from the heifers. He was enjoying his job, one for which he was bred. He's a cow horse, and a good one.

The dun gelding was Zans Genuine Roper. By Genuine Doc, he was out of a daughter of the old Stuart Ranch mare Seven S Margarita. Now, that was a mare.

"Margarita," as she was called, had 15 foals before her death in 2005. Four of those picked up quite a record in various roping events, capturing all kinds of AQHA titles. One foal, Genuine Redbud, was a world champion and the 1995 World Championship Show Superhorse. But Margarita is not alone. There's a whole herd of good mares at Stuart Ranch. Let's just say it's a "super" band of mares.

The Ranch

Stuart Ranch traces its history back to 1868, when Robert Clay Freeny acquired some property near Caddo, Oklahoma, in the southern part of the state. His great grandson, R.T. "Bob" Stuart Jr., eventually inherited the property, and in 1991, expanded the holdings by purchasing about 24,000 acres 120 miles west of Caddo, near Waurika. That made a total of 40,000 acres, and allowed Stuart to change up his operation. However, the new acquisition not only offered the opportunity for change—it necessitated it. The Caddo property is rough and brushy, and requires a different management system than does the Waurika operation, which is more prairie. That's the situation Bob Stuart was facing when he decided to hire his daughter, Terry Forst, to help him make some decisions on how to run the two ranches.

Terry had grown up with a very real interest in livestock and ranching, and with those interests at heart, had acquired an animal science degree from Oklahoma State University and a ranch management certificate from the program at Texas Christian University. When Bob hired Terry, she was running her own consulting business, so she was prepared to step right in. She took a look, made some recommendations, and the change began. But those changes worked, and soon Terry's dad began to depend on her more and more. In the mid 1990s, she moved to the ranch at Waurika, and became an integral part of the operation.

"Dad had run into some problems with the cow herd," Terry says, "so he hired me to see if it would be feasible to operate the Waurika ranch in conjunction with the one at Caddo.

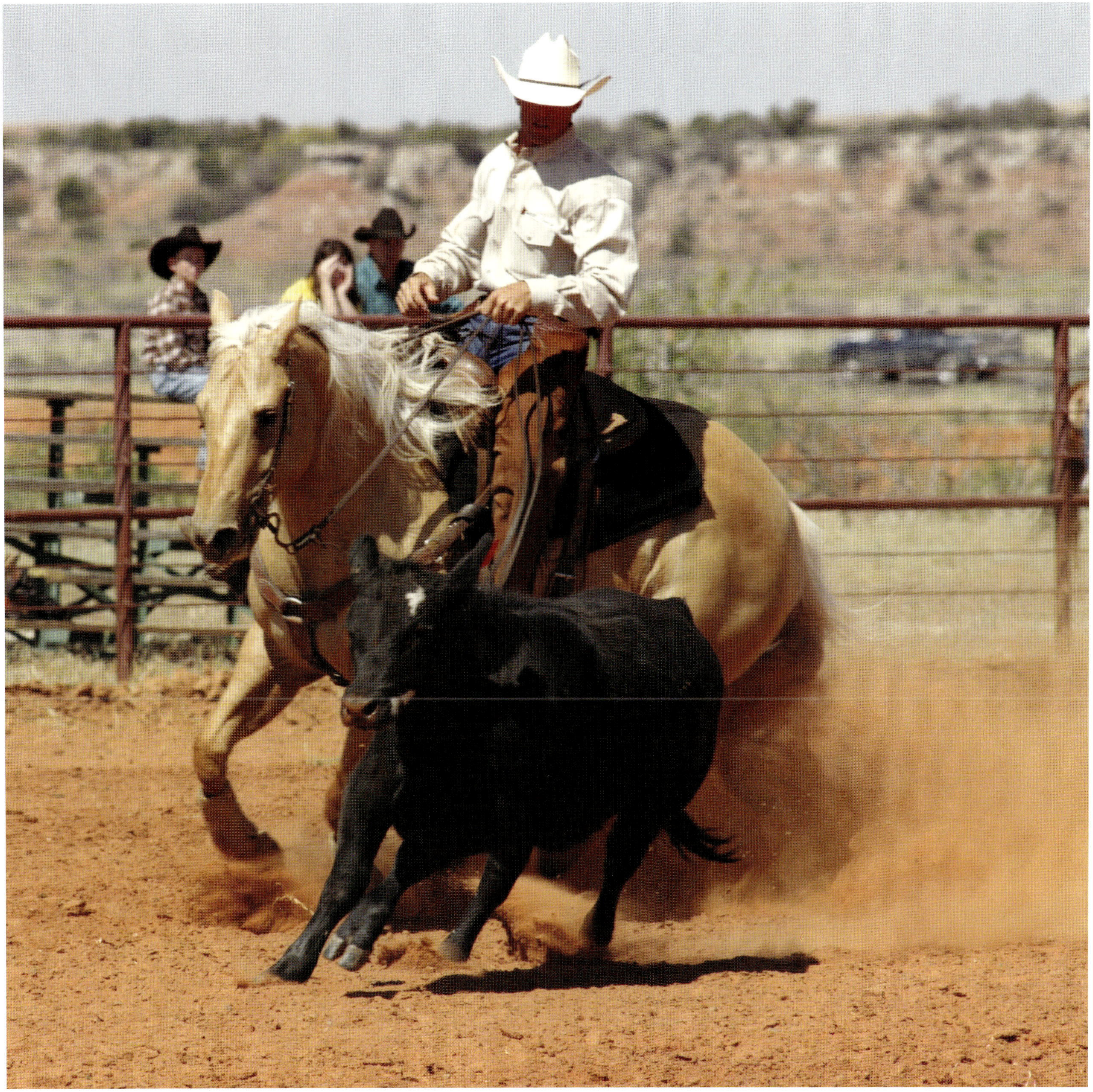

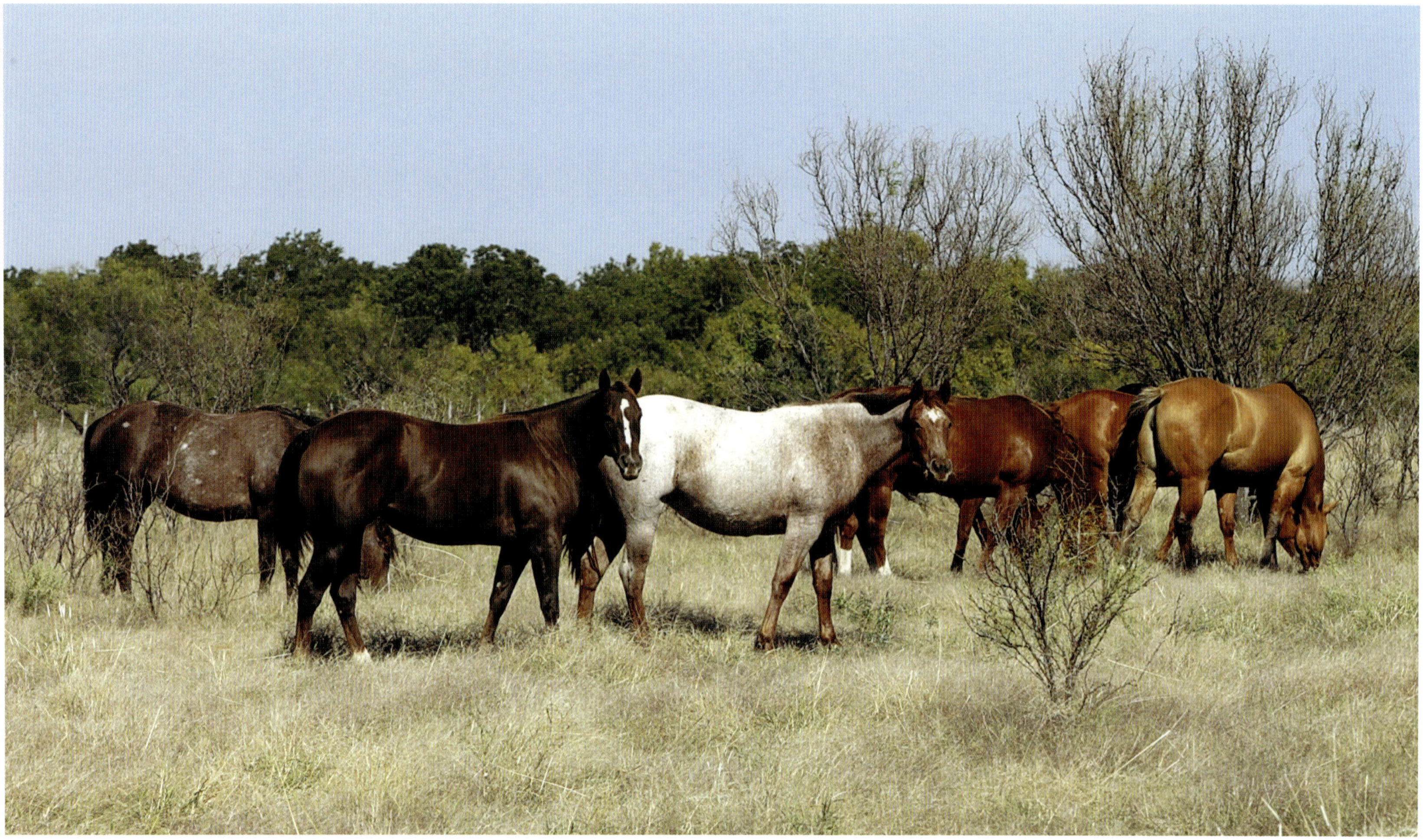

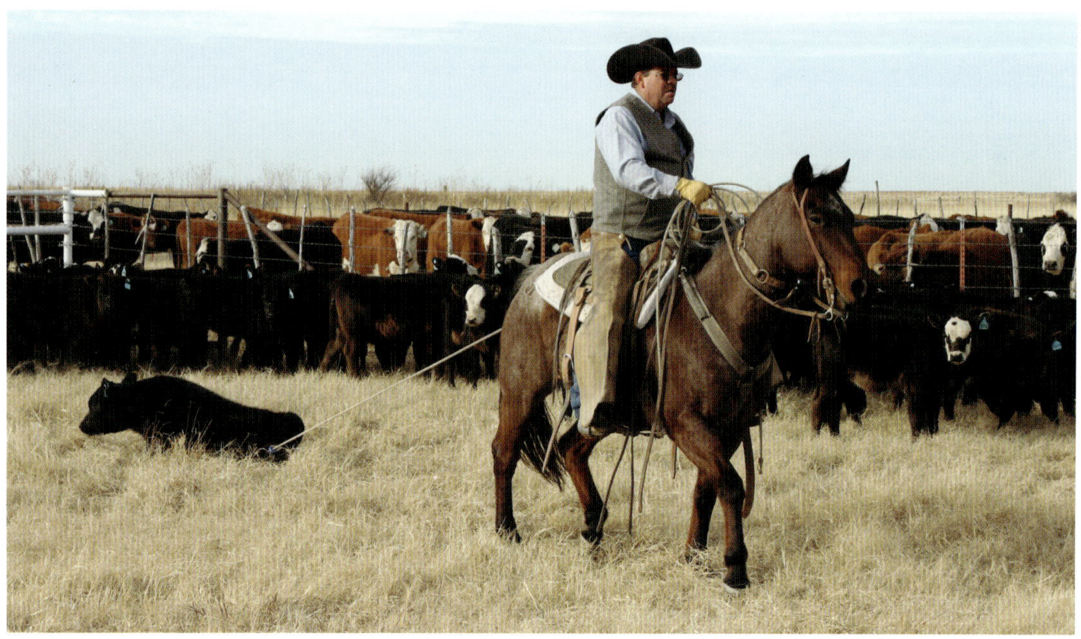

"Initially, he didn't want any help with the horses," she continues, "but that kind of evolved as we went along. We were pasture breeding the stallions at that time, and we started having problems with Zanaday (Seven S Zanaday). He wasn't getting a lot of his mares settled. So we decided to collect him and breed him artificially, and we got every mare in foal. Turned out he was culling his mares. He wouldn't cut them out of his herd; he just wouldn't breed them. And it was some of our really nice mares, like Margarita."

After that, Terry moved to the ranch and her responsibilities increased. Stuart Ranch had some of the best ranch horses anywhere—they had been honored with AQHA's Best Remuda Award in 1995—and foals that they raised and sold were being shown, and were really making a name for themselves, and for the ranch. Terry decided that maybe they should try to show a few themselves. In 1999, she hired Chris Littlefield to help them with the horses.

Raised in Texas, Chris had been a cowboy all of his adult life. He liked what he saw at Stuart Ranch. At the same time, Terry saw an opportunity in Chris and in his ability to help them with their program. Terry felt that where their horses fit the best was in the roping arena, and that fit Chris perfectly. But as the program developed, they began to look at a more versatile horse, one that could do more than rope.

"The horses we had were a little old fashioned," Terry says. "They were a little heavy fronted, and had some things that needed to be changed to fit today's style of horse. So we started looking at that, but at the same time, we didn't want to compromise the things we think are very important, like structure, bone and athletic ability. Our horses have always been able to go all day, whether they are dragging calves or sorting bulls, and we wanted to keep that and improve some of the other areas."

The Stuarts' goal was to have a ranch horse that could do everything necessary on the ranch, but still compete in roping and working cow horse. They were already breeding some mares to some outside stallions, including Carol Rose's Genuine Doc and Shining Spark, and Dick Pieper's Playgun. In 2000, Terry's dad, Bob, with the advice of Terry and Chris, bought a 3-year-old son of Playgun by the name of Real Gun. The trio thought the young stallion, out of a War Leo-bred mare, might be able to start them in the direction they needed to go.

Chris was really high on Real Gun because he had been a big fan of Playgun. Trained to be a cutting horse, Playgun had won more than $170,000 in the event, but Chris says, "The first time I saw him, I was thinking, 'Let's rope calves and tie steers on him.' He was bred to be a cutter, and he did it very well, but I was just thinking what a nice calf horse he would be."

Chris had the same thoughts about Real Gun, and as soon as Stuarts acquired him, Chris started showing him in Ranch Horse Cutting Association and Ranch Horse Association of America events with some tremendous success. Plans to start developing the horse operation were set in motion.

Bob, however, died in September 2001, and everything then was left up to Terry.

"Dad was well aware of where we wanted to take the ranch," Terry says. "We had everything in place when he died. The indoor arena was under construction, and our overall plan was underway. If he had just lived another two years, he would have been able to see how everything was going to work."

Everything is working, and the selection of Real Gun has helped make it so. Under Chris' direction, Real Gun has won a number of titles in both AQHA and National Reined Cow Horse competition. In 2004, he won the AQHA World Show Superhorse title, making two superhorses that Stuart Ranch has either bred or owned.

The Mares

But it's that super band of mares that forms the backbone of Stuart Ranch's horse program, going all the way back to a 1961 mare called Miss T Stuart.

Stuart Ranch at that time owned a stallion called Son O Leo, and the cross of him on "Miss T" began a legacy of good horses that extends yet today. Among Miss T's 20 foals are AQHA Champions Seven S Sunset and Seven S Valentine. Seven S Taffey was also an AQHA Champion, but because she earned more than 50 points in western pleasure, was designated as a Superior western pleasure horse. Miss T was also the dam of Superior halter horse Seven S Cinnamon, and she foaled Seven S Suzanna, who was Superior in both heading and heeling. And then there was Seven S Sunshine, whose foals earned more than 800 AQHA points. And that brings us to Seven S Margarita.

Oh, yes, Margarita. Not only did she produce world champion and Superhorse Genuine Redbud, she was also the dam of Genuine Hombre, who has Superior titles in reining, working cow horse, heading, heeling and tie-down roping, among other awards. And she produced Genuine Jezabelle, who's been to the World Show in both heading and heeling, and is the dam of Seven S Genuine Zan, World Show qualifier in heading, heeling and working cow horse. And to mention one more, there was Seven S Genuine Miss, who has Superiors in heading and heeling, and, under Chris, was the high-point versatility ranch horse the first year the event was an AQHA-approved class.

Yes, Stuart Ranch horses have made a name for themselves, and it's that legacy that Terry is determined to continue. Although there are no more Son O Leo mares in the broodmare band, there are some daughters of Port Command, a stallion by Commander King that Stuarts owned, and some Seven S Zanaday daughters. Seven S Zanaday was a son of Zan Par Bar that Stuarts bred and showed to a world championship in heading.

Daughters of their matings to Shining Spark are in the broodmare band, and they have purchased a couple of Peppy San Badger daughters, as well as one by Dual Pep and one by Grays Starlight.

"We want Real Gun to have a chance at some mares that could really get his career started," Terry says. "But we have to go look for these matches and really do our homework. We want to keep the genetics fresh and our program current, but we like size and substance in our mares. That way, if the foals don't make it in cutting or reined cow horse, you can still go ride them all day on a ranch and do whatever you need to do."

The Cattle

Unlike many of the Best Remuda ranches, Stuart Ranch is involved in the show arena with its horse operation. However, it is a ranch first, and with 2,500 mother cows, ranching is a full-time job that requires a number of horses.

The cow herd consists of Herefords and black baldies, and probably the biggest part of the business is producing replacement commercial heifers for sale, using Angus and Hereford bulls. The cows are split up, with about half of them calving in the fall and half in the spring. Spring calves go to wheat pasture when they are weaned the following fall, and the fall calves go straight to the feedlot when they are weaned in the spring.

"We utilize both ranches with our cow herd," Terry says. "At the Waurika ranch, we raise our cows, get them up to 4 or 5 years old, and get a couple of calves out of them. Then when they are big, mature, strong cows, we send them to the Caddo ranch, and they will last until they are 11 or 12, some of them even to 13. The two ranches are completely opposite, but they complement each other. We can do some things over there that we can't do here, and vice versa."

The horses on Stuart Ranch are bred to work cattle, and that's what they are used for, whether it's in the arena or in the pasture. Some of those at home in the pasture are full sisters and brothers to those that are gone to a reined cow horse show, and they might be just as good.

Terry says that even those that are destined for the show arena get their start in the pastures, working cattle.

"The colts always go to the pasture," she says. "The 2- and 3-year-olds will always go. We might not take the finished horses as much, because they don't need to go. But what we're doing with our horses is what ranch horses are bred to do. They're working cows, regardless of where they are."

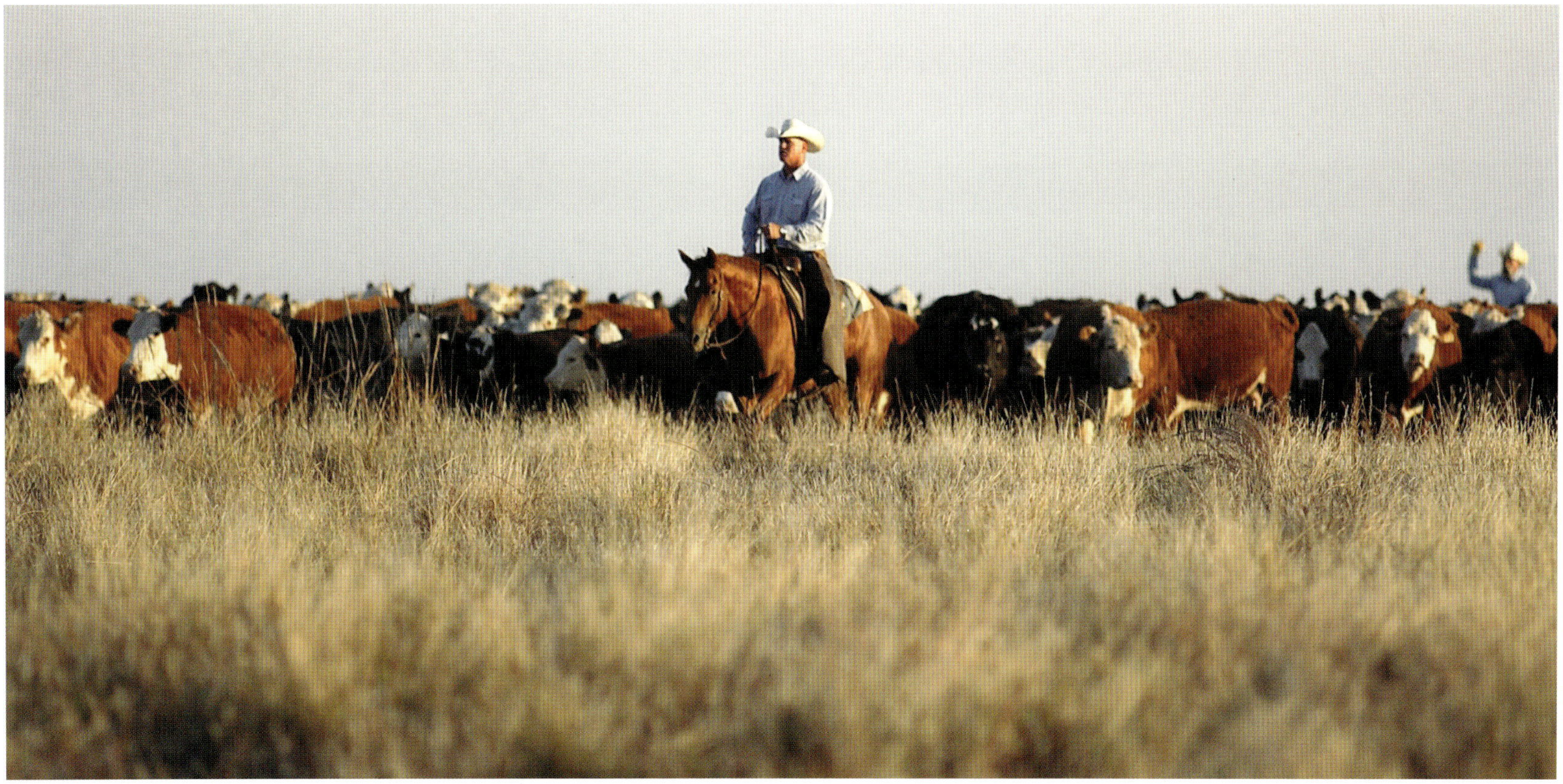

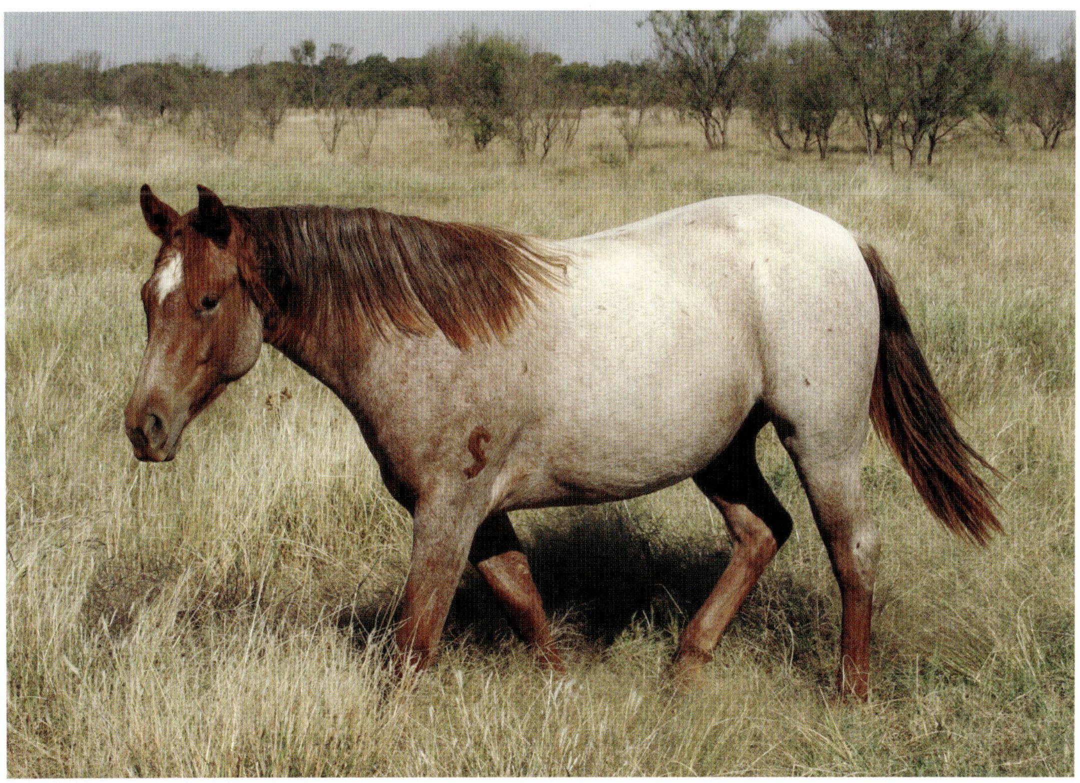

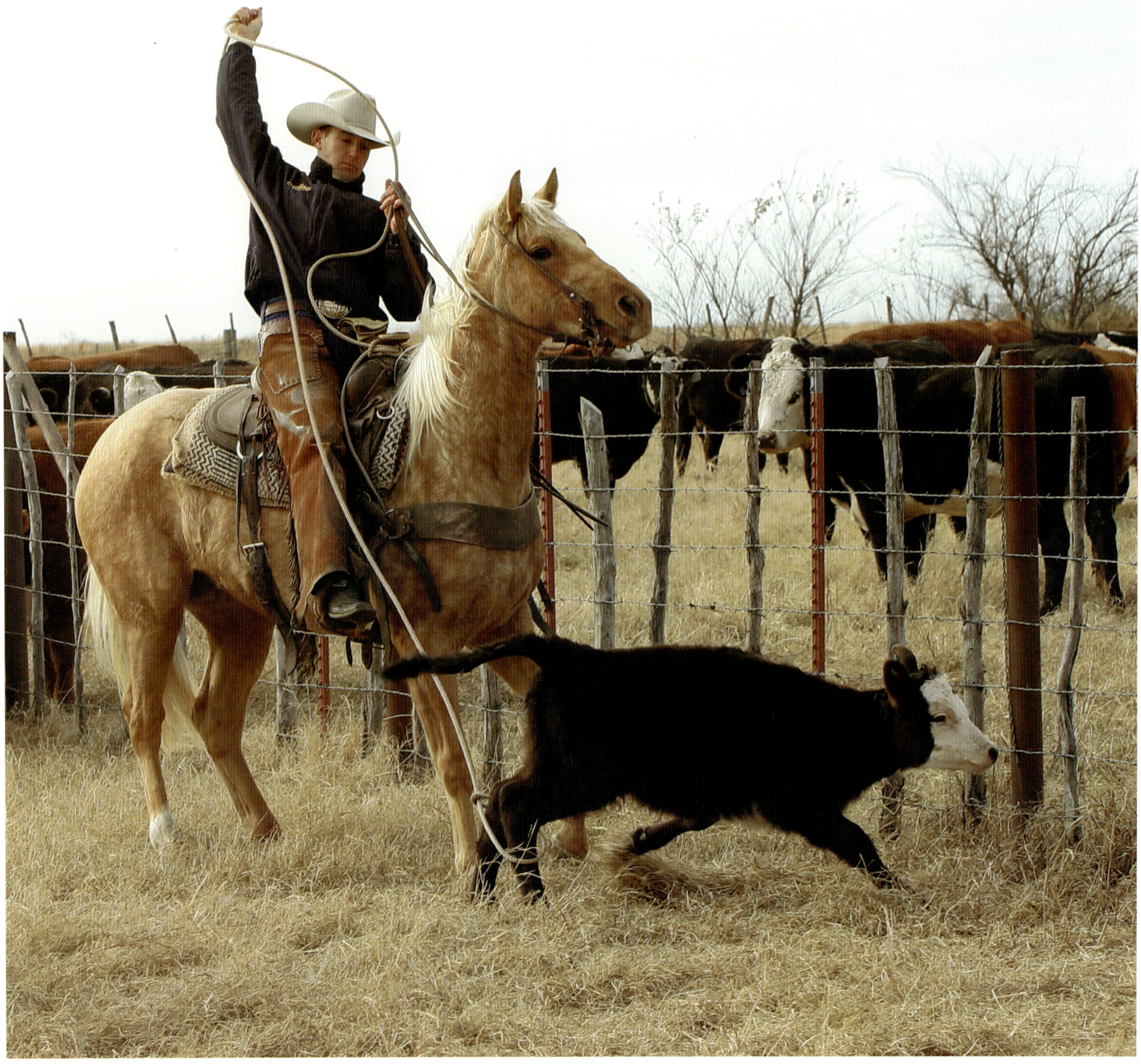

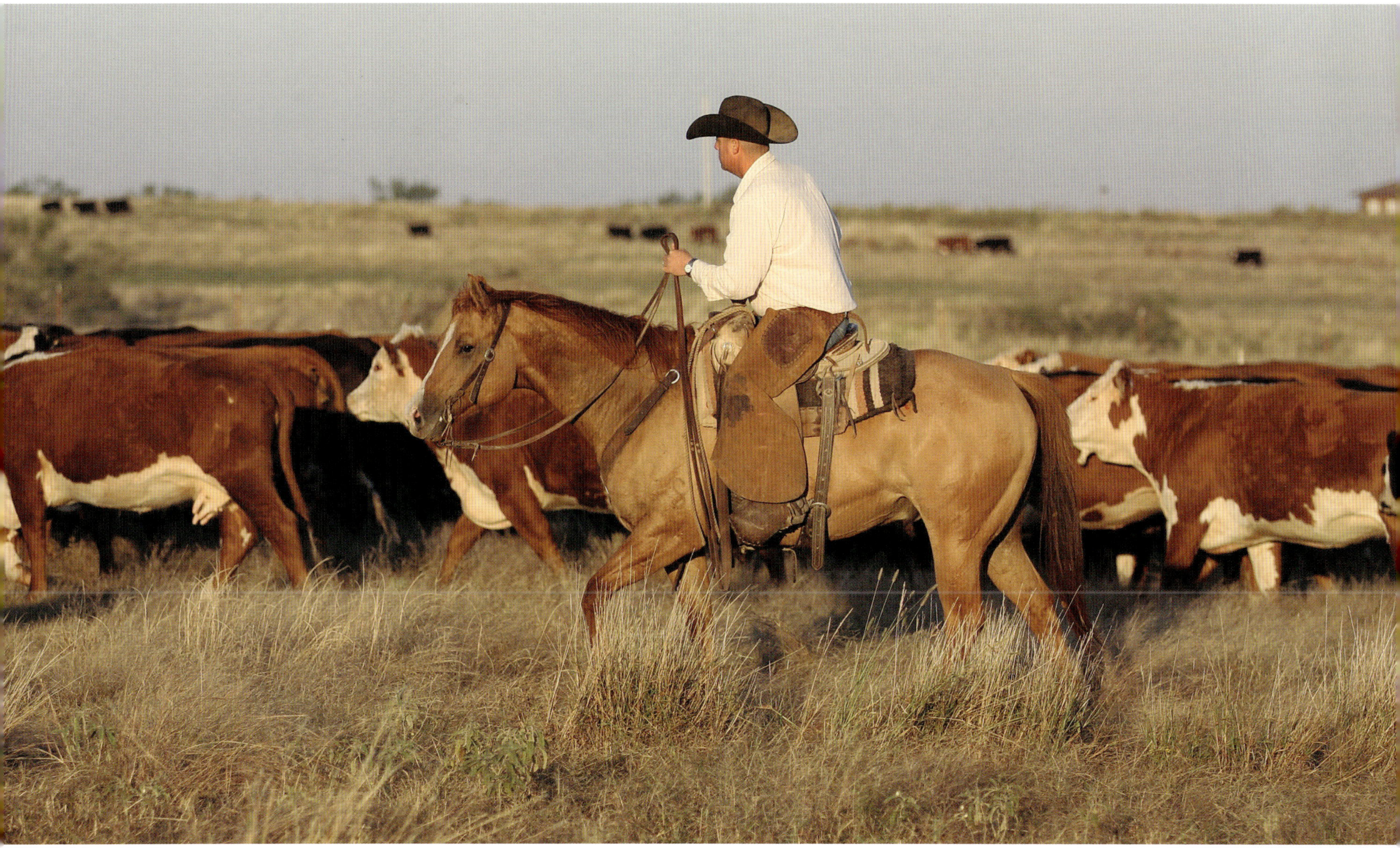

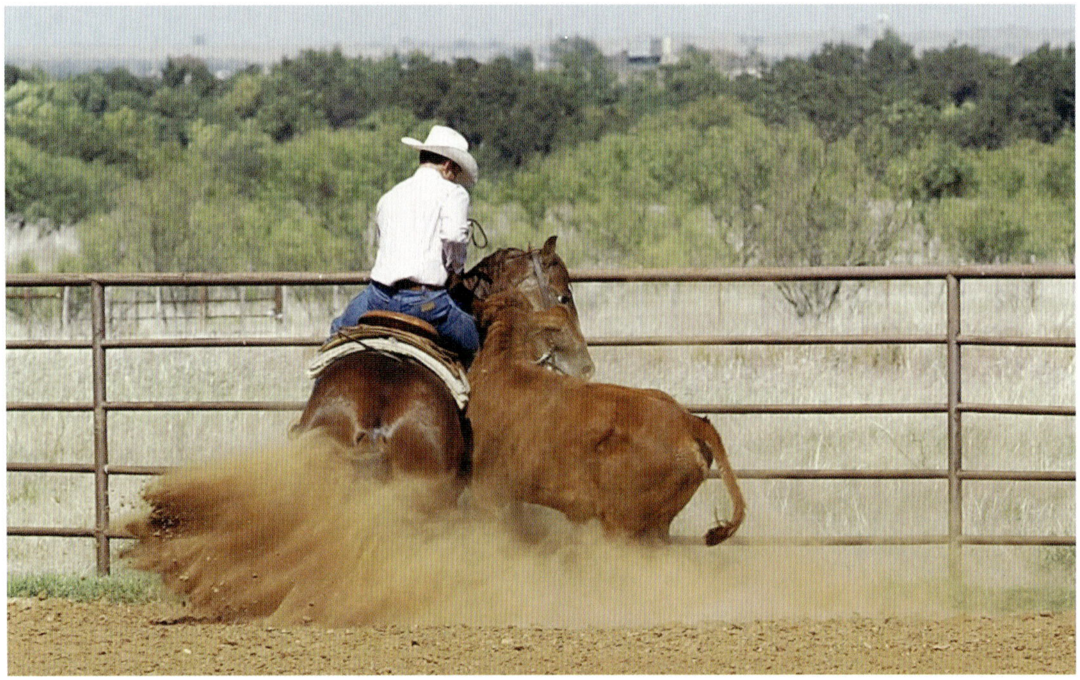

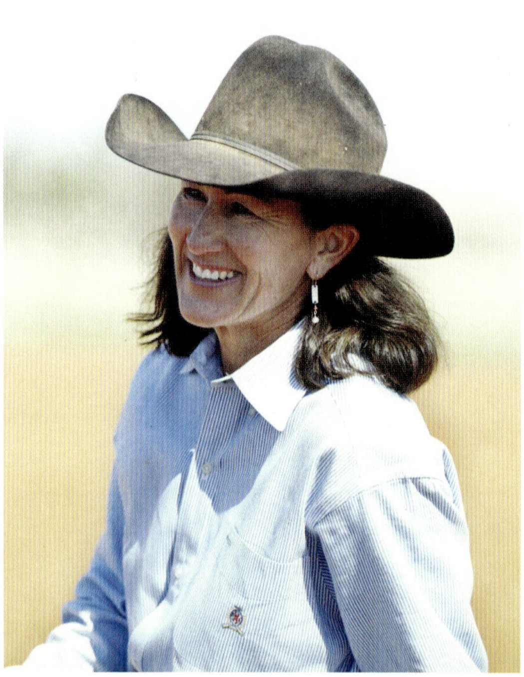

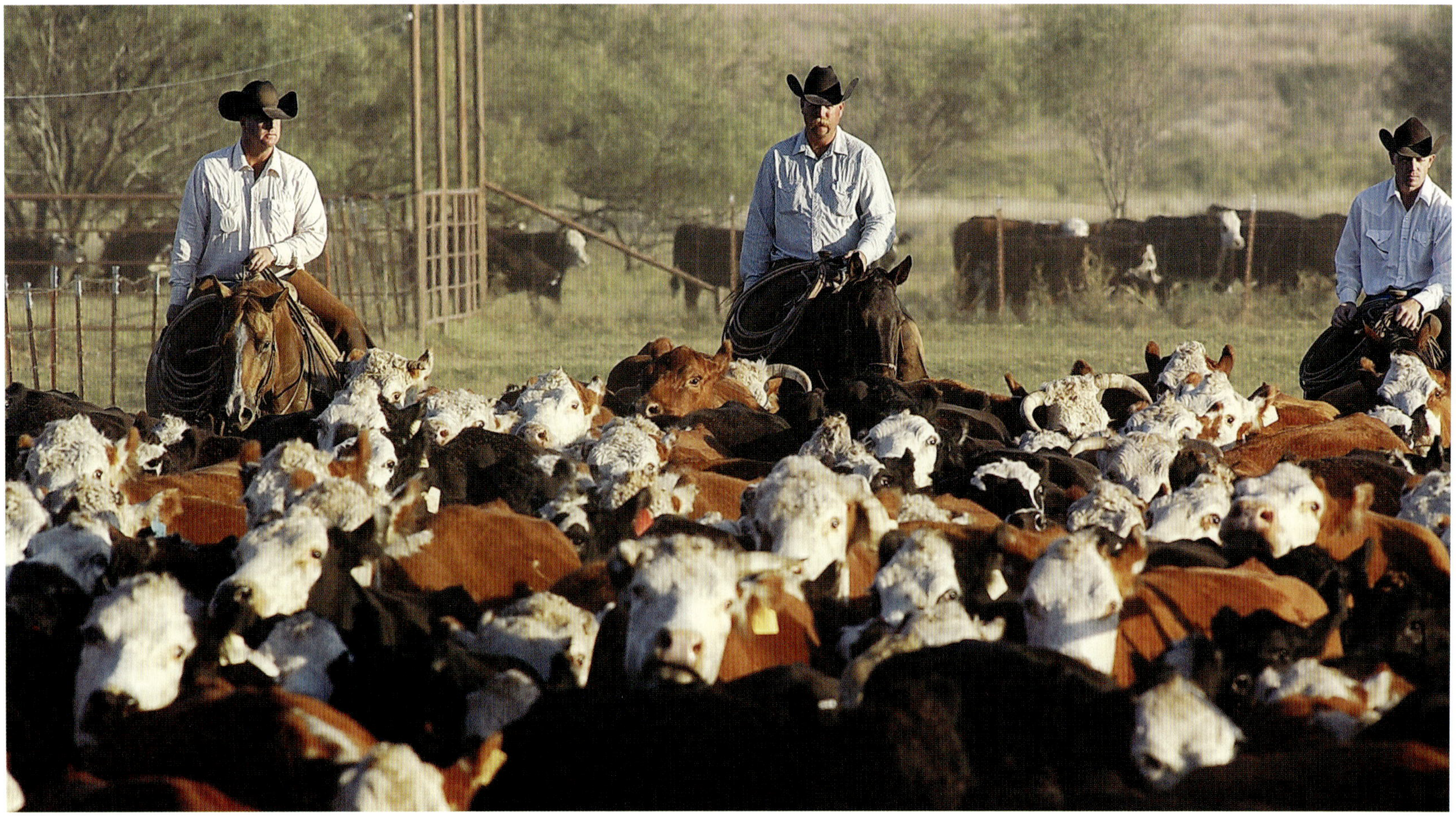

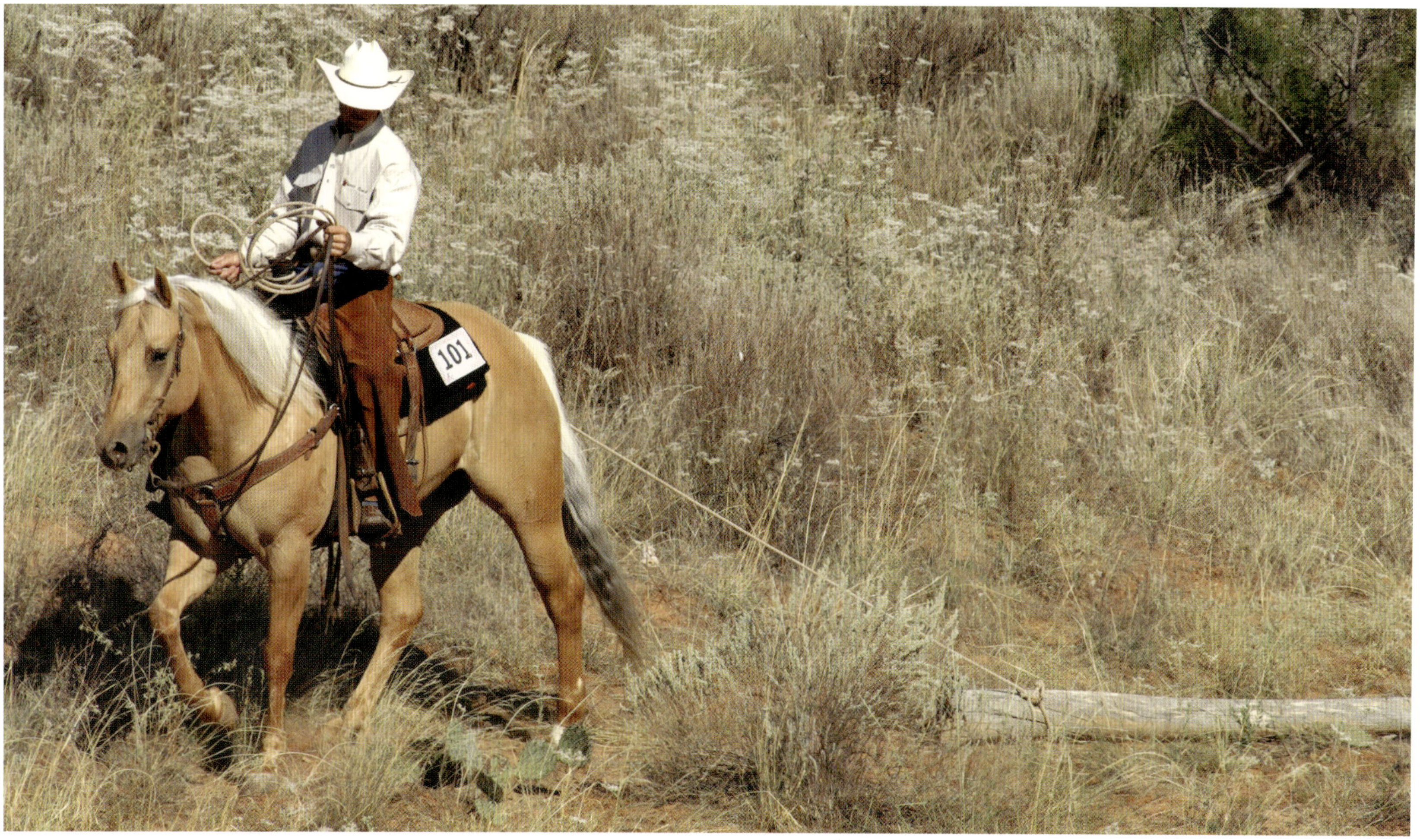
101

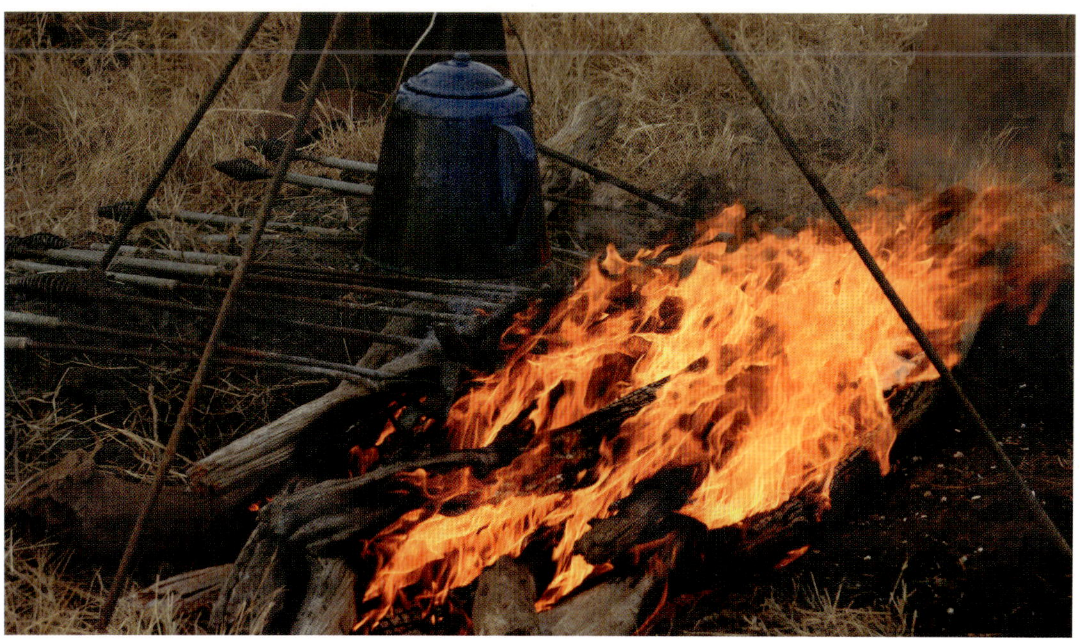

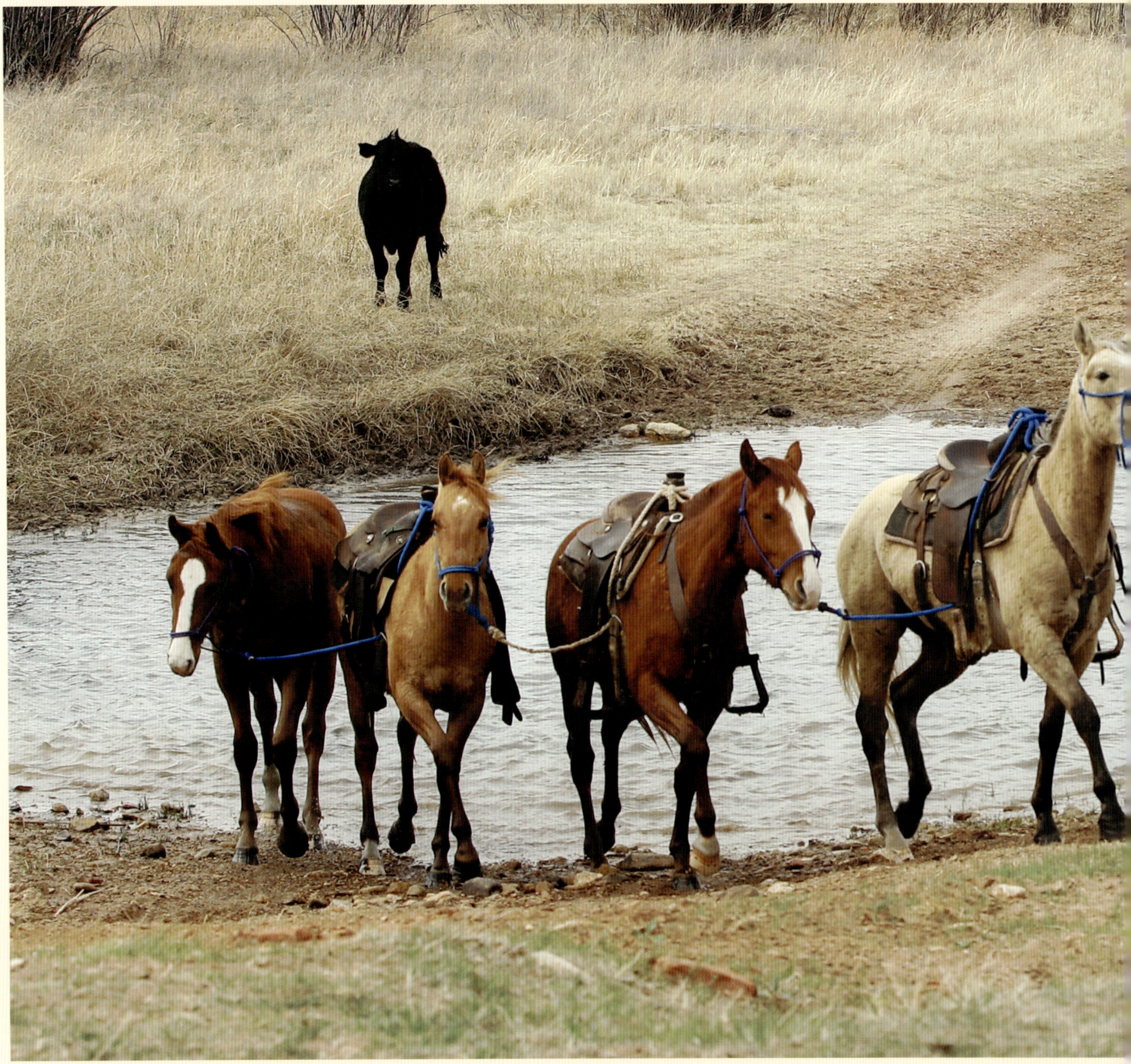

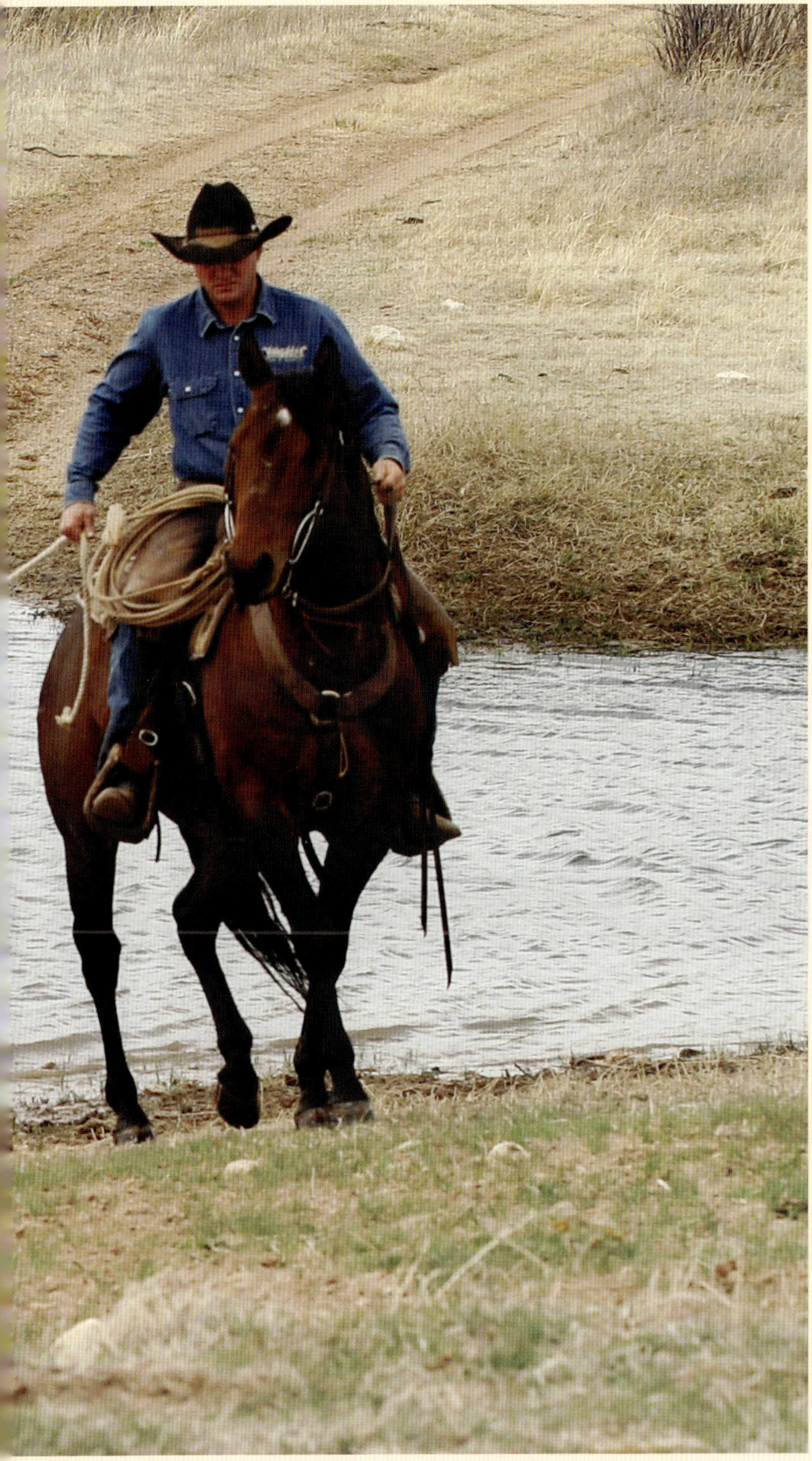

Bar B Ranch

Beaver, Oklahoma

1996

Cow Horse Country

The Bar B Ranch in the Oklahoma Panhandle has been raising good horses for more than 100 years.

Otto Barby got his first taste of cowboying in the Oklahoma Panhandle in 1879, at the ripe old age of 13. During the next few years, he worked for some of the biggest ranches in existence at the time. Ten years later, after a dispute over a land claim in central Oklahoma, he returned to the Panhandle and made it his home.

"Someone had jumped his claim," says Stanley Barby, Otto's grandson, "so rather than get into a gun fight over it, he just left and came back to the Panhandle."

But Otto ended up with a claim anyway. Through some negotiation and trading, he purchased a relinquishment—where someone had filed on a claim and then turned it back to the government—on the banks of the Beaver River in what is now Beaver County, Oklahoma.

"He then married the neighbor girl," says Stanley. "Over the next several years he acquired more and more land, and gradually built his ranch.

"He was an agent for a Kansas City bank," Stanley continues, "and during the Depression, there were lots of people who couldn't pay their notes. He never foreclosed on anyone, but when someone would starve out and leave, and just turn everything back over to the bank, he would assume their note. Then, he formed a partnership with his three sons, and they ended up with about 150,000 acres here in the Panhandle."

Stanley, through his father, who was one of Otto's sons, inherited a portion of that, including the original homestead. Today, Stanley and wife Junetta, along with their son and daughter-in-law—Rodney and Robin Barby—and daughter and son-in-law—Debbie and Jim Hilton—ranch in Beaver County in the Oklahoma Panhandle. As a matter of fact, Stanley lives in a house located at the original Bar B Ranch headquarters.

Otto Barby and Sons was running a cow/calf operation back in the beginning, and with as much country as it was operating, it took a lot of horses. Probably the first big move they made in the horse business was to make a deal with the Army and get in on the remount stallion breeding program. Many ranches in the West participated in this same program, and it consisted of the Army furnishing the various ranches one or more Thoroughbred stallions to breed to the ranch's mares. The Army, then, would pick a certain amount of the colt crop to use as future cavalry mounts.

"They had to keep those colts until they were 4 years old," Stanley says. "Then the Army would come out and sort them, and take what they wanted. The ranch got to keep the rest. So they rode what the government rejected."

Those rejects weren't all bad, though, and in the 1940s, the Barbys paid a visit to Walter Merrick, who was pretty proud of a Quarter Horse stallion he owned named Midnight Jr. They bought a son of Midnight Jr named Proud Johnie, brought him home and put him on the mares they had raised by the remount stallions, and they were in the Quarter Horse business.

They still are, and just like when Otto Barby and Sons was in business, raising cow horses is the primary goal. As a matter of fact, every colt raised on the ranch gets his start as a cow horse.

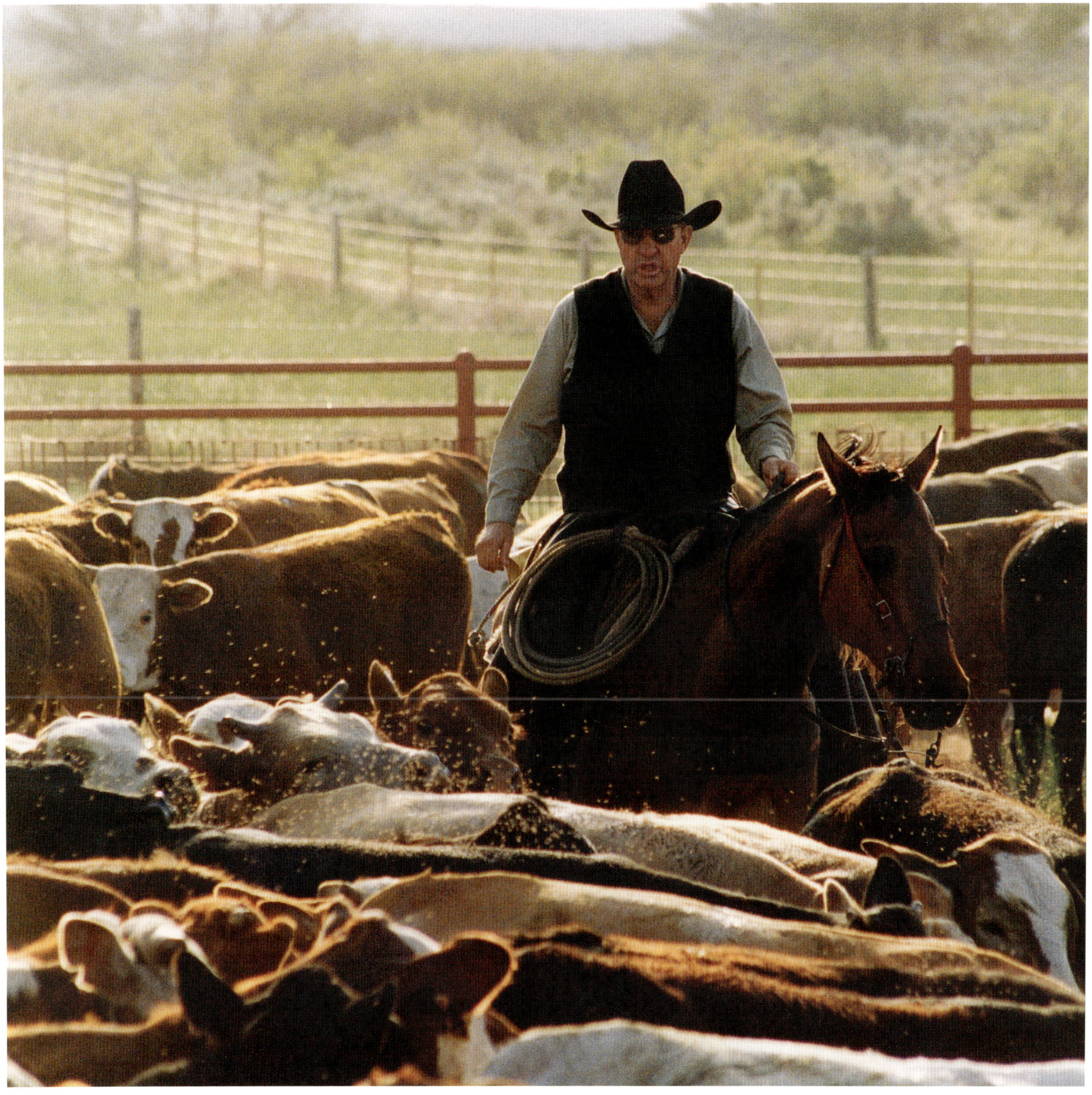

"We start all our horses in the round pen," Stanley says, "get them to where we can rope something on them, and then take them outside. That's where we really get them broke. We do a lot of doctoring in the pastures, which means the boys do a lot of roping on these young horses. They get jerked around, and they find out what it is to have to chase one of these yearlings. Then Rodney goes through them, and the ones that look like they have the ability are taken to the arena and we start team roping on them.

"We don't necessarily try to finish one," Stanley continues. "We just get them going and sell them as good prospects."

Rodney adds, "We have sold some pretty nice finished horses, but it takes a lot longer, and you have to haul them. We're better off if we rope on them 30 days, just get them started, and then let someone buy them. We get them to where they are really gentle, and working well in the box, and then someone can come get them and do what they want to do on them."

One of the outlets for their horses is with cutting horse trainer Shannon Hall, who rides them and markets them as turnback horses.

"Of course, these are not finished cutting horses," Rodney continues, "but they handle well, and you can take them then and go any direction you want to on them."

Not all the horses raised are sold, however. The Barbys are in the cattle business, and the cowboys' needs on the ranch are taken care of first. Each of them has five to six horses in his string.

Although Bar B Ranch started out as a cow/calf operation, the business today has evolved into strictly stockers. In a good year, the Barbys will handle about 20,000 head, getting cattle in and shipping every month. Calves arrive weighing from 400 to 600 pounds. They are then turned out on either grass or wheat—depending on the time of the year and how much rainfall the ranch has had. After the cattle reach about 800 pounds, they are shipped to a feedyard.

"We quit the cow operation kinda slowly," Stanley says, "and worked our way into the yearling operation as we began to buy some more land and put everything together like it is today. The yearlings fit us a little better. You can definitely prepare for drought a little easier—in a dry year, you just don't get in as many stockers. When you have a cow/calf operation, you have to figure out what you're going to do with those cows when you don't have enough grass. The yearlings give us a little more flexibility and enable us to limit our risks.

"We rotate those cattle around, depending upon their weight, and that, along with the doctoring, means lots of horseback work."

The Barbys breed about 50 mares a year to five different stallions. Their broodmare band primarily reflects the breeding of Playgun, Rocket Wrangler, Grey Badger, Peppy San Badger, Joe Hancock and Driftwood, but there are also some mares by the well known Tanquery Gin and Jackie Bee, and some by The Calico Prince, who's by a Thoroughbred son of Native Dancer out of a Rocket Wrangler mare. The Calico Prince has AQHA points in halter, working cow horse, tie-down roping, team roping and barrel racing.

"We think Playgun puts some cow in our horses," Rodney says, "while Rocket Wrangler gives them the endurance, some stretch in the forearm and some stride. Grey Badger and Peppy San Badger put the grit to them, and the Hancock and Driftwood breeding gets us some big bone."

The stallions they are using today include Bar B Red Kiowa, who's by a Poco Tivio-bred horse and out of a daughter of Kiowa Thad. Stanley raised Kiowa Thad, who's by a son of Zantanon H and out of a daughter of Kiowa Dusty, who Stanley showed very successfully back in the 1970s in both cutting and reining. They are also breeding Bar B Top Gun, a 4-year-old stallion by Real Gun out of a Zans Misty Gold mare, and JR Driftwood Spice, a 2-year-old double-bred Driftwood stallion that they are trying on a few mares.

Those stallions are all owned by the Barbys, but they have also leased a couple of stallions: Jaysons Dunit, by a son of Hollywood Dun It, and Four Figure, a Haythorn Ranch stallion that goes back to Joe Hancock and Driftwood.

Lonestars Pistolero, who's by Playgun and out of a Colonel Freckles-bred mare, and the 3-year-old Joes Country Tune, a double-bred Two Eyed Jack stallion, are both owned by the Barbys but are being used by neighbors on a share deal—Barbys get some of the foals. Then there's Rattlin Ron, who's by the Peppy San-bred horse Gray Badgers Tip and out of a mare that goes back to

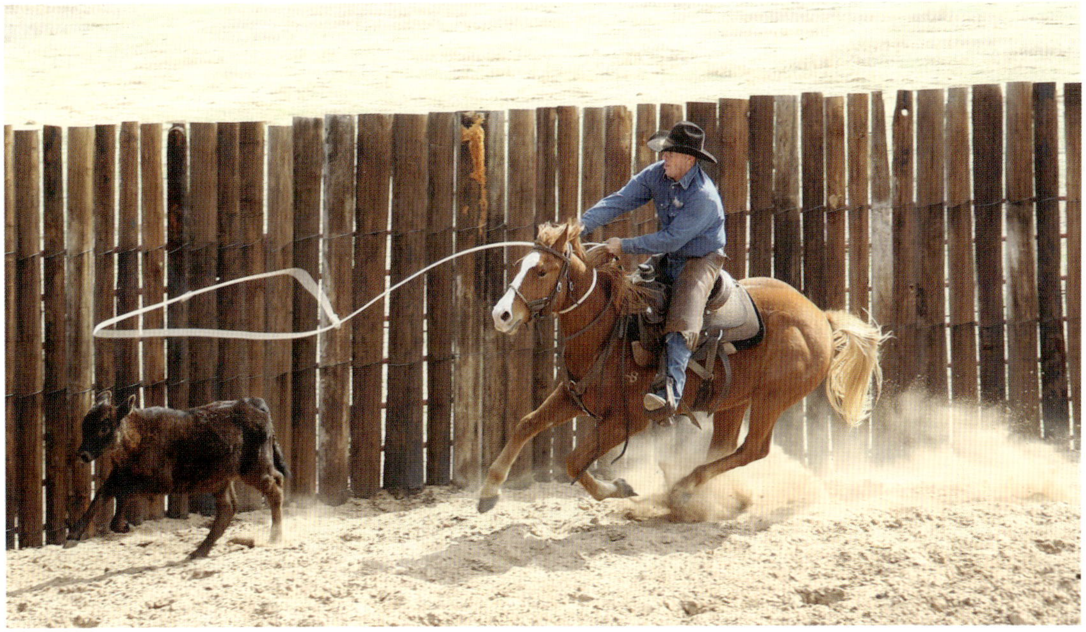

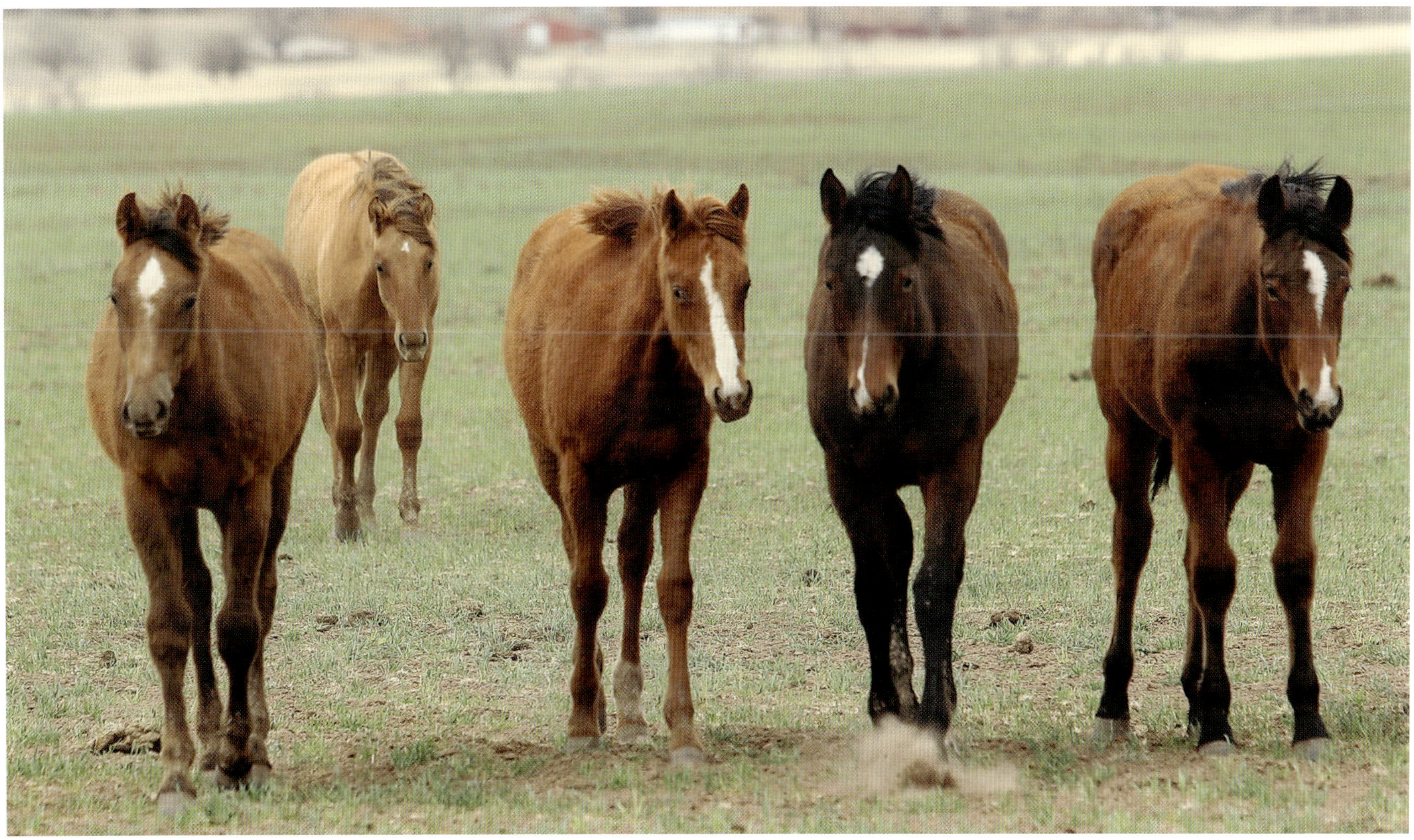

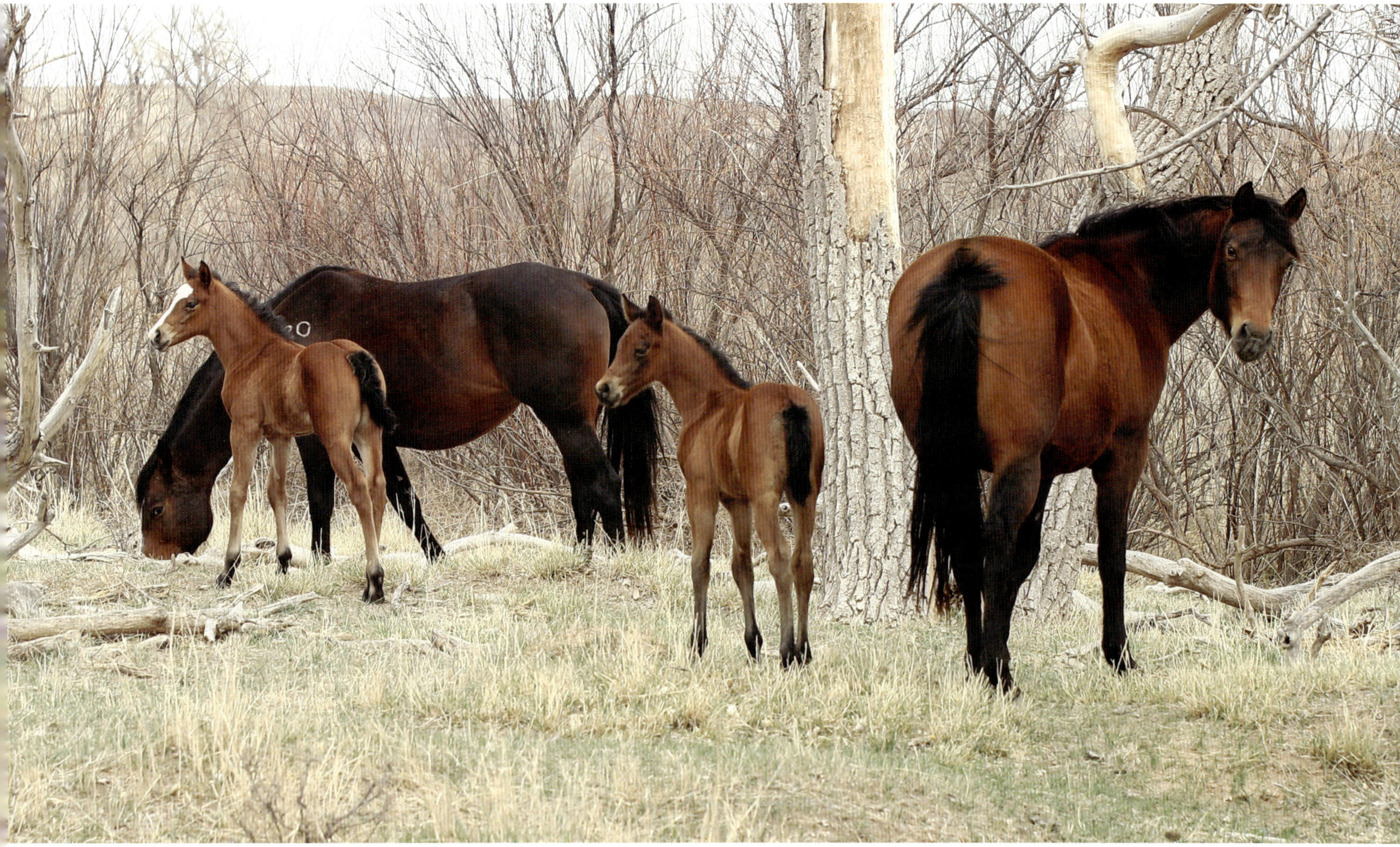

Kiowa Dusty, and who was loaned to Redlands Community College at El Reno, Oklahoma.

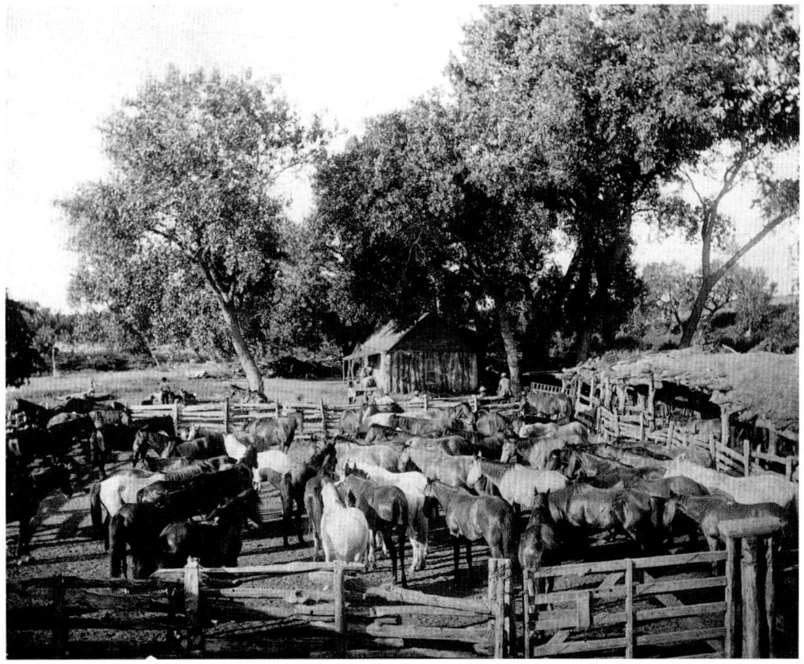

"On those stallions that we loan out, we have them if we need them, but we can farm them out if we don't," Rodney says. "We do that instead of selling them and then later on wishing that we still had them. And after breeding season, we have them here on the ranch to ride if we need them while we're working cattle.

"We try to get our horses anywhere from 15 to 15.2 hands," Rodney continues, "and we try to hit about 1,250 pounds. It seems like everyone wants a 15-hand horse. He's big enough to take a jerk, but he's not too big to get on."

Each of the stallions is turned out with a band of mares about the third week of April and picked up in early July. The foals are weaned about the first of October, but that's not the first time they are handled.

"We work with the colts quite a bit while they're little," Rodney says. "We get each set of mares and colts in during the summer, work with them for two or three days, and then turn them back out and go on to the next set. By October, you can sure walk up to them and touch them, do whatever you want to do."

"We think we are on the right track with our mares right now," Stanley adds. "The colts we're getting have good backs, good legs, good feet and a good attitude. Rodney works on that attitude, because he's the one who has to get on them in the round pen."

The colts are usually started during the holidays at the end of the year. Rodney says that the action from the cattle buyers tapers off from around Thanksgiving until after New Years', which gives them plenty of time to ride young horses. Most of the colts are about 1½ years old when they are first ridden, but they are worked for only two to three weeks, and then turned back out. That process continues periodically until their training really begins.

"It seems like that periodic rest really helps their minds respond," Rodney says. "It attaches what you're trying to teach them. We think that's better than just pounding them for three months straight, where they get run down."

Although the Barbys have sufficient numbers of horses that they choose to ride only geldings for the cow work, the fillies are started and ridden for about 30 days before Rodney decides which ones to keep for broodmares and which to sell.

Rodney is also bringing back a tradition that his dad started back in the 1970s, and that is showing some horses. Stanley showed several horses in cutting, reining and the various roping events. Today, Rodney feels that they need to start doing that again in order to better market their horses.

"At most of the sales that we've been to, people ask about points," Rodney says. "It seems to have a big effect on both price and demand, so I'm going to try to show three of our studs and probably some of our geldings, too."

Rodney rodeoed through college, and still competes in tie-down roping. It's that event that he plans on entering at the horse shows. His long-range plans are to qualify three of the stallions for the AQHA World Show.

A lot has changed since Otto first rode across the sagebrush-covered sand hills of Beaver County. But a lot hasn't changed, either. There are still Barbys running cattle, riding good horses. And there likely will be for a long time to come.

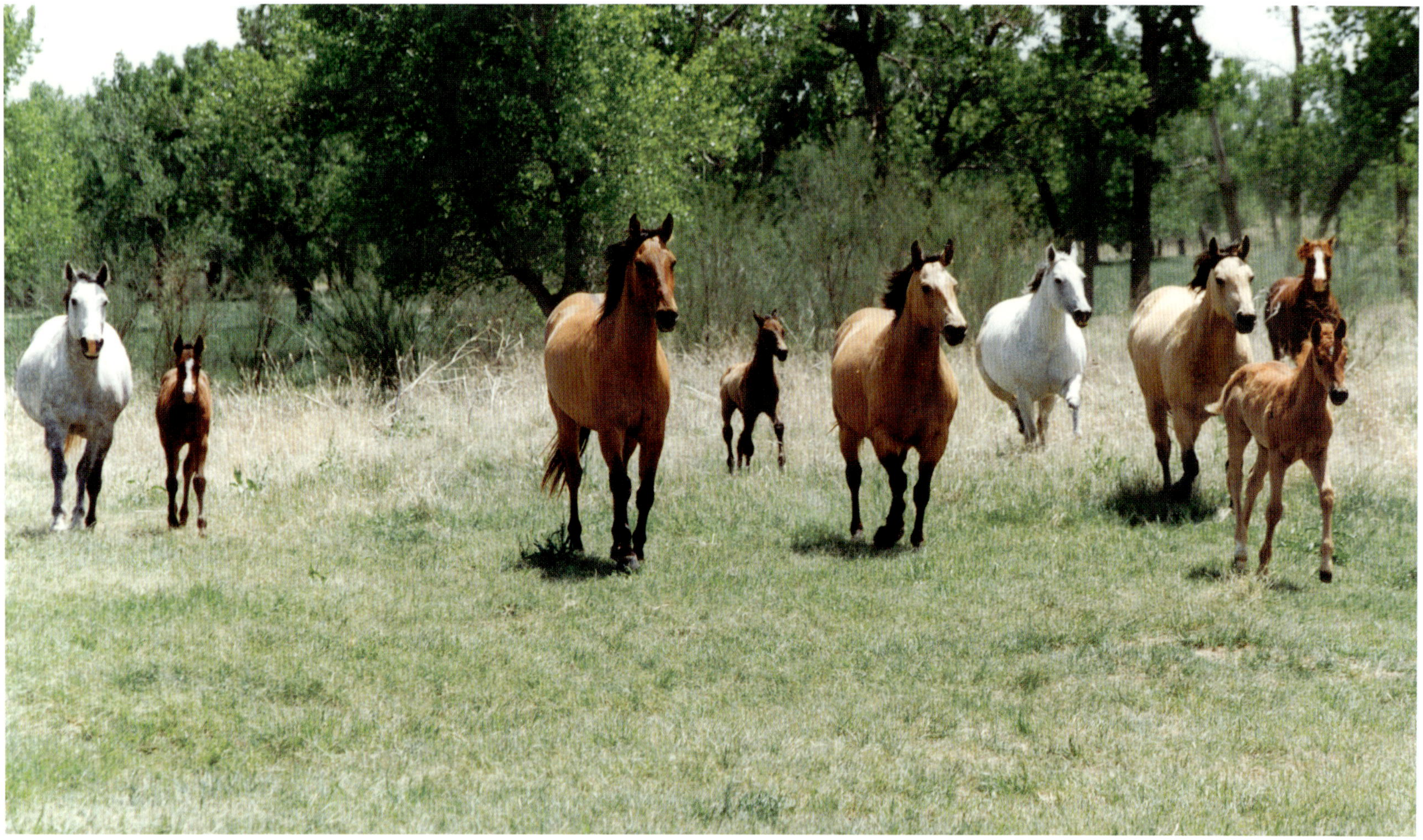

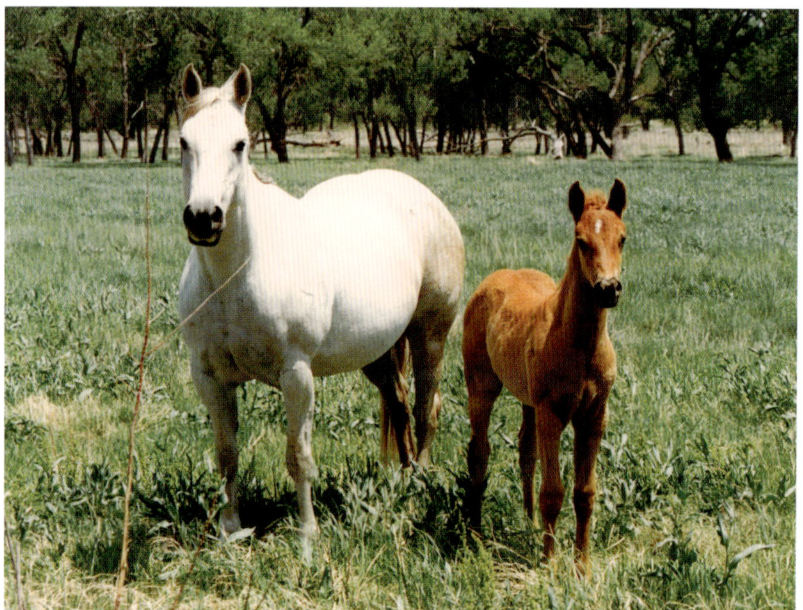

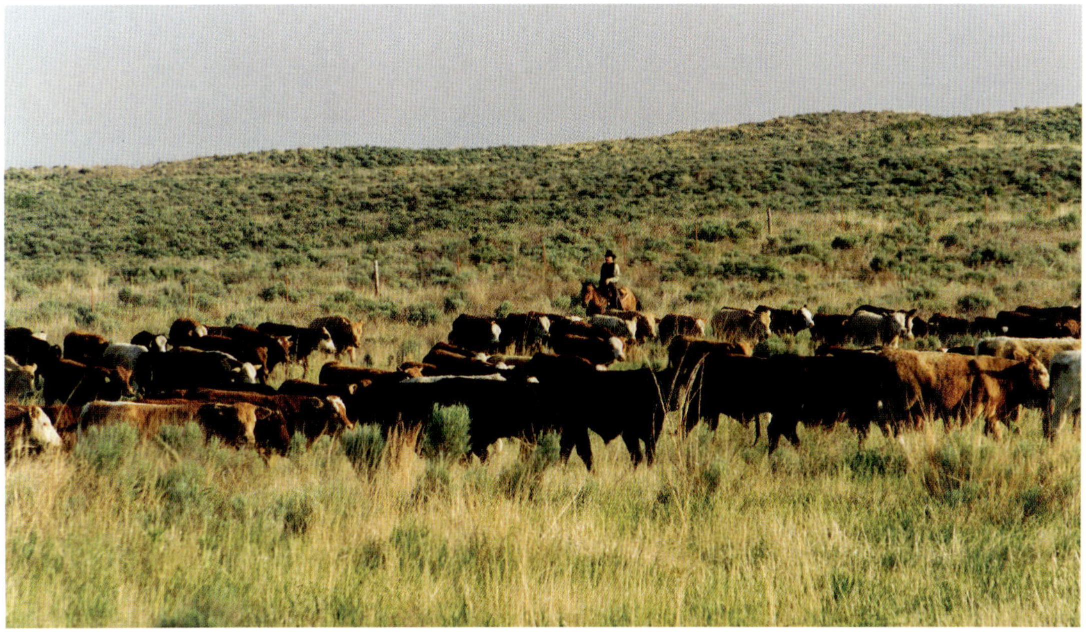

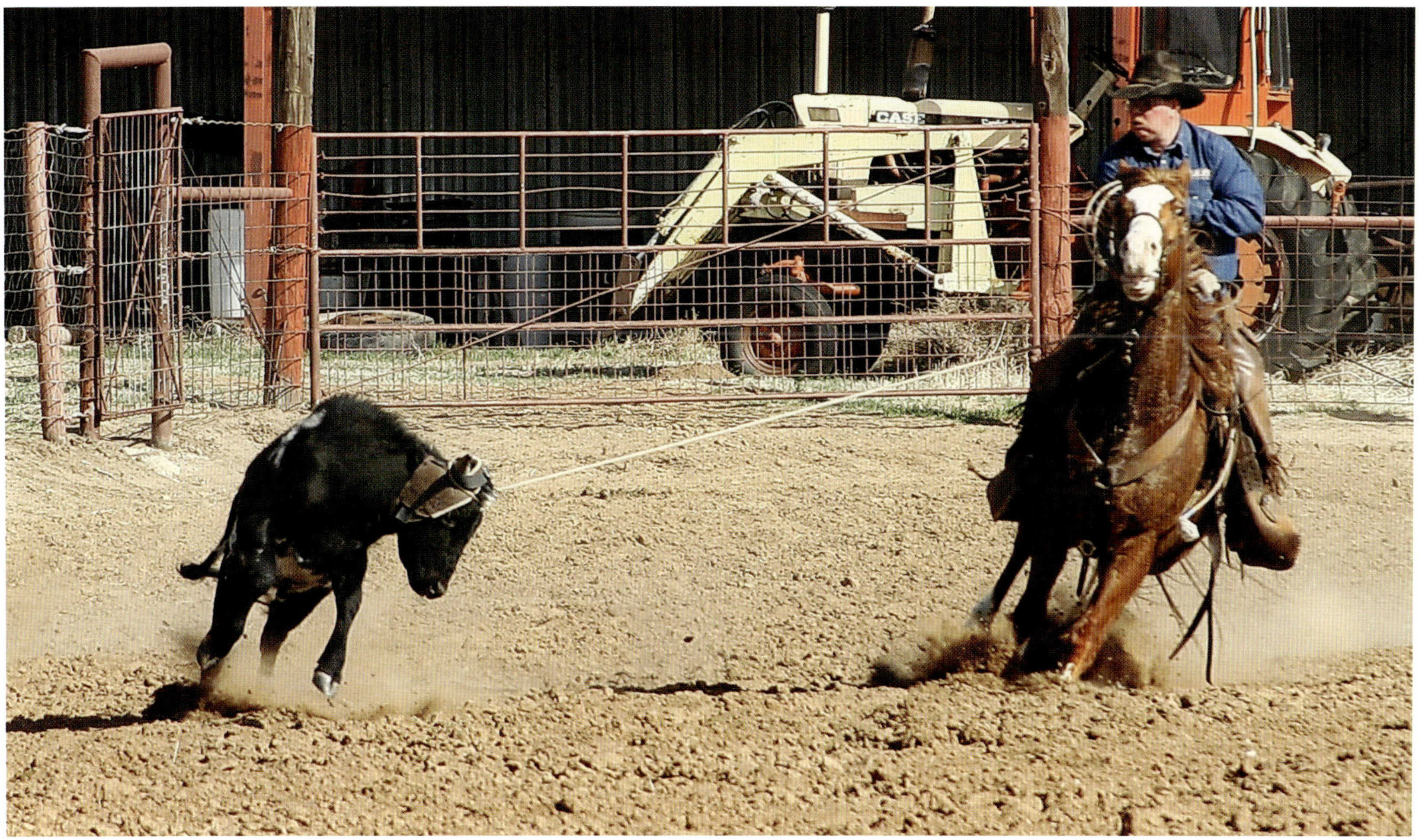
CASE

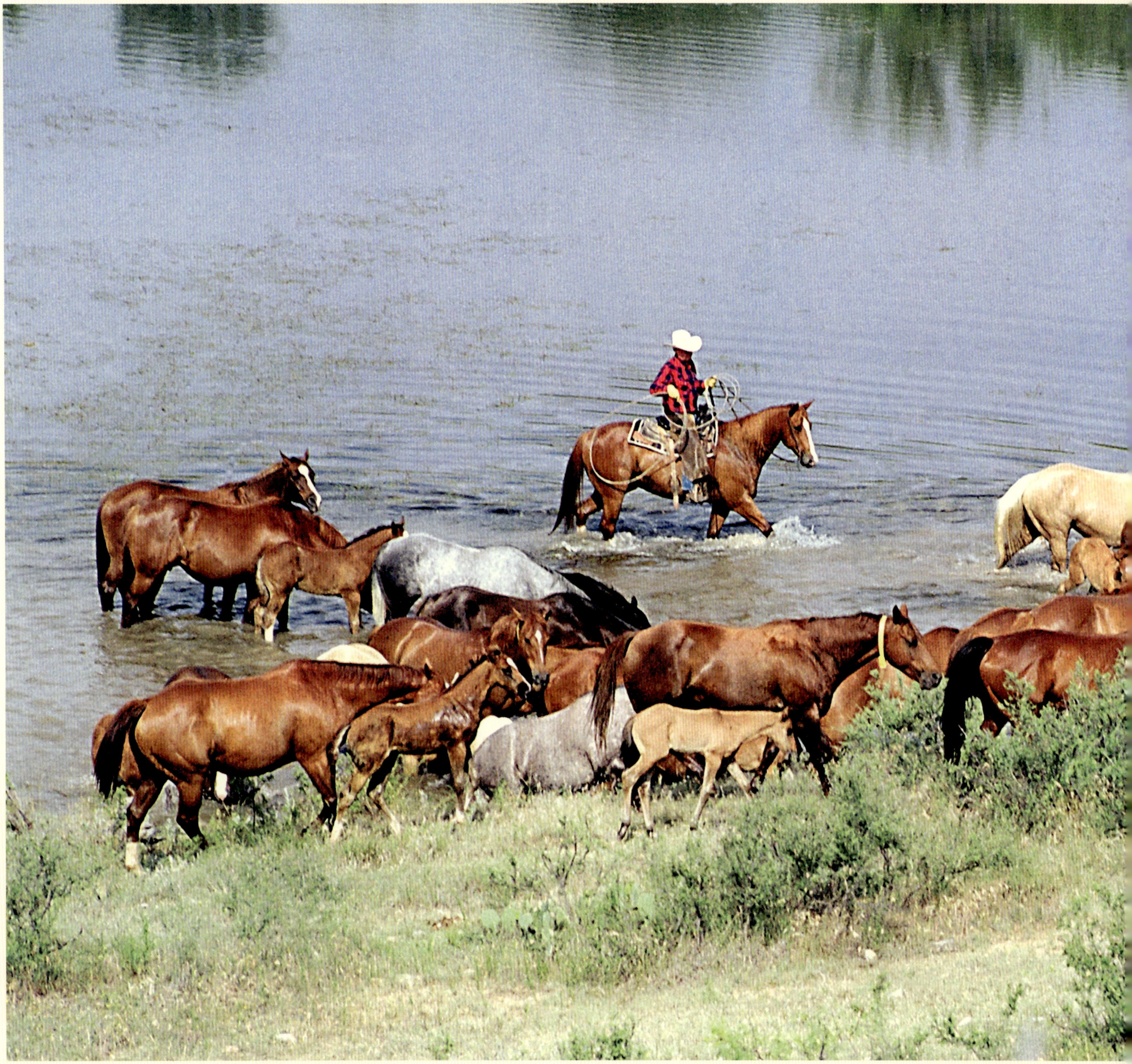

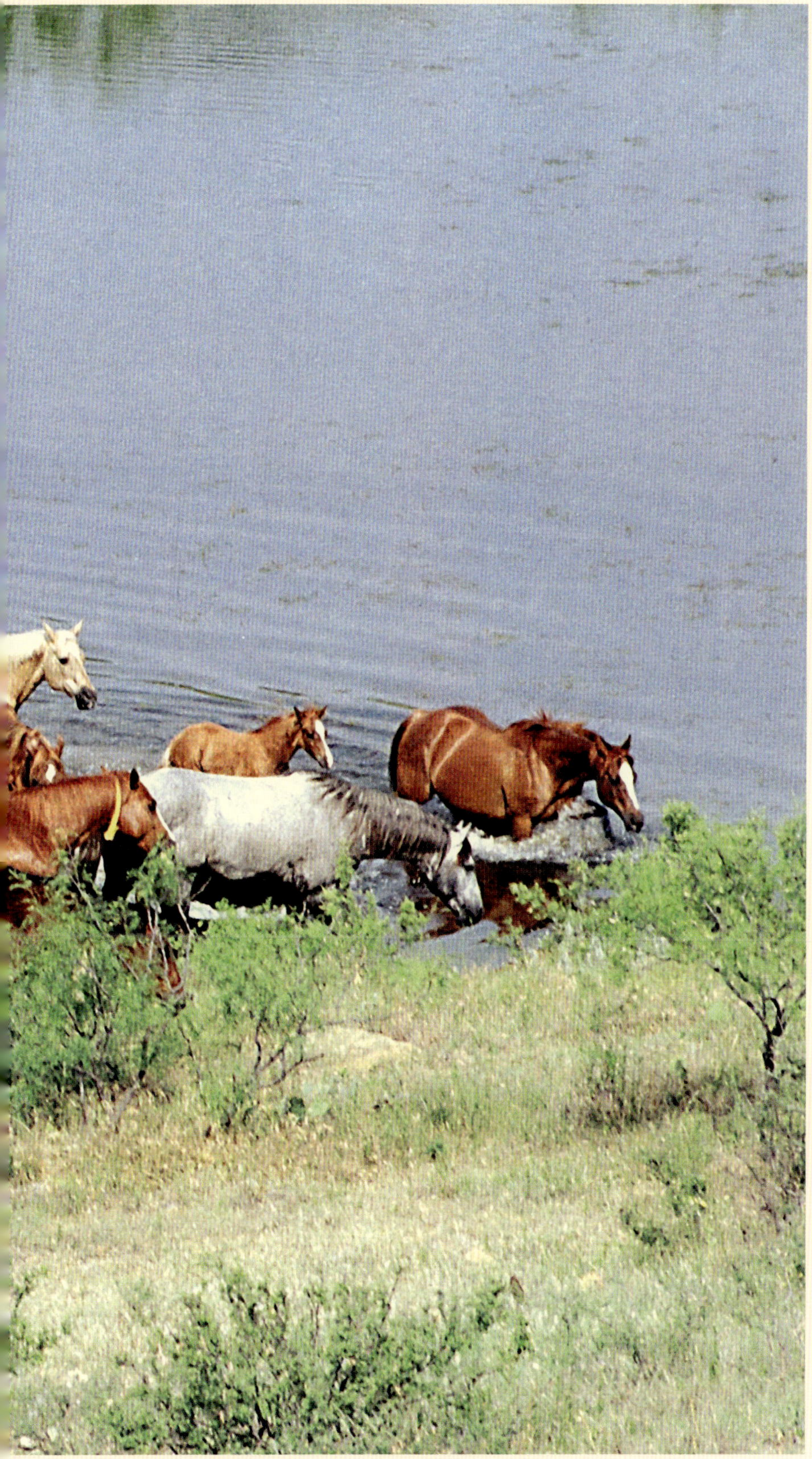

R.A. BROWN RANCHES

Throckmorton, Texas

1997

Ranch Horses

Texas' R.A. Brown Ranches earned the Remuda Award for raising some of the best cow horses in the country.

When R.A. (Rob) Brown Jr.'s grandfather established his Throckmorton, Texas, ranch in 1903, one of the first things he did was to make sure he had some top horses. He had been in the cattle business since 1895, and he knew the value of a good cow horse.

When R.A. Brown Sr. inherited the ranch, he followed that same philosophy. Brown recognized early the importance of a good horse, and he made sure his family and cowboys had plenty of them.

R.A. Brown Sr. died when his son, Rob, was only 29, but the value of a good horse was already instilled in the young man. It's something Rob has never forgotten, and it's something he makes sure his children know today. R.A. Brown Ranches is a family operation that encompasses more than 77,000 acres and approximately 2,500 head of cows. That takes good horses.

Rob and his wife, Peggy, share ownership in the ranch with their sons, Rob A. and Donnell, and daughters Betsy and Marianne. All four of the children are married, with children of their own, and all of the families are part of the ranch. However, only two of the Brown children are currently involved in the ranch's management—Donnell and his wife, Kelli, and Betsy and her husband, Jody Bellah. Rob A. and his wife, Talley, have their own ranch in the Texas Panhandle, and Marianne's husband, Todd McCartney, is in business for himself in the agriculture industry. But all the families are involved, and they all know and appreciate the heritage of the ranch and its history with horses.

Rob says that his father raised him to be a cowboy, and that he grew up horseback, learning the ways of horses and cattle. He says that once, while he was in his teens, he and his father went to a neighboring ranch to help work cattle for a friend of his father's, Roy Parks, who later became AQHA President.

"It was a 10-section pasture without any cross fences," Rob says. "Since I didn't know the pasture, Mr. Parks sent me around the outside. I could see the fence at all times as I was going around. That was 40 miles. We started at daylight, and we had that pasture gathered by 10:30 that morning. Then I started heeling calves and dragging them to the branding fire, and I never did change horses."

It took a special horse for that kind of work, and that's the kind the Browns became known for. Horses that were to reside on the R.A. Brown Ranch during the next several years are, to this day, still in the bloodlines of many ranch horses throughout the West.

"One of the first horses I remember us having was one we called Black Hancock," Rob says. "That was in the early '40s. He got a hip knocked down when they were halter breaking him at the [Four] Sixes. Somebody else ended up with him after that, and he got fistula of the withers. Daddy gave $50 and a hundred bushels of oats for him. He got the fistula cleared up, and was in the process of registering him when the horse got tetanus and died. Daddy only got one or two colt crops out of him."

Rob says his dad had a number of good horses, including one called Tubal Blake, which had the Blake and Steeldust breeding of the early days. But he thinks the one that is really the foundation

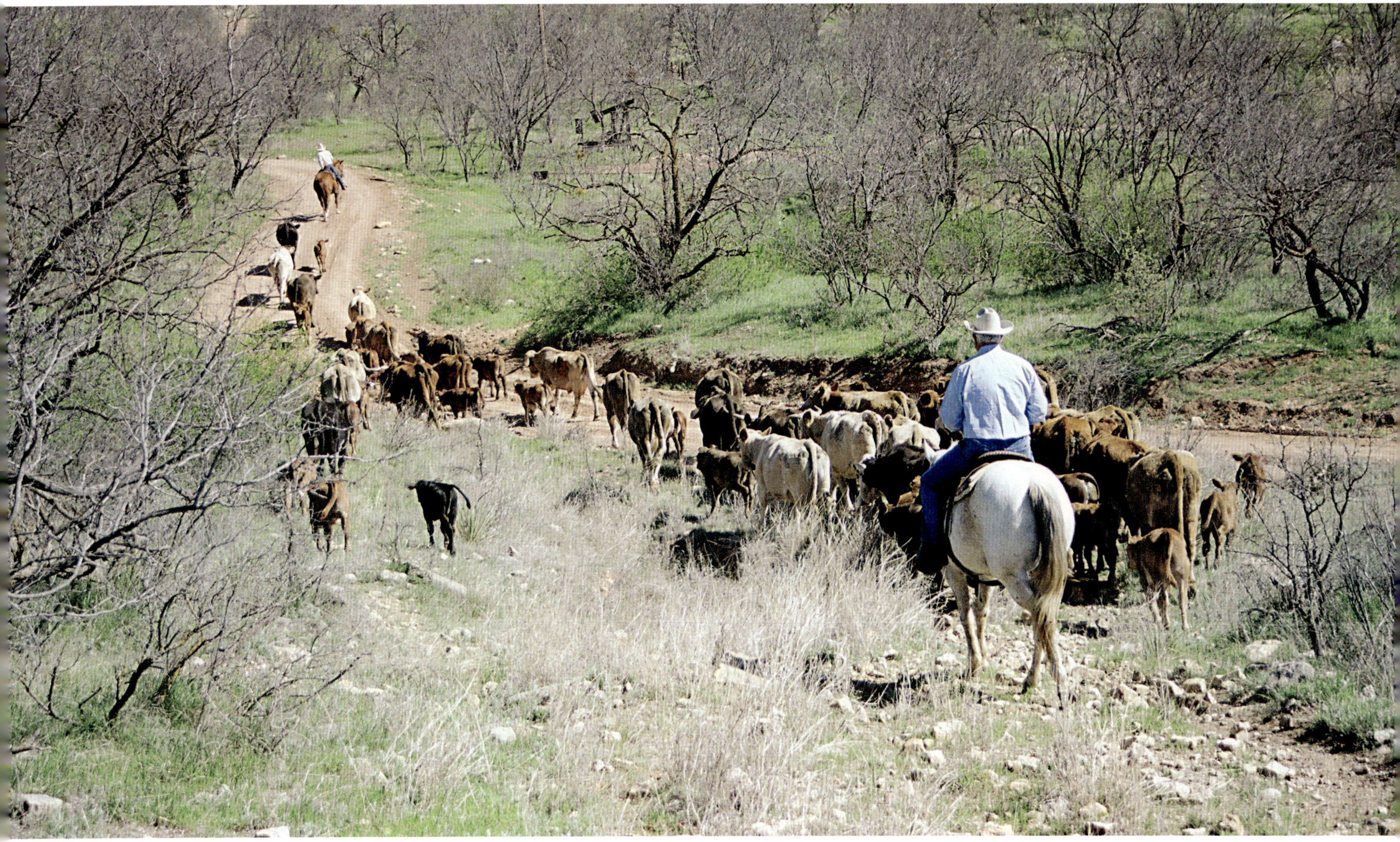

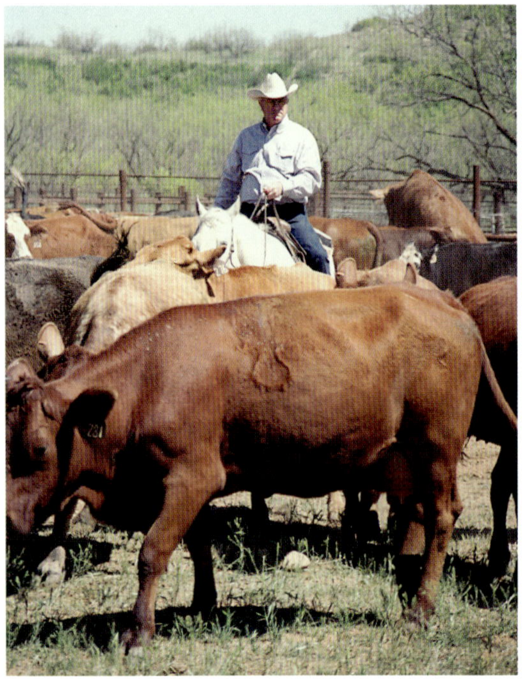

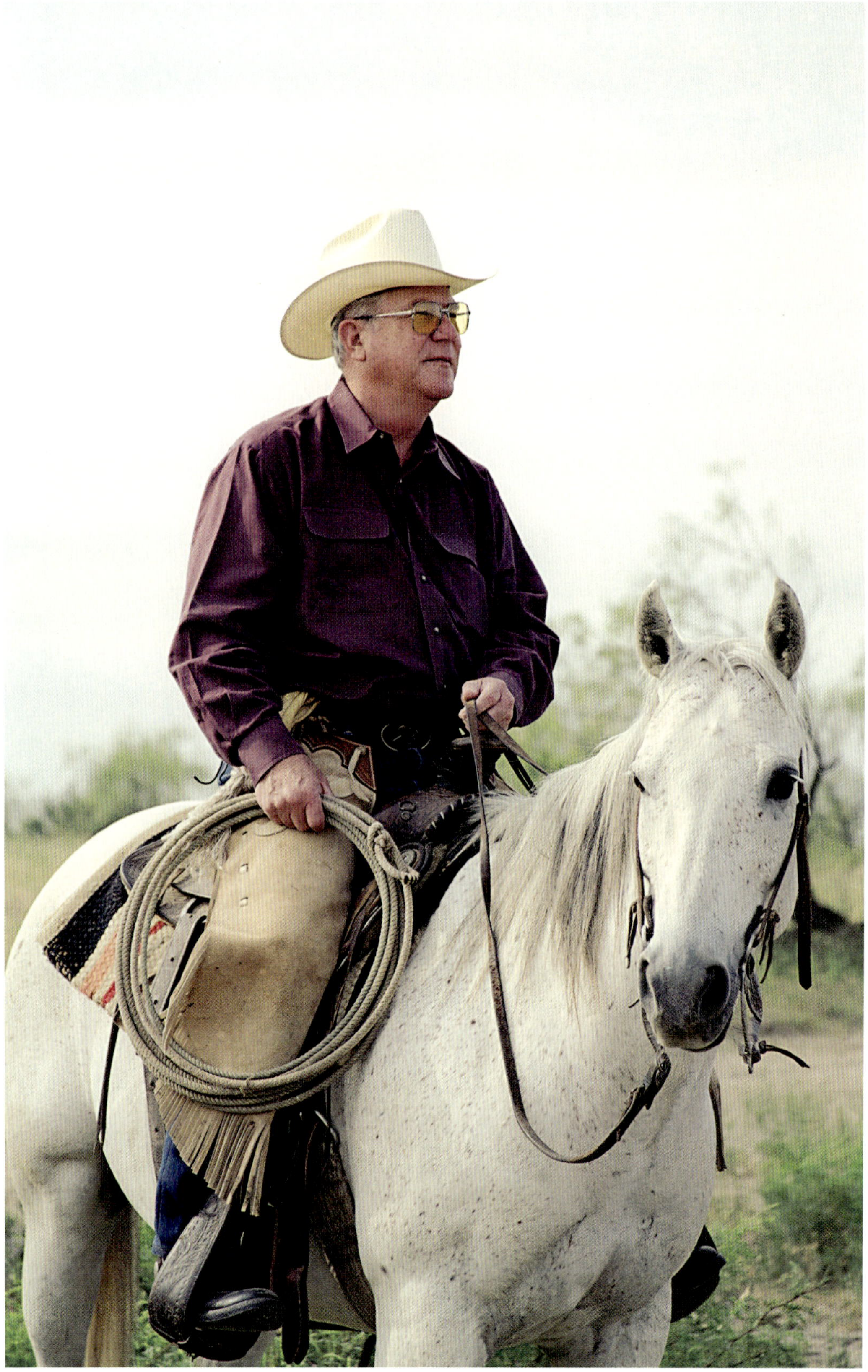

of their breeding program today was a horse called Blue Gold, which they got from the Four Sixes Ranch.

Rob's dad was a friend of Miss Anne Burnett Tandy, who owned the Four Sixes, and of Johnny Schwartz, the Four Sixes manager. "When Johnny had a heart attack, Dad volunteered to go to the Wyoming Hereford Ranch and pick out the Four Sixes' Hereford bulls that year. He also bought the stud Blue Rock for them."

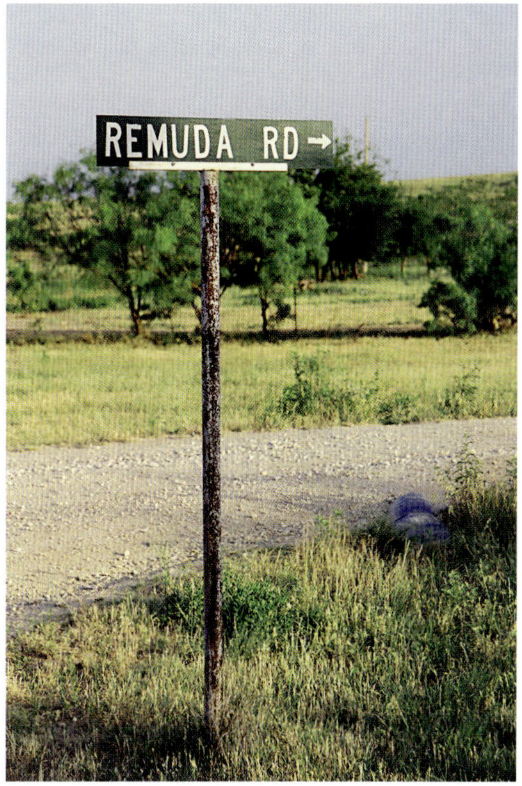

Blue Rock was by a Thoroughbred horse called Lost Cause by Eternal out of a Yellow Wolf mare. Yellow Wolf was by Old Joe Bailey. As a way of repaying Rob's dad for his favor, Miss Anne told ranch foreman George Humphreys to pick out the top stud prospect from Blue Rock's first colt crop and give it to the Browns. They called him Blue Gold. Blue Gold's dam was by Hollywood Gold, another top Four Sixes stallion, and out of a daughter of King O'Neill.

"That old Blue Rock line of horses had a lot of quality, really good heads and necks," Rob says. "We already had another Old Joe Bailey-bred stud that was getting some good, big, stout horses that had some cow and a lot of stamina, a lot of heart, but they were pretty coarse and needed some refinement. Blue Gold nicked really well on them."

Next came the Eddie horses. Eddie was a stallion that was foaled in 1940 by the Thoroughbred Raffles. He was being used for cutting when the Haythorns in Nebraska bought him, and they felt he would do their program a lot of good. Rob's dad sent six mares to Haythorns to breed to Eddie back in the '50s, and the fillies that resulted from that mating were subsequently added to the Brown's broodmare band. Rob says, "We got a little band of those Eddie mares started, and crossed about 20 of them on Blue Gold. With those colts, you could do anything you wanted to do horseback."

All of the aforementioned horses are still in the bloodlines of R.A. Brown ranch horses, and their pedigrees were strengthened a few years ago by the addition of a gray stallion called Hesa Eddie Hancock, who, for a few years, was the main stallion on the ranch. He traces back to both Eddie and Blue Gold.

"We've got 35 broodmares," Rob says, "and 22 of them are daughters of Hesa Eddie Hancock. We're crossing them on a stallion we have now, Gold Peppy Freckles, who's a grandson of Hollywood Gold, and goes back to both Little Peppy (Peppy San Badger) and Colonel Freckles on the bottom. We've also got a son of Hesa Eddie Hancock that we're breeding to a few of our own mares as well as some outside mares. I think you can see we're really high on that horse."

Hesa Eddie Hancock now belongs to Rob's son, Rob A., who's using the stallion on his own ranch.

Rob says, "We've always had horses that had a lot of stamina because we ranch in a big, dry country, and we don't change horses at dinner. We may ride a horse 30 or 40 miles a day, and think nothing of it. We've kept a little more bone, a little more size under our horses, but there's a reason for it.

"We do feel like the ideal horse should be a super cutting horse," he continues, "but we want him big enough that if we need to go catch something, even if it's pretty big, we can catch it and handle it."

Most of the cowboys on the ranch have three or four good horses in their strings. When they take a young horse to ride, they usually sell an older horse.

"We sell some older horses," Brown says, "but we will never sell a horse out from under a cowboy. We give him 25 percent of what the horse brings, and the refusal of the sale of any of his mounts. In other words, he has to agree to the price. Even if the price suits me, if it doesn't suit him, I won't sell the horse.

"But I think that's why we have some of the best cowboys in the country. They've got the pride of riding some sur'nuff good horses. They know they're not going to have that horse sold out from under them. They can ride him as long as they want to."

Rob pasture breeds all his mares. He turns the stallions in with the mares about April 1, and picks them up around the first of July.

"We get an 85 to 90 percent foal crop every year this way," he says. "We try to breed them when the grass comes in the spring. Most of our colts are born in March and April. That's Mother Nature's way to make it happen. We've got some mares that have foaled within the same week for 15 years. If a mare doesn't breed readily, we don't keep her, kinda like our cow outfit. She's got to be making a buck if she's going to stay around here.

"We also breed our 2-year-old fillies; we don't hold them until they're 3. They may not grow out and be as big a mare, but we think they are certainly big enough and sexually mature enough to breed and raise a colt. We used to break all our fillies, but we've gotten so involved with our registered cattle operation that we don't have time anymore. But we'll try to keep some of the first colts out of those fillies to see how they ride, make sure they're going to be what we want.

"Practically every mare on this ranch was raised here, as was her mother, her grandmother and her great-grandmother. We've brought in very few outside mares over the years."

Brown says, "We're pretty aggressive animal breeders, in both our cattle and our horses, and it's worked for us for a long time. My grandfather started it, and my kids are staying with it. We need these horses, so we're going to try to make good ones."

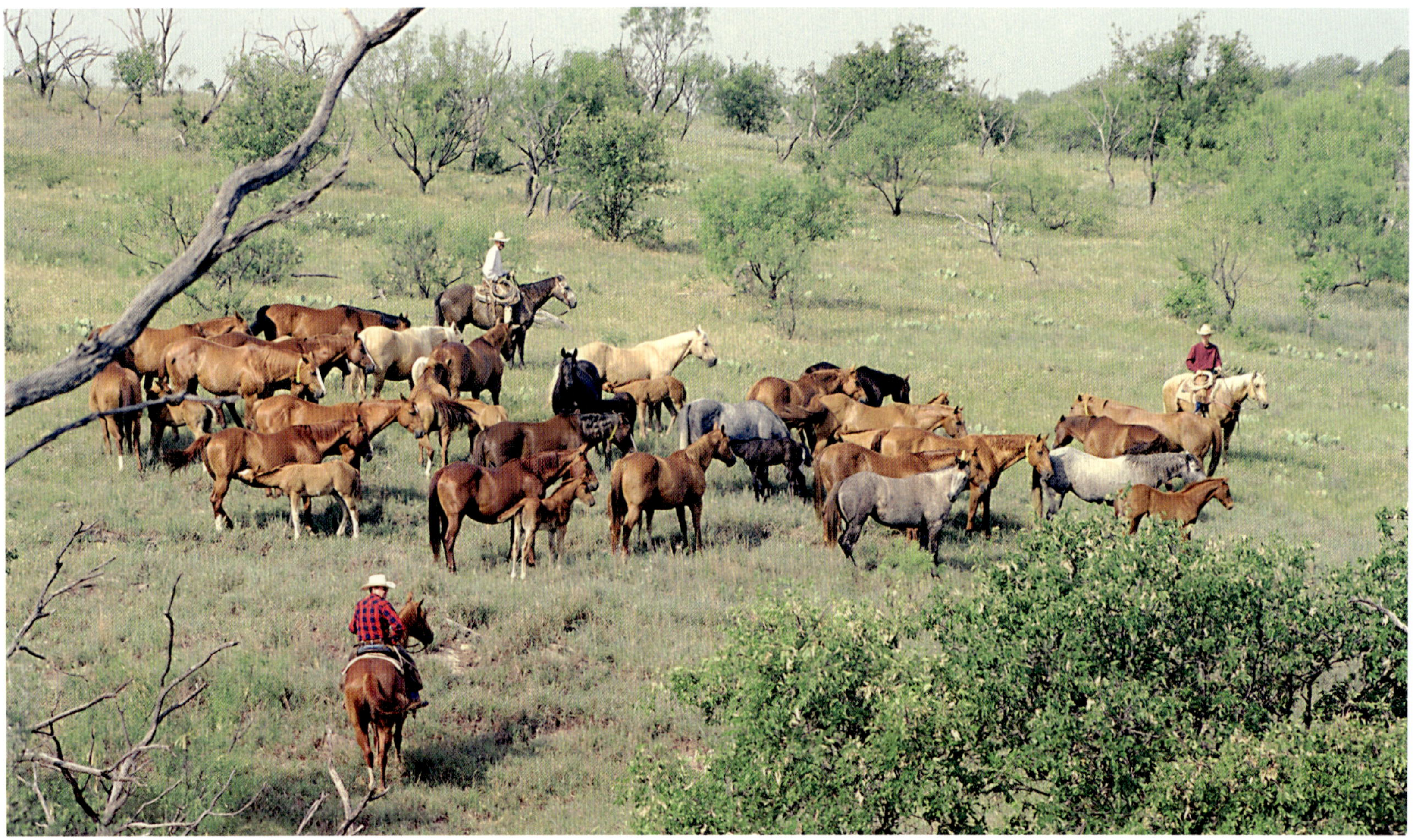

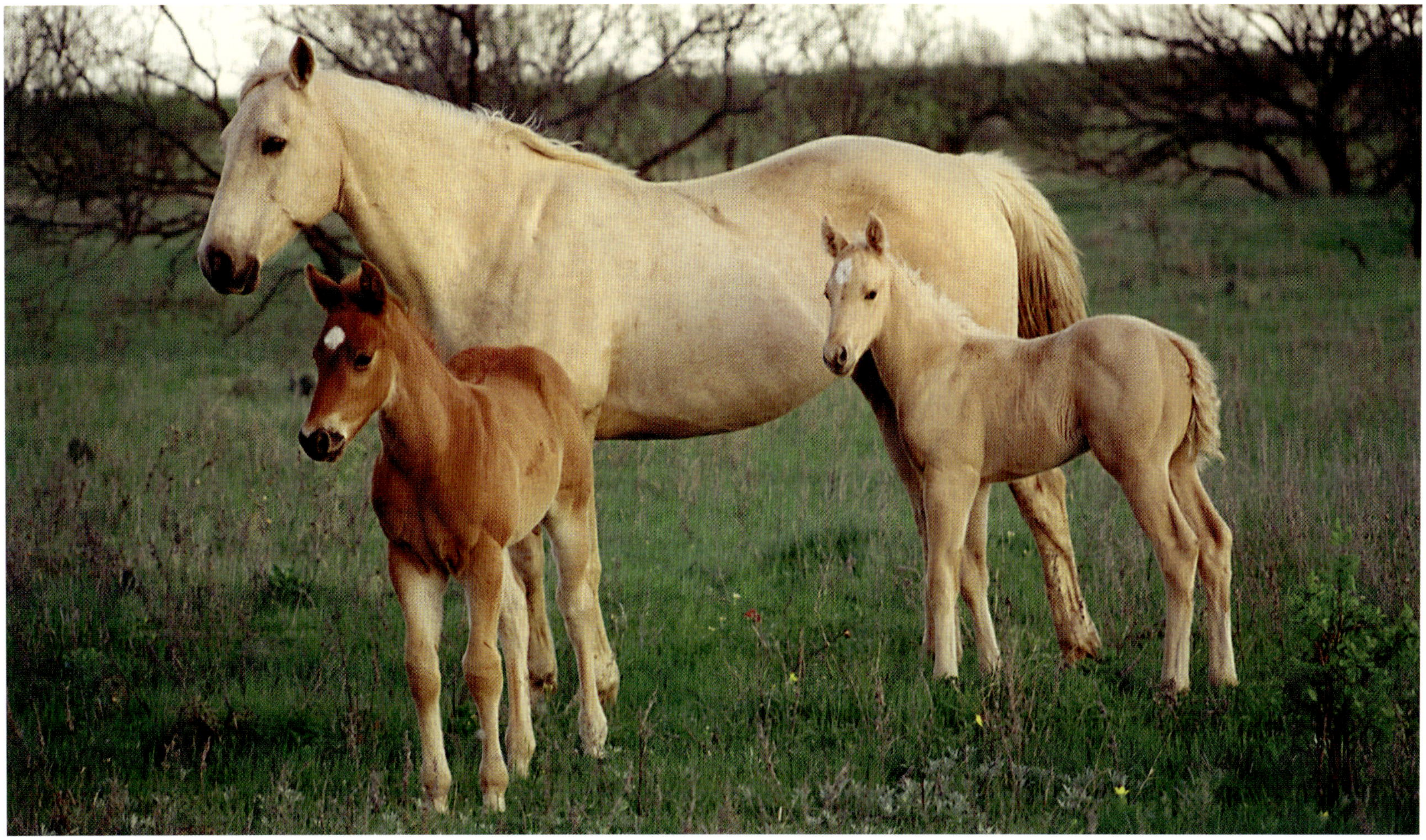

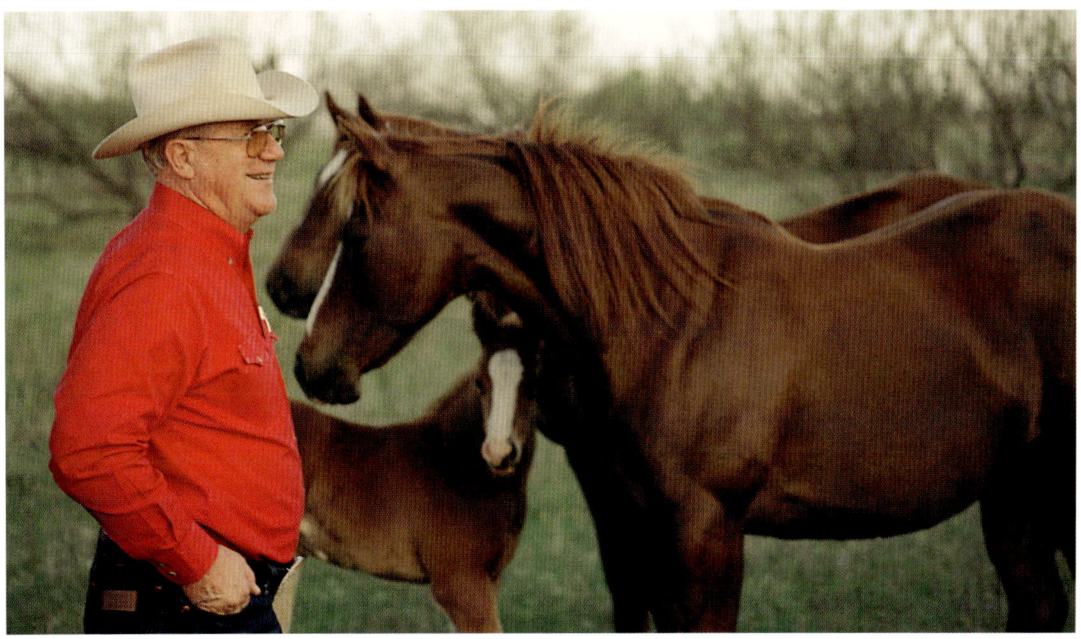

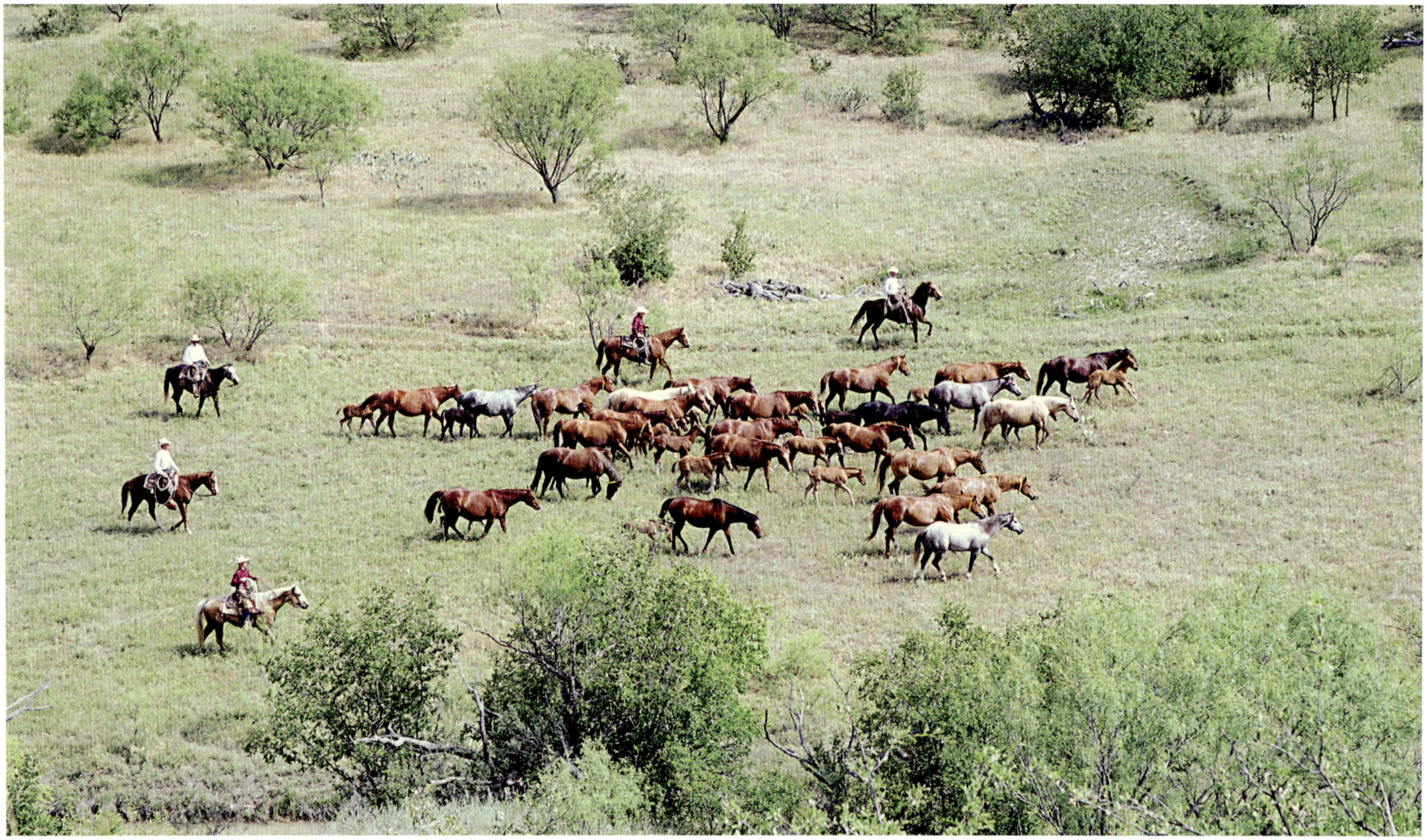

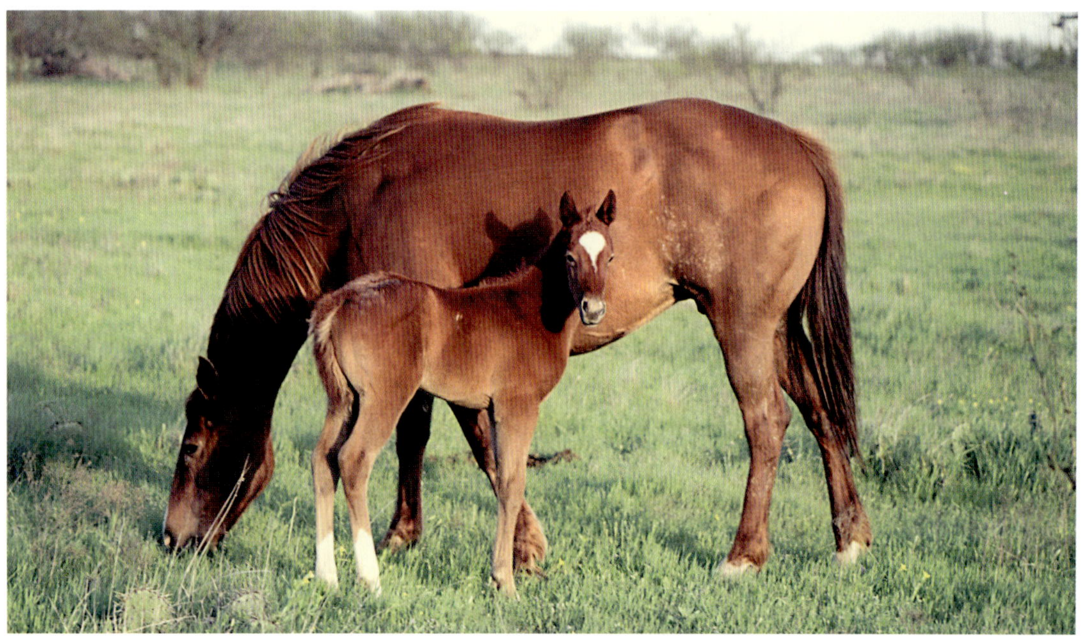

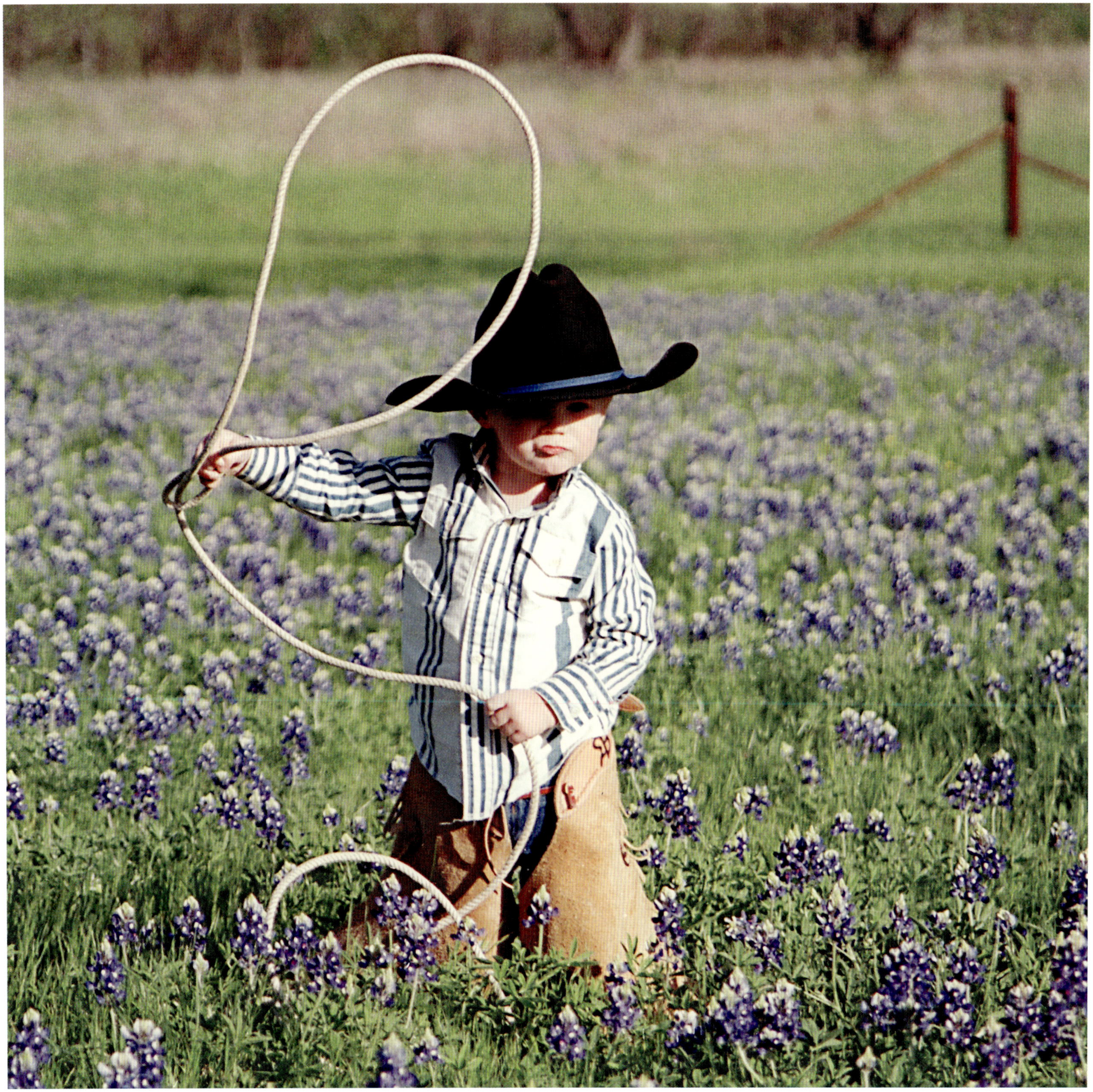

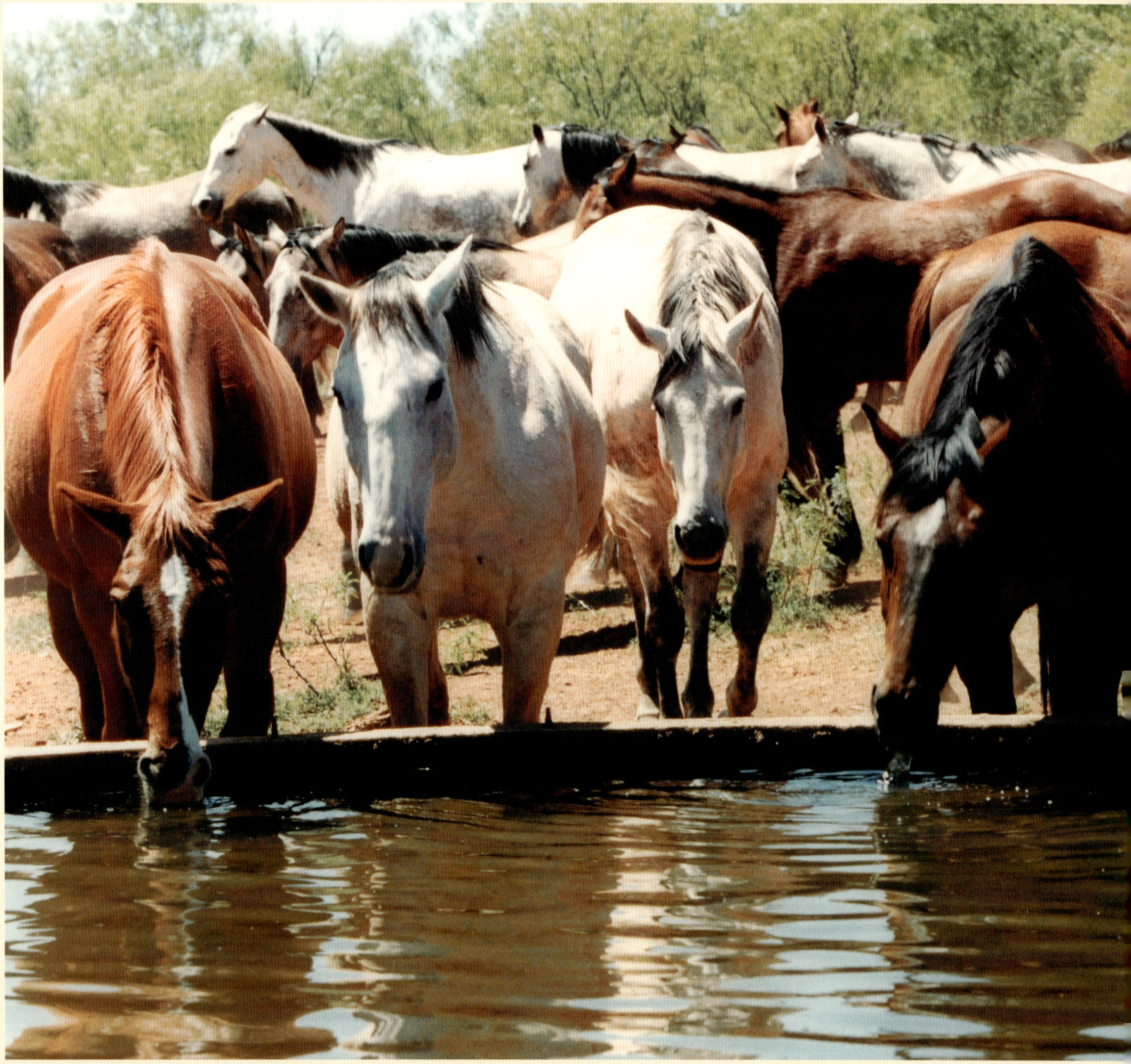

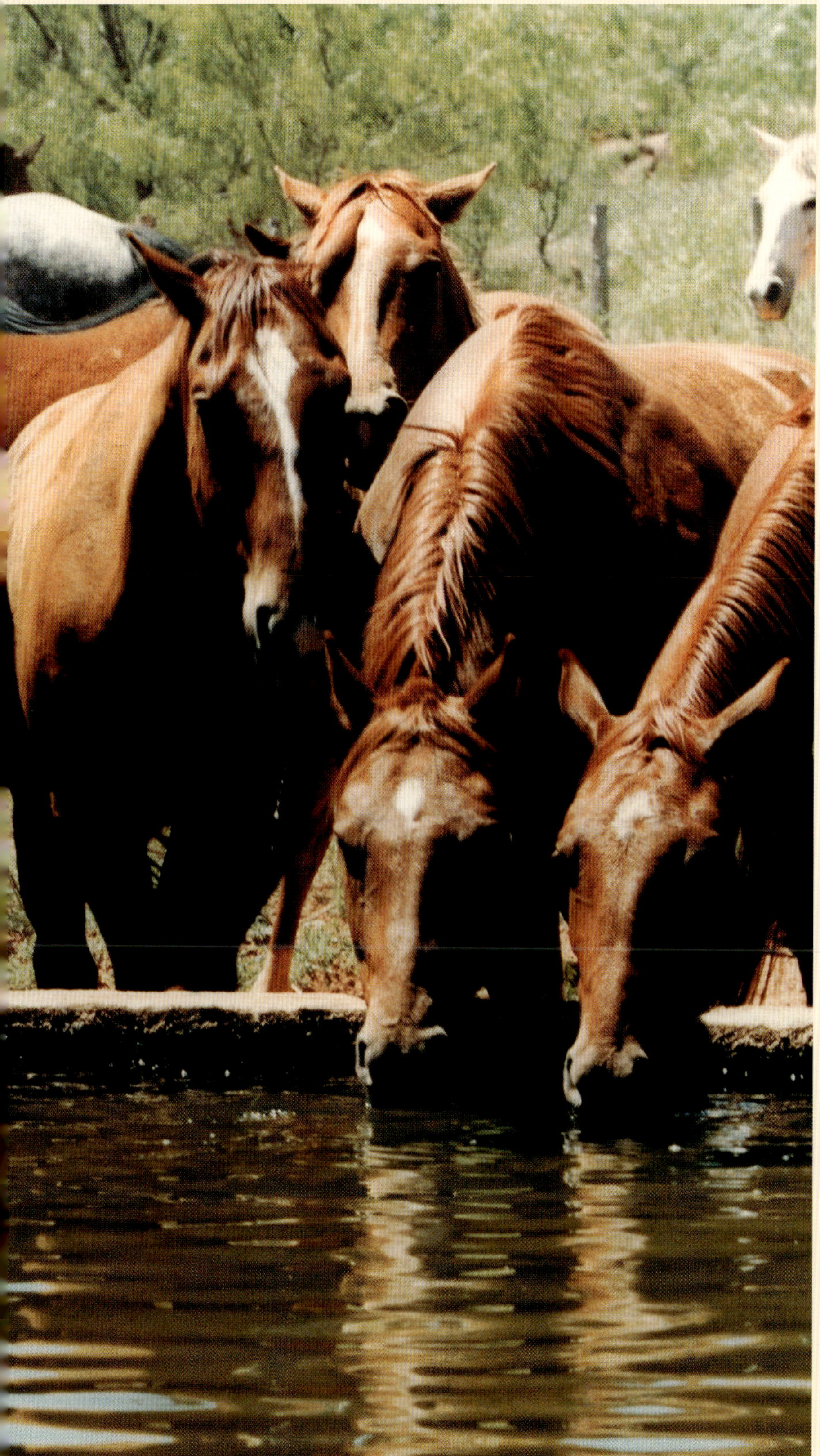

Pitchfork Ranch

Guthrie, Texas

1998

Necessities

The 125-year-old Pitchfork Ranch breeds good horses by necessity; no modern convenience can take their place in rugged terrain.

In 1881, Dan Gardner was looking for a partner. He and Col. J.S. Godwin had contracted to buy a herd of longhorn cattle, but due to family problems, Godwin wanted out of the deal. Gardner wasn't strong enough financially to handle it all himself, so he began searching for help.

At the last minute, Gardner remembered his Mississippi boyhood friend, Eugene F. Williams, who lived in St. Louis. When approached, Williams must have liked the idea of becoming a Texas rancher. He took a train to Fort Worth, and he and Gardner bought out Godwin's interest in the herd. Williams then returned to St. Louis, leaving Gardner to manage their operation.

Gardner took delivery of 2,600 head of Longhorns, all carrying a pitchfork brand. Along with the cattle came 70 horses. The cattle and horses were on land that ranged through Dickens and King counties in central West Texas, and Gardner made a deal for the land, too.

Over the next few years, to help with business expenses, Gardner and Williams took in other partners and incorporated the ranch as the Pitchfork Land and Cattle Company. Each of the partners was named a director, but Gardner and Williams continued to call the shots. Under their direction, the Pitchfork became one of the most respected ranches in the state. Today, it is one of the largest and oldest ranches in American history.

Williams died in 1900 and Gardner in 1928, but by then Williams' two sons, Eugene and Gates, had already picked up their father's interests and worked closely with Gardner. Eugene's son, Gene, followed in the footsteps of his father and uncle, and today Gene Williams is chairman of the board of Pitchfork Land and Cattle Company. From his St. Louis office, he works closely with current general manager Bob Moorhouse.

Bob, an AQHA Director, is only the seventh manager of the 125-year-old ranch. He oversees the entire operation, which includes 165,000 acres at the headquarters near Guthrie, Texas, 10,000-acre and 3,000-acre places at nearby Truscott and Benjamin, and still another 3,300 acres in the Kansas Flint Hills. Together, the four ranches run about 6,000 head of cattle—either cows or yearlings, depending upon how each ranch is stocked—and 55 broodmares. Approximately 100 geldings make up the ranch's saddle horse remuda.

"We make use of all the modern conveniences we can," Moorhouse says. "We have two-way radios in our pickups, and we use stock trailers to haul our horses to different parts of the ranch. However, we have not found anything to replace the horse on this ranch."

The Pitchfork raises all its saddle horses. Mares are divided into different bands, and a stallion is turned out with each band. The stallion battery includes Pay Forty Four by Preferred Pay out of a Genuine Doc mare; Tejons New Star by Grays Starlight out of a Doc O'Lena mare; Special Gun by Playgun out of a Doc's Prescription mare; and the 3-year-old MH Dyno Peppy by Little Dyno by Smart Little Lena out of a daughter of Peppy San Badger. In addition, Cat Silver, by High Brow Cat out of a Peppy San Badger daughter, is owned by the ranch, but stands at the nearby Four Sixes Ranch. Cat Silver qualified for the senior cutting at both the 2003 and 2004 AQHA World Championship Shows, and has earned money in NCHA cutting competition. Tejons New Star has also earned NCHA money, and Special Gun is being

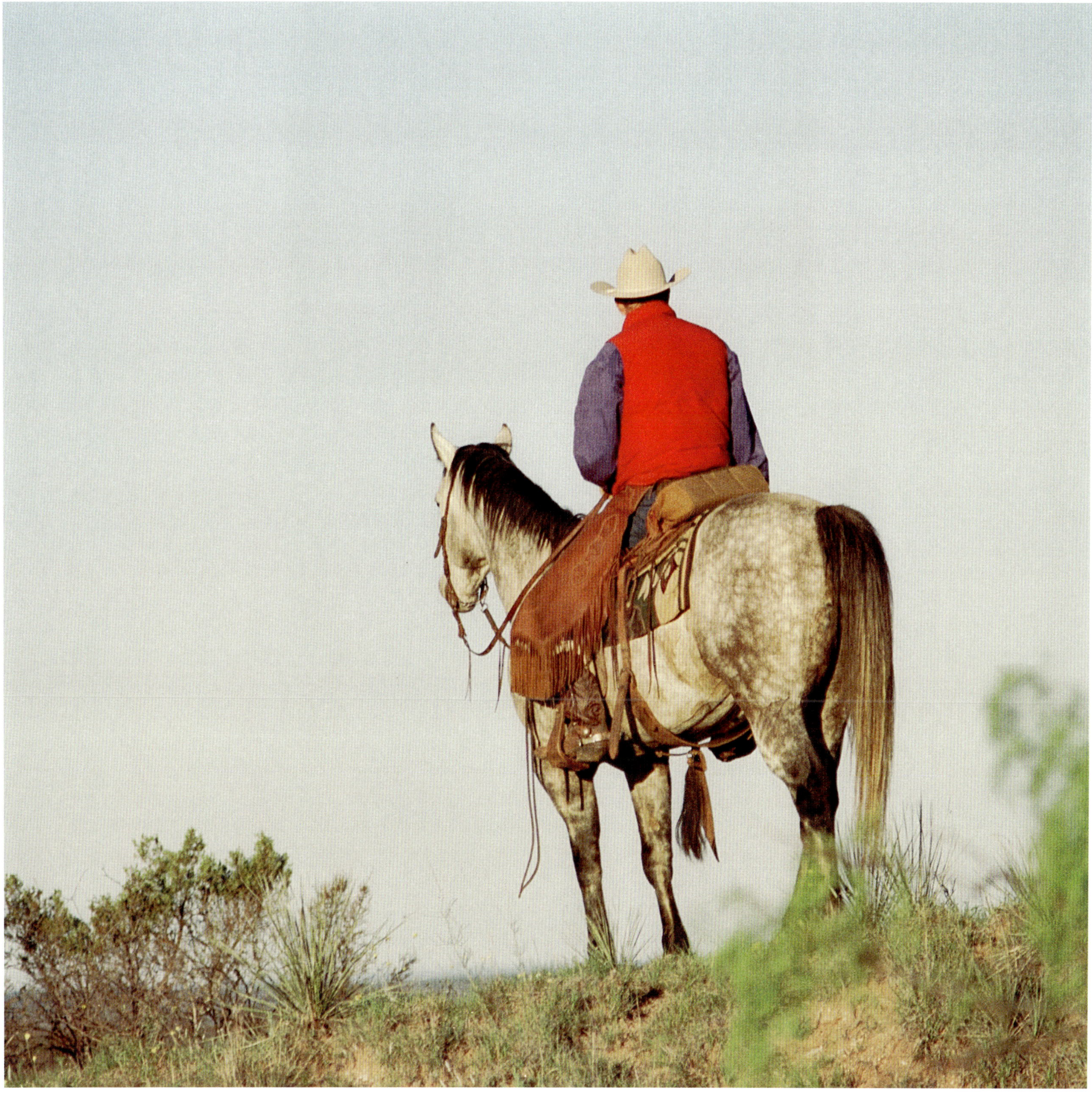

shown extensively in Ranch Cutting Horse Association competition.

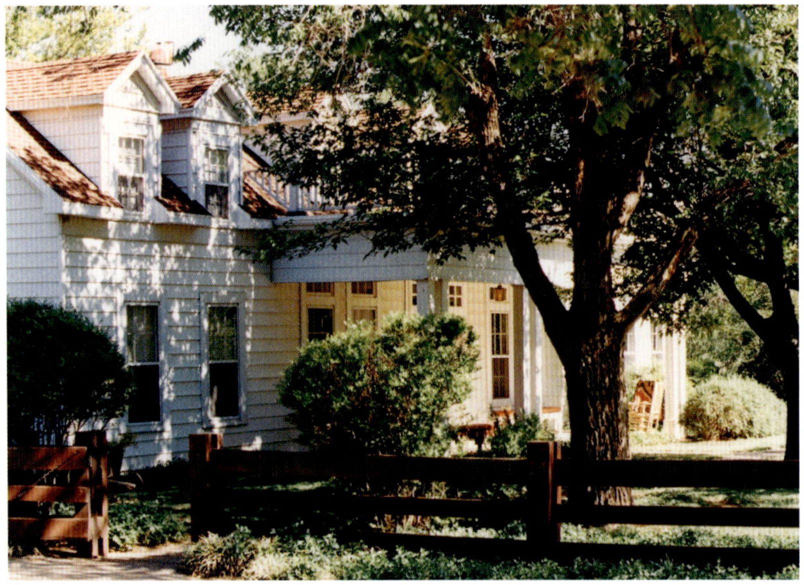

Foals are halter broken and branded in July, and then they're turned back with the mares and not weaned until fall. Each foal receives four brands: the signature pitchfork brand on his left shoulder; a number on his left buttock denoting his sire; a number on his left jaw identifying his mare family; and a number on the right jaw indicating foaling year. With this series of brands, someone can recognize the horse at any time in his life, and tell his age and how he's bred.

The cowboys ride only geldings. The fillies are either sold or used as replacements in the broodmare band. The Pitchfork holds a production sale with the neighboring Four Sixes, Tongue River and Beggs ranches each fall, and they consign horses to the annual Best of the Remuda sale, which is open only to winners of the award. Bob says they sell some geldings as well as mares, but, "I don't want to sell geldings away from the cowboys when they need them to do their job."

Currently, 10 full-time cowboys work on the ranch. Each is allowed 10 horses in his "string." In other words, each cowboy has 10 horses that he can call his and use to perform his work.

Each cowboy adds a 2-year-old to his string annually. Bob says, "To me, it is very, very important to get a good colt each year."

The cowboys start riding their 2-year-olds right after they finish branding, usually in late spring or early summer, and the horses are slowly introduced to cow work. As 3-year-olds, they are ridden more, and by the time the horses are 4, they are fully dependable as cow horses.

"The Pitchfork is known for its good horses," Bob says. "Cowboys have been known to stay on the ranch instead of taking another job because they didn't want to leave their horses."

Some cowboys live with their families in various "camps" around the ranch. The residences are not actually camps, but consist of a house near a set of corrals and a barn. A cowboy who lives in a camp keeps his string of horses in a small pasture or "trap" near his house, and cares for the portion of the ranch in his area.

Other cowboys live in houses at the ranch headquarters, and the single cowboys live in the bunkhouse there. All of these men eat breakfast and dinner (lunch) at the cookhouse, where Bob and wagon boss James Gholson give orders for the day. The cowboys who live in the bunkhouse also take supper at the cookhouse, while the married men eat with their families.

At branding time in the spring, or weaning in the fall, each cowboy takes his horses and joins the other cowboys on the ranch at a designated location, where they all work together, moving across the ranch until all the cattle have been worked. Bob sends the chuckwagon, which becomes the cowboys' home for most of the week. The chuckwagon cook prepares three meals a day over a wood fire, and the cowboys eat all their meals under the wagon "fly" or tarp.

They stay with the wagon Monday through Saturday, sleeping in range tepees or tents. This eliminates the need to drive to work each day. By daylight, the cowboys have finished breakfast and are ready to catch their horses. The wagon boss, or foreman, ropes each cowboy's mount.

Much of the Pitchfork Ranch is made up of extremely rough terrain. Small, steep canyons and cedar and thick mesquite trees make crossing some pastures by any method but horseback impossible. The terrain reinforces Bob's claim that there will always be horses on the ranch because the rough pastures require a unique way of gathering cattle.

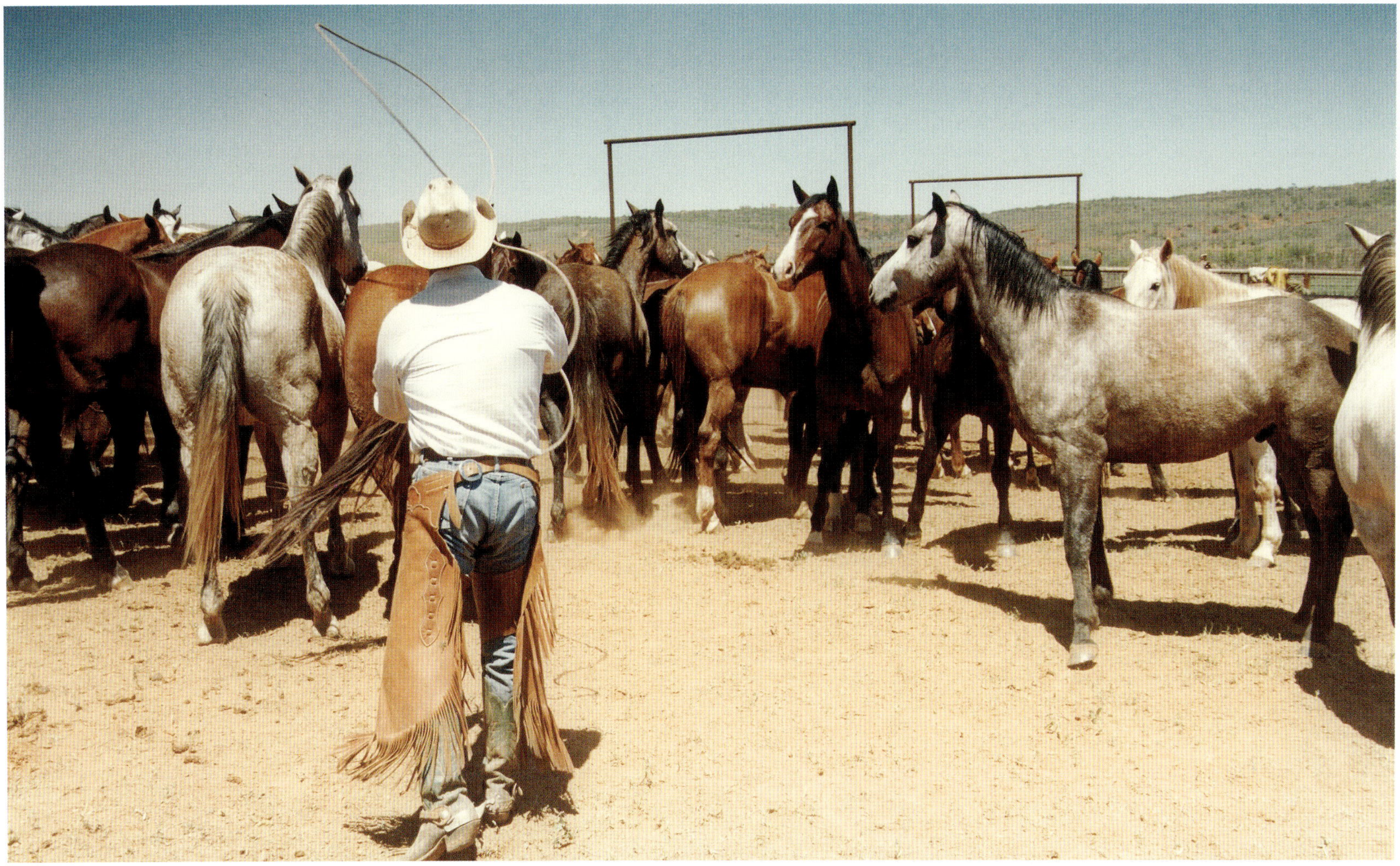

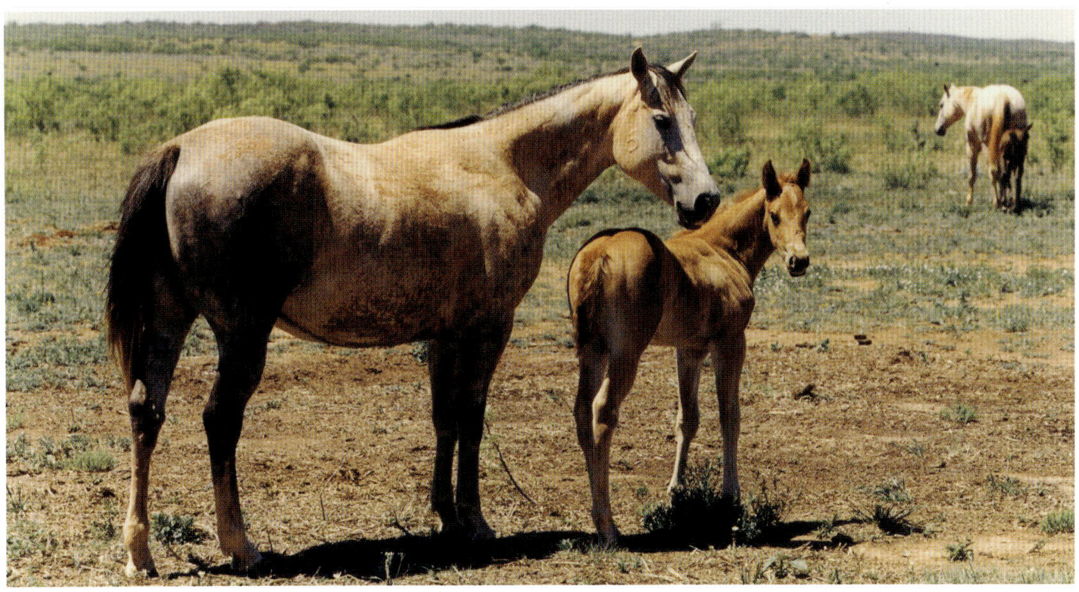

As a drive begins, the cowboys ride single file in a long trot across the pasture, following the wagon boss, who periodically stops and drops off a man. Extra cowboys are hired when the pastures are gathered for branding or shipping, so by the time each cowboy is assigned a spot, there might be up to 12 to 14 men, 200 to 300 yards apart, in a line 1½ miles long. The number of riders depends upon the size of the pasture and the density of the brush. In a real brushy pasture, where a cowboy might not even be able to see the man next to him, the wagon boss whoops to the closest cowboy, and the call is passed down the line until the last cowboy knows it's time to move out.

As the line moves across the pasture, the cattle are pushed to a designated point and penned. Calves are branded, or weaned, depending on the time of the year and the operation underway.

On Saturday afternoon, most of the cowboys return to their homes and their families, but they're back at the wagon by Sunday afternoon. Branding usually takes about a month, and weaning in the fall takes four to five weeks.

The work is hard, and the days are long, but the cowboys work on the ranch by choice. As one of them puts it, "It's good to get home in the evening, take a shower and wash the dust off, but then you just can't wait to get back out there and catch another horse."

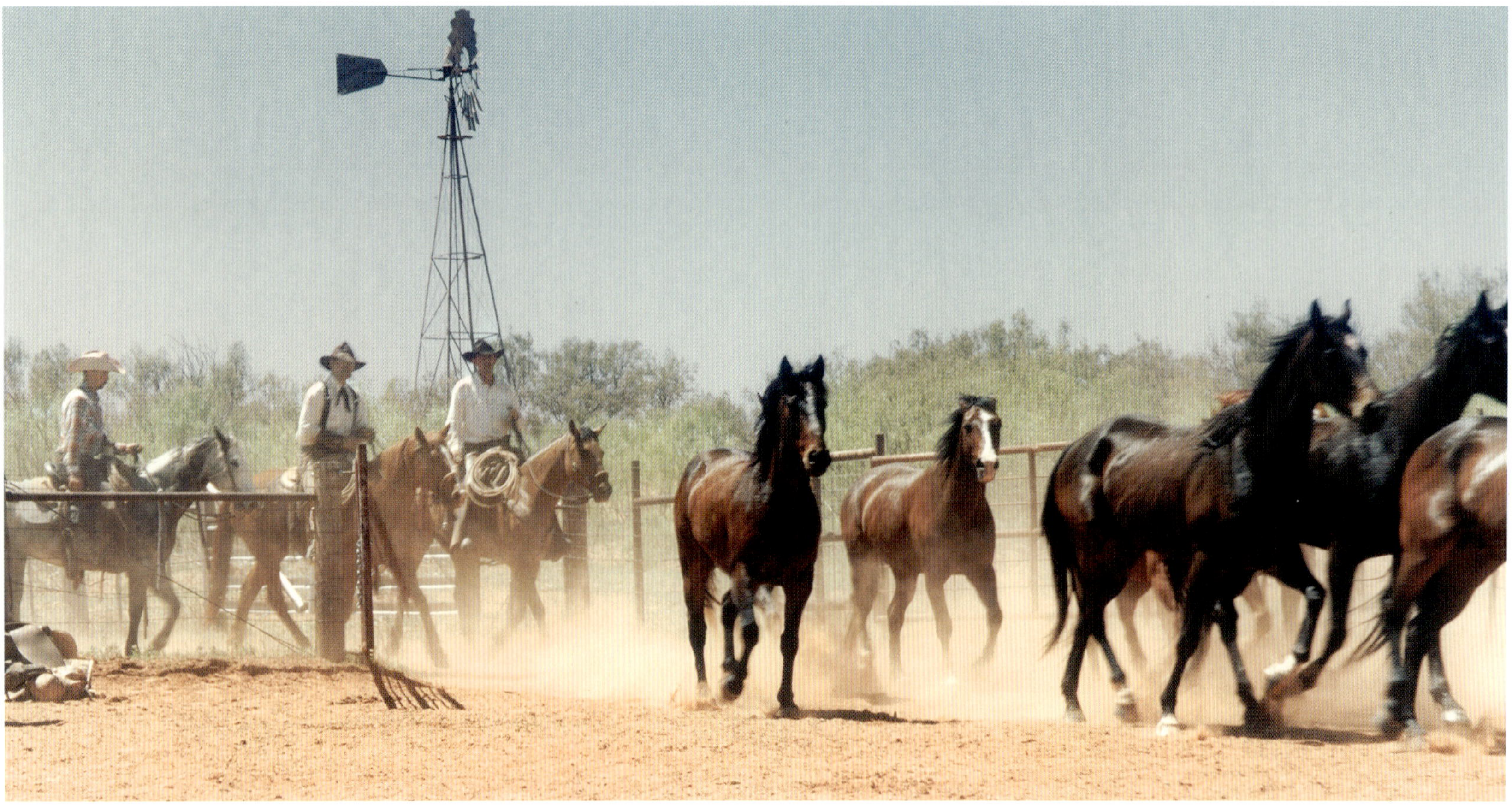

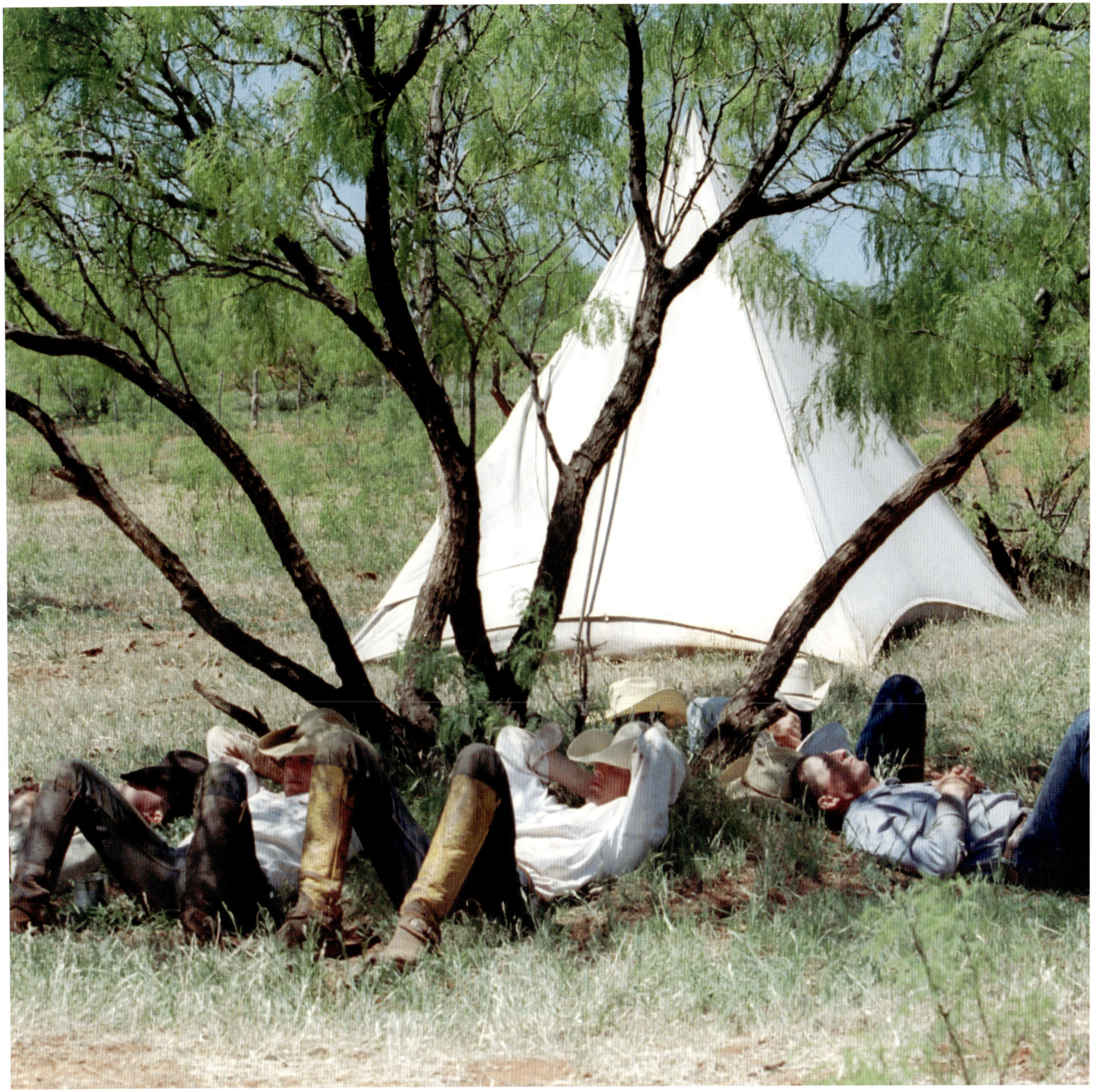

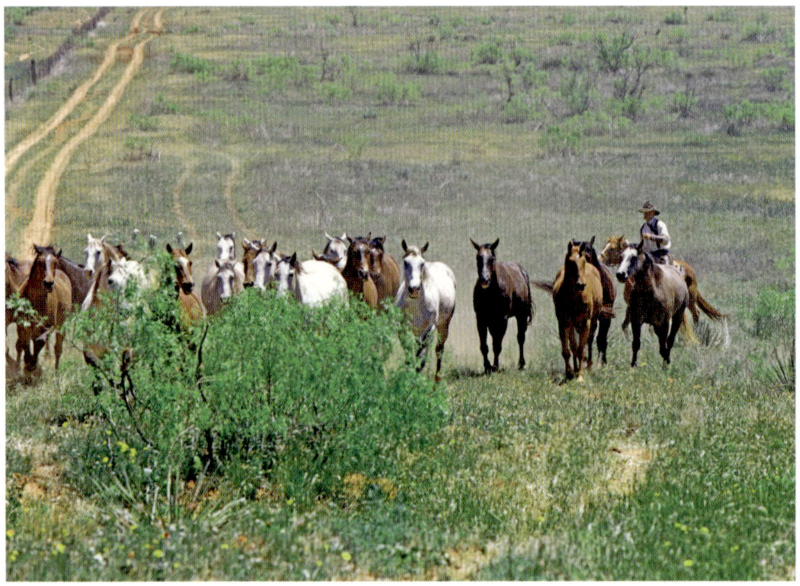

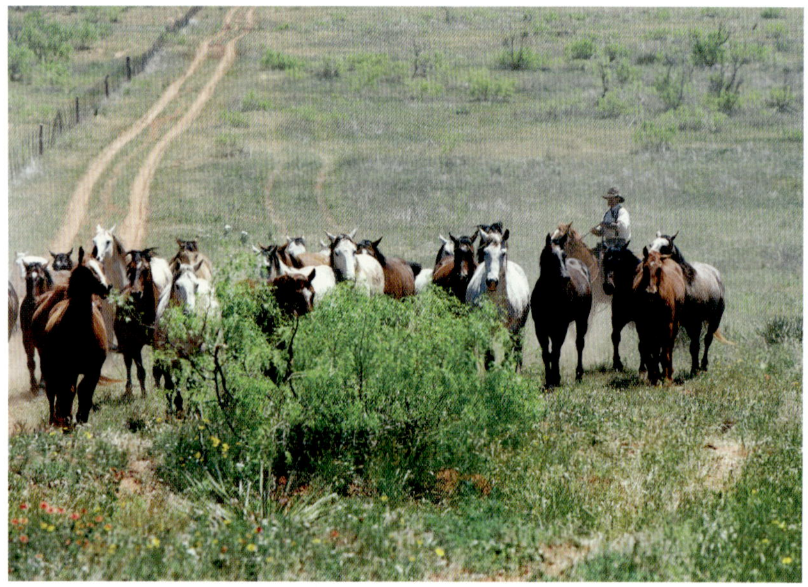

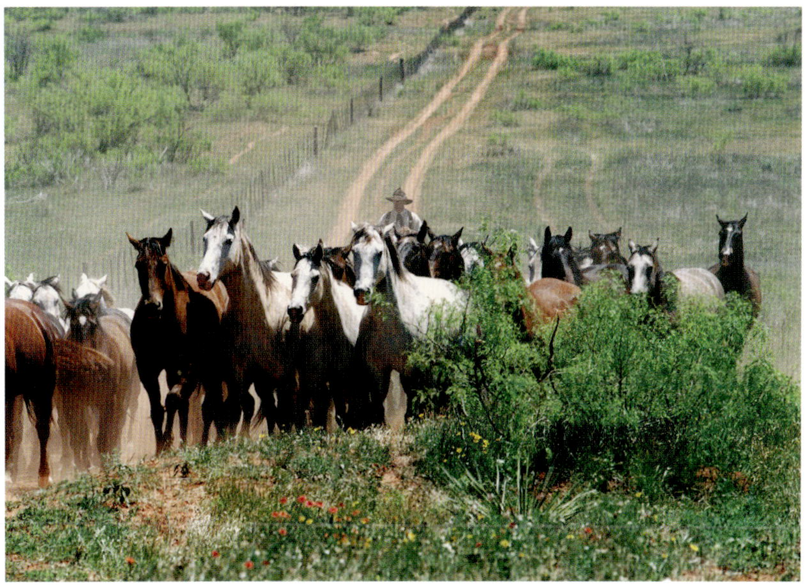

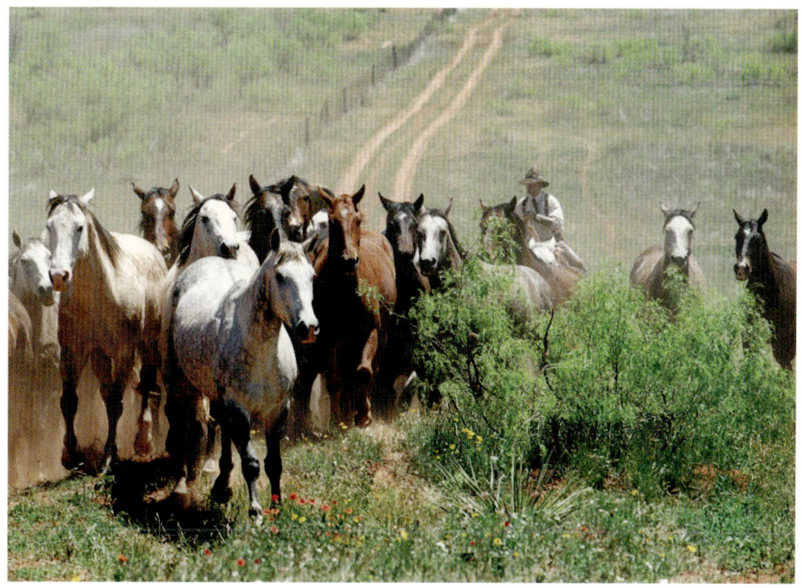

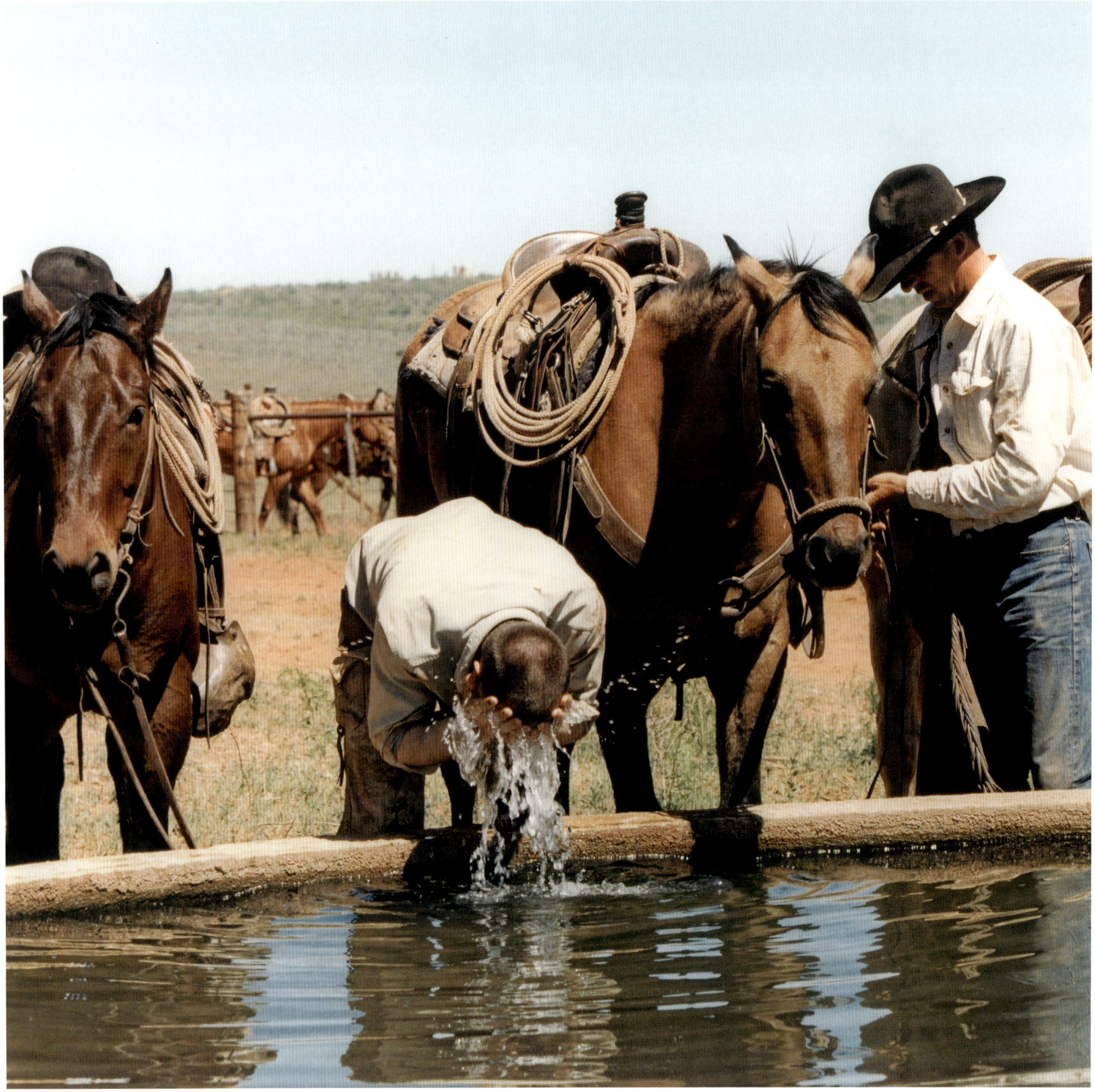

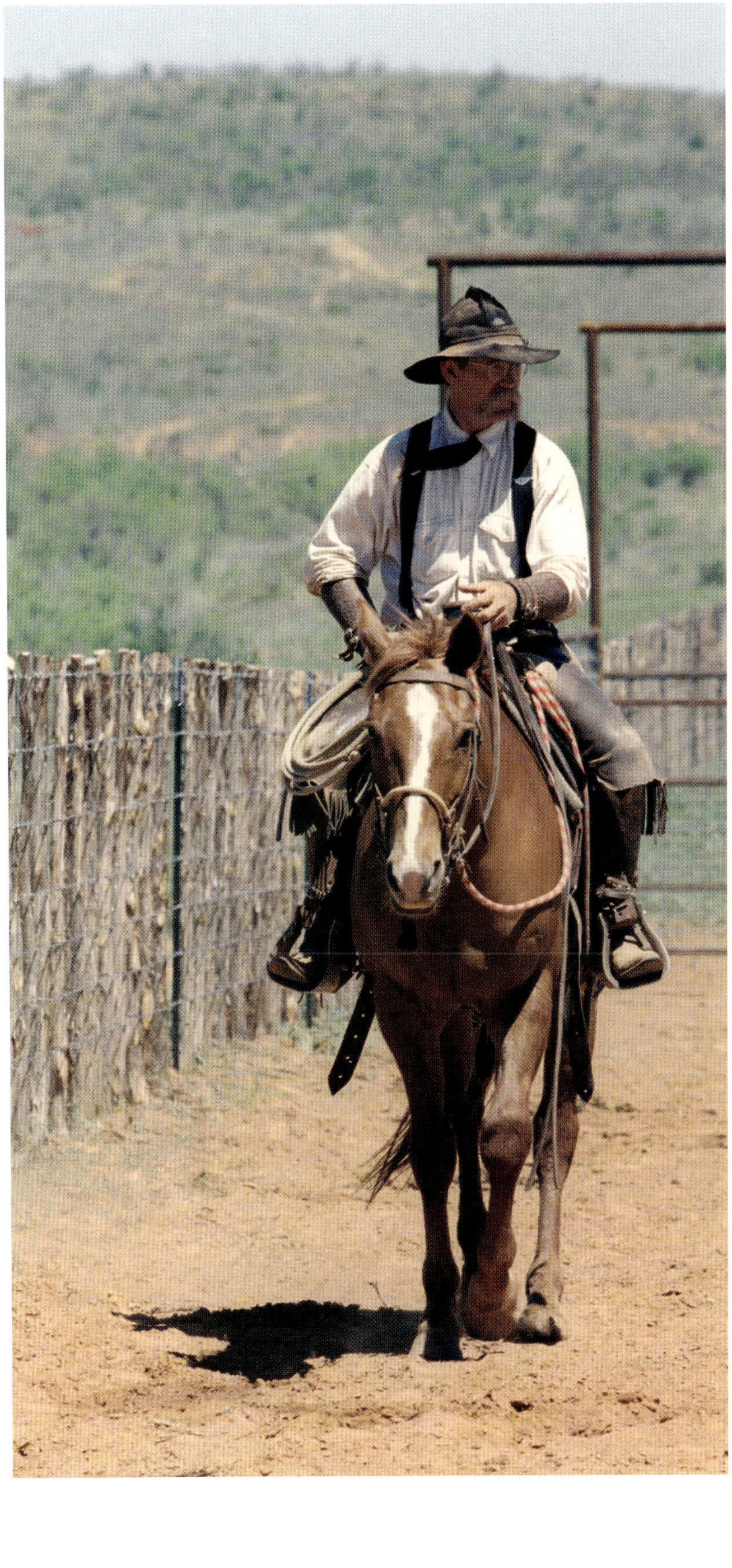

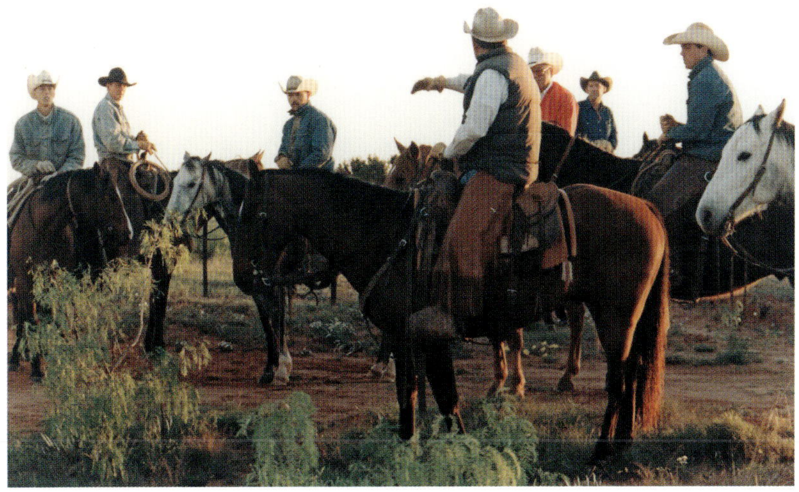

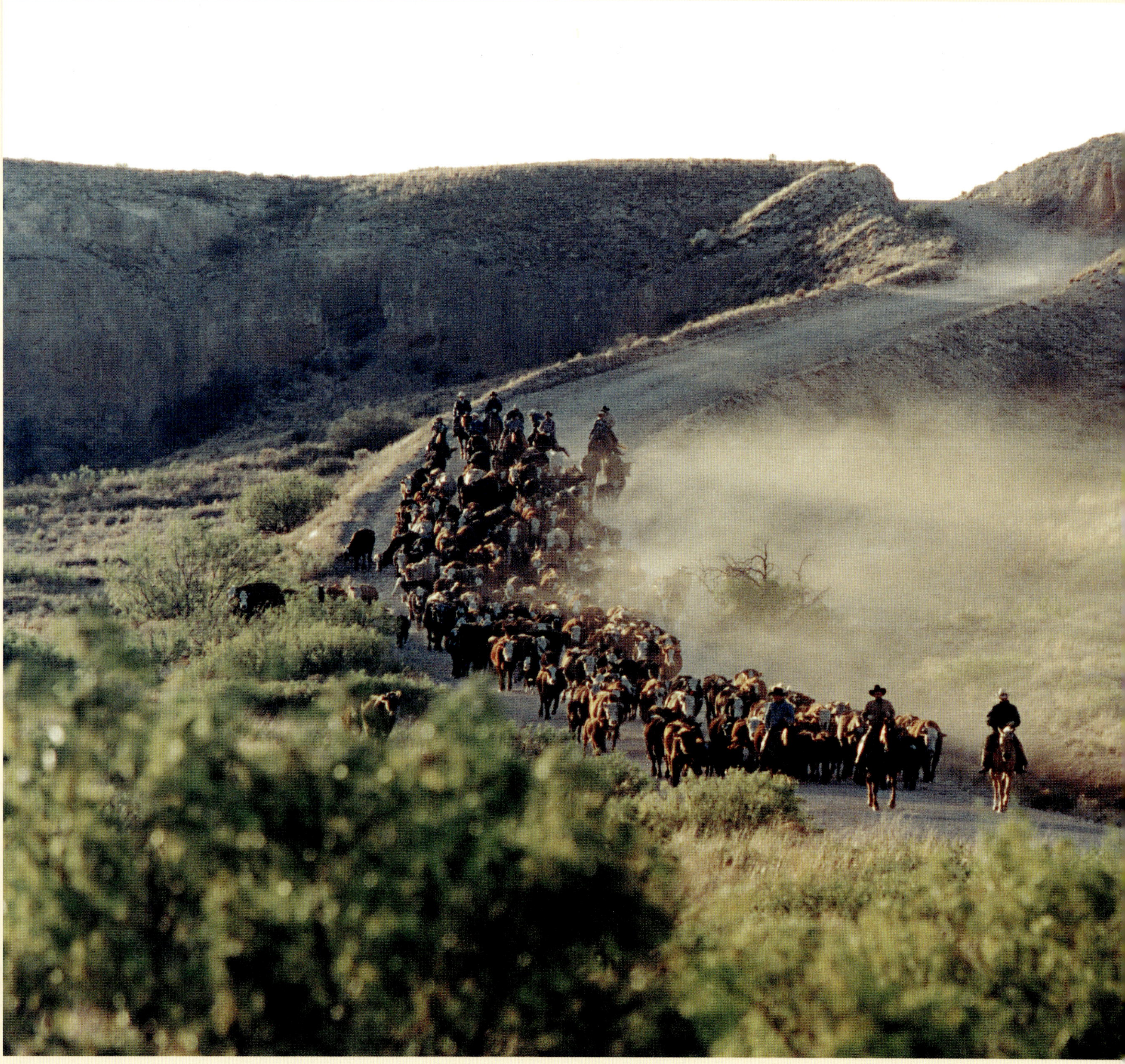

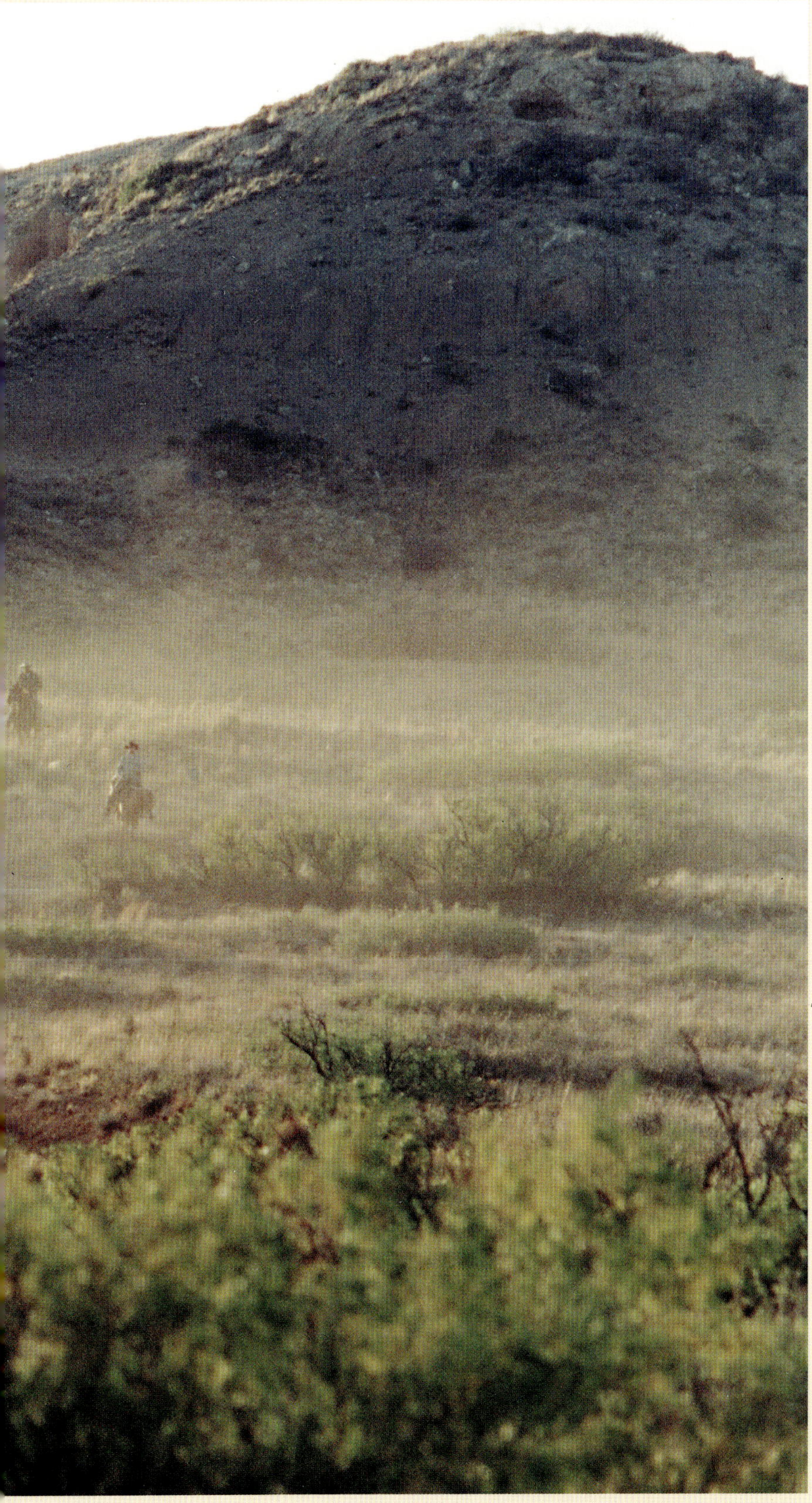

Bogle Ltd.

Dexter, New Mexico

1999

Good Horses, Good Cowboys

It's no accident that Bogle Ltd. was named the winner of the 1999 Best Remuda Award. They planned for it.

Bogle Ltd. ranches lie in the southeast corner of New Mexico, in an area centered around Roswell, reaching east to near Tatum and Lovington, south to Artesia, and north and west toward Corona. The property encompasses more than 840 square miles, better than half a million acres. Much of the land is plagued by drought, dotted with mesquite and cactus, and scorched by the hot summer sun. It's where Hal Bogle established his ranching interests when he arrived in New Mexico from Tennessee in 1917, and it's where his grandsons continue to ranch today. It's also where some of the best ranch horses in the country reside.

Daylight is still more than two hours away when Wayland Newberry hollers at his houseguest.

"Jim, you awake?"

"Yeah," I holler back, knowing full well that he realizes I'm lying. I also know that he's probably grinning to himself as he steps out on the back porch to smoke a cigarette and wait for me to get dressed. A glance at the clock glowing in the dark room tells me it's 3:50 a.m.

I hustle into my clothes, grab my camera bag and meet him on the back porch. By 4:20 we are at the corrals and Wayland is catching a horse. At 4:30 we're having breakfast in the headquarters kitchen, and by 4:40 the horses are loaded and we are headed for another part of the ranch.

Wayland Newberry is foreman of the Turkey Track Ranch, one of three divisions that make up Bogle Ltd. He's worked on the ranch for more than 30 years, and has been foreman since 1988. The Turkey Track Ranch makes up the largest portion of the Bogle holdings—640 sections, more than 400,000 acres.

Forty-five minutes later we join up with some other cowboys, including Stuart and Donald Bogle, two of the ranch's owners, and the day's work is planned. Light on the eastern horizon says dawn is not far off now.

Stuart and Donald Bogle and their brother, Scott, are managers of what was once known as the Hal Bogle Estate but today is incorporated as Bogle Ltd. Hal Bogle was their grandfather, and the three brothers inherited responsibility for the ranch from their father, Bill.

Headquartered at Dexter, New Mexico, a small town just south of Roswell, the estate is made up of three separate ranches plus a farming operation and a feedlot. Stuart manages the Turkey Track, which runs from the Pecos River east about 30 miles. Scott takes care of the Four Lakes Ranch, a 50-section (32,000 acres) operation between the towns of Caprock and Tatum, and Donald runs the Corona Ranch, 150 sections (96,000 acres) between Roswell and Corona. All three ranches are cow/calf operations. The ranches' broodmares are primarily on the Turkey Track and the Four Lakes.

Also included in Donald's responsibility is the feedlot. All three brothers pitch in and help with the farm.

"The farm is about 5,000 acres," Stuart says, "mostly hay, corn and wheat. We sell all the corn and most of the hay to the dairies around Dexter, but we chop some of the wheat into silage and we'll plant some sorghum silage to use at the feedlot."

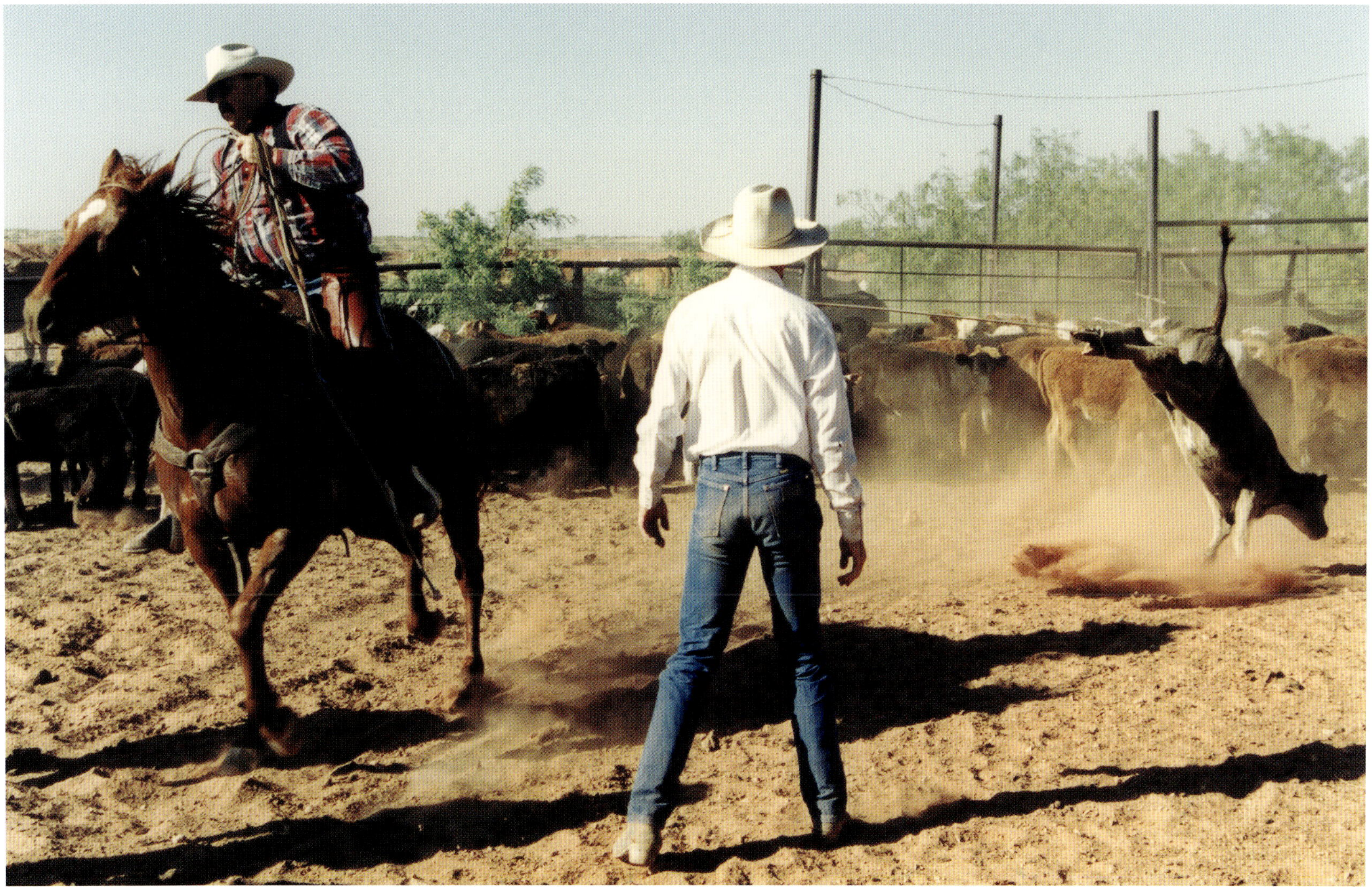

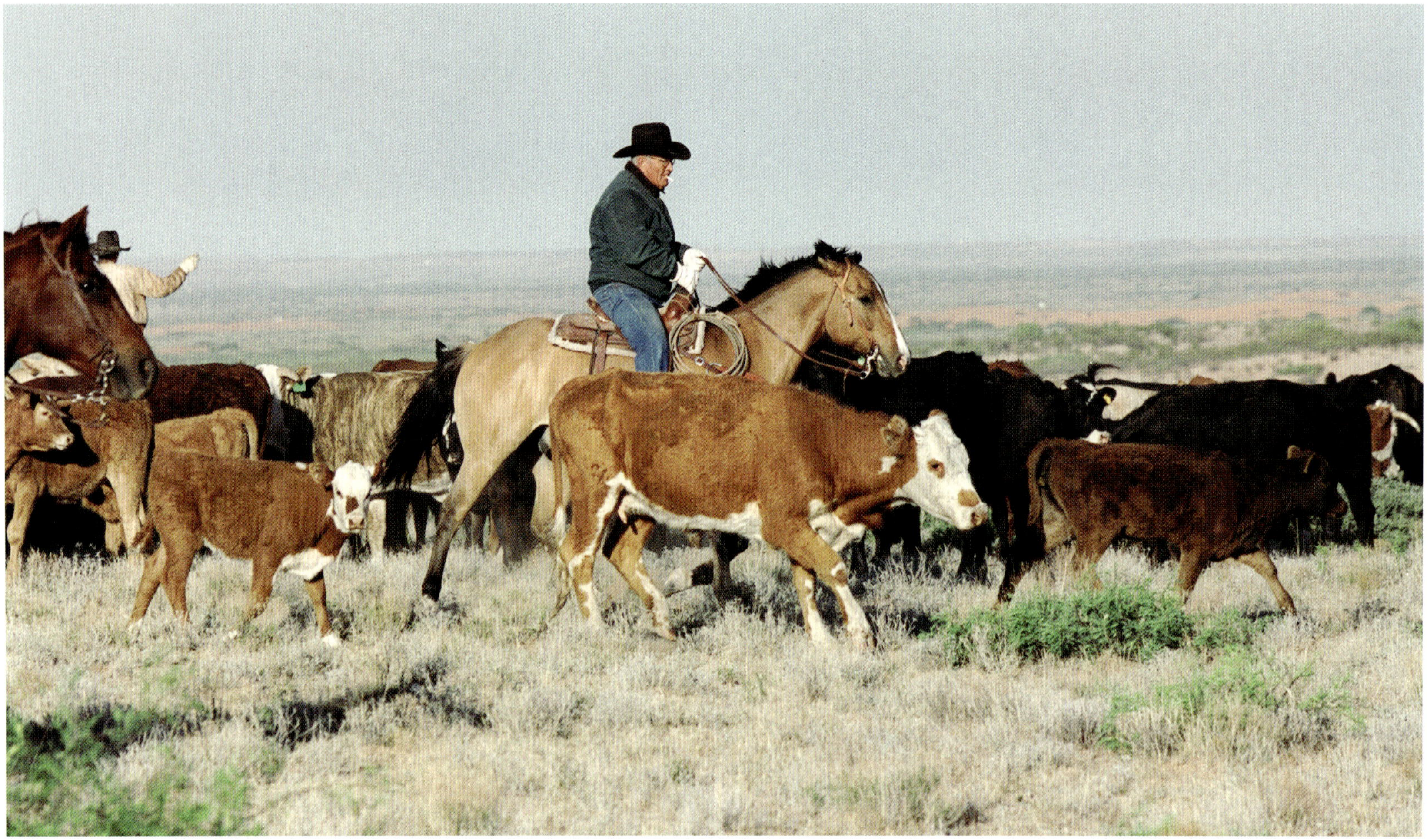

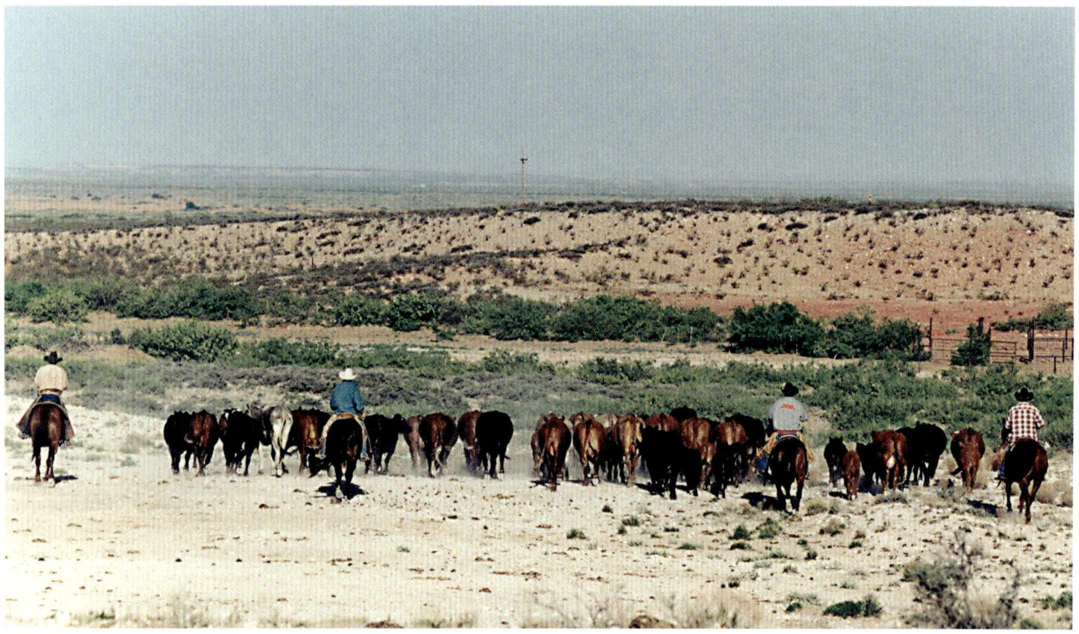

Cattle on the ranches number between 4,000 and 5,000 head, depending upon pasture conditions. For the most part they are crossbred cows with the base breed being Angus. There is one herd of straight-bred Angus cows, from which they raise bulls for the rest of the ranches, and they are using some Braford bulls. Mother cows make up the base of the operation, but they will occasionally run yearlings, too.

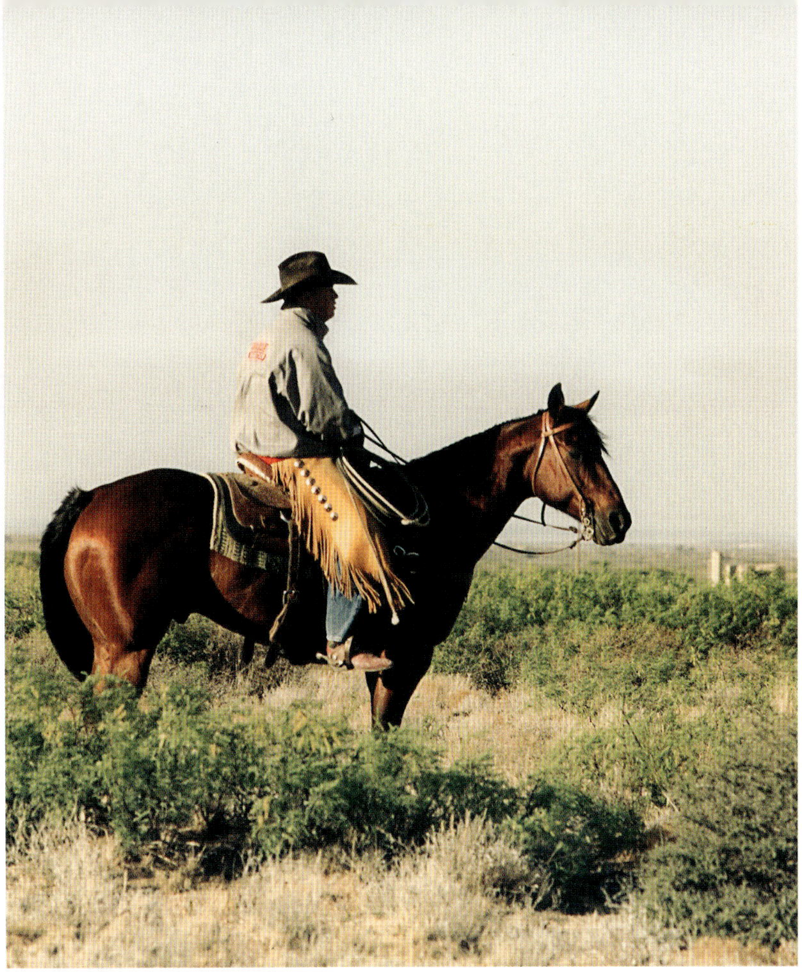

"Corona and Four Lakes are good yearling ranches, but at the Turkey Track, water is too far apart for it to be good yearling country," Stuart says. "We run yearlings on those ranches when we get some good rains."

This was not the year for yearlings.

The sun is not yet up, but all the cowboys are horseback now. This morning they are moving the cattle from a pasture in which the grass is exhausted, into one that contains some oak shinnery brush that the cattle will eat. It should be enough to last them two weeks—two weeks that that particular herd won't have to be fed.

For all practical purposes, it hasn't rained for a year in south-eastern New Mexico. An already harsh ranch country—annual stocking rate is just six cows per section—has been made even harsher by the drought. An occasional shower this past spring has given hope, but the clouds have always moved on, usually dropping a half-inch or less of moisture. Most of the dirt tanks are dry, and for water, the cattle have to depend upon the windmills that bring it from far below the ground. The almost never-ending wind is a Godsend—it keeps the windmills turning.

Horses are the only way to work the cattle on the Bogle ranches. The terrain, especially on the Turkey Track, is too rough for any kind of motorized device. Hal Bogle's first Quarter Horse stallion was a horse named Bogle. Foaled in 1939, he was by San Siemon and was brought to the ranch by world champion cowboy Bob Crosby of Roswell. Crosby was at a roping when he found the stallion in Stonewall, Oklahoma, on Dick Truitt's place. Crosby called Hal Bogle, who wired Truitt the money for the horse, and Crosby brought him to New Mexico. Bogle then became the base of the ranch's Quarter Horse breeding program.

Bogle was followed by such sires as Midnight Red B by Midnight, Flashy Waggoner by Waggoner Snip, and others whose bloodlines go back to King, Old Sorrel, Begger Boy (TB), Red Buck, Chicaro Bill, Redman and Little Joe. In later years the Bogles used Jet Twist Deck by Easy Jet and Peppy Motor Scooter by Mr San Peppy. Many of these sires are still reflected in the bloodlines of the ranch's broodmares.

Recently they used Zan Parr Jet by Zan Parr Bar and Docen Colonel by Colonel Freckles, but both of those stallions are gone now. Today they have Ray Ray Leo, a son of War Leo out of a daughter of Jackie Bee; Lil Peppys Sugar Bar by Peppy San Badger; and C J Goldrush by C J Sugar by Son O Sugar out of a daughter of Smooth Herman. They also have a son of Playgun, Play To The Max, but, only 3 years old, he's in training and is not being bred yet. The stallions that are being bred are turned out with their own bands of 15-20 mares in late March or early April, and picked up in July.

"Docen Colonel was really good for us," Stuart says. "He was out of a daughter of Doc Bar, but he was a pretty good sized horse, and he threw a little more size than a lot of horses that are bred that way. His foals have a really good disposition, and they are pretty athletic."

Ray Ray Leo is 30 years old, but still being used on a few mares. Lil Peppys Sugar Bar is a big stout horse with $7,700 in earnings from the National Reining Horse Association, and Donald Bogle is showing C J Goldrush in AQHA versatility ranch horse shows and in Ranch Horse Association of America competition.

With the exception of an occasional stallion prospect, the colts are all gelded and placed into the ranch remuda for the cowboys to ride. The better fillies are returned to the broodmare band, and the others are sold.

Stuart says, "Sometimes we'll break the fillies and ride them a little, but we'll definitely mess with them enough to be sure they're the kind we want to keep. We'll breed the fillies the first time when they are 3."

The Bogles try to keep their mares on decent pastures, but they are also supplemented throughout the year, and, of course, wormed and vaccinated.

Also, Stuart will carry two or three of his best mares to outside stallions each year. The primary goal for the horse breeding program is to keep good horses under good cowboys. As a matter of fact, that's always been their plan.

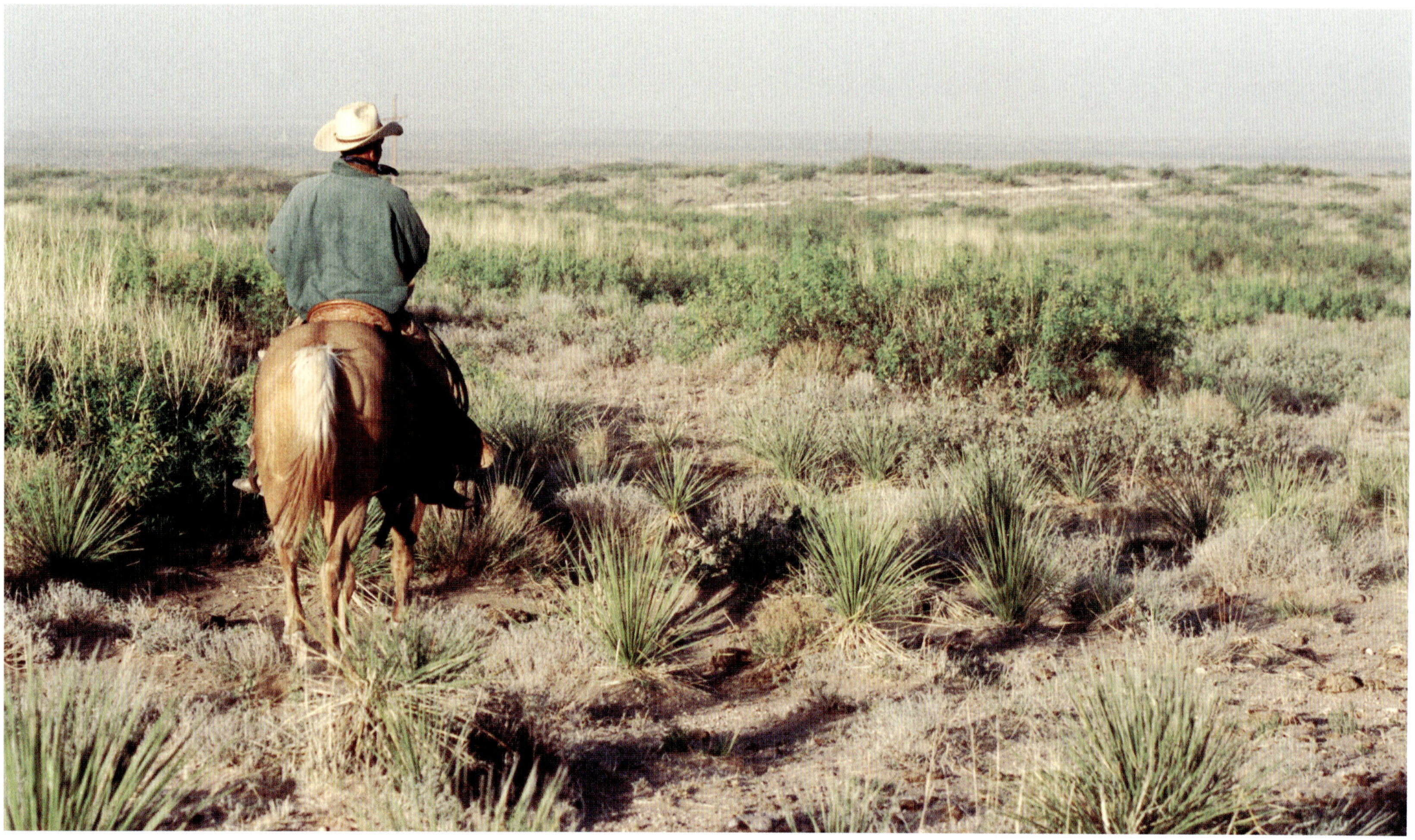

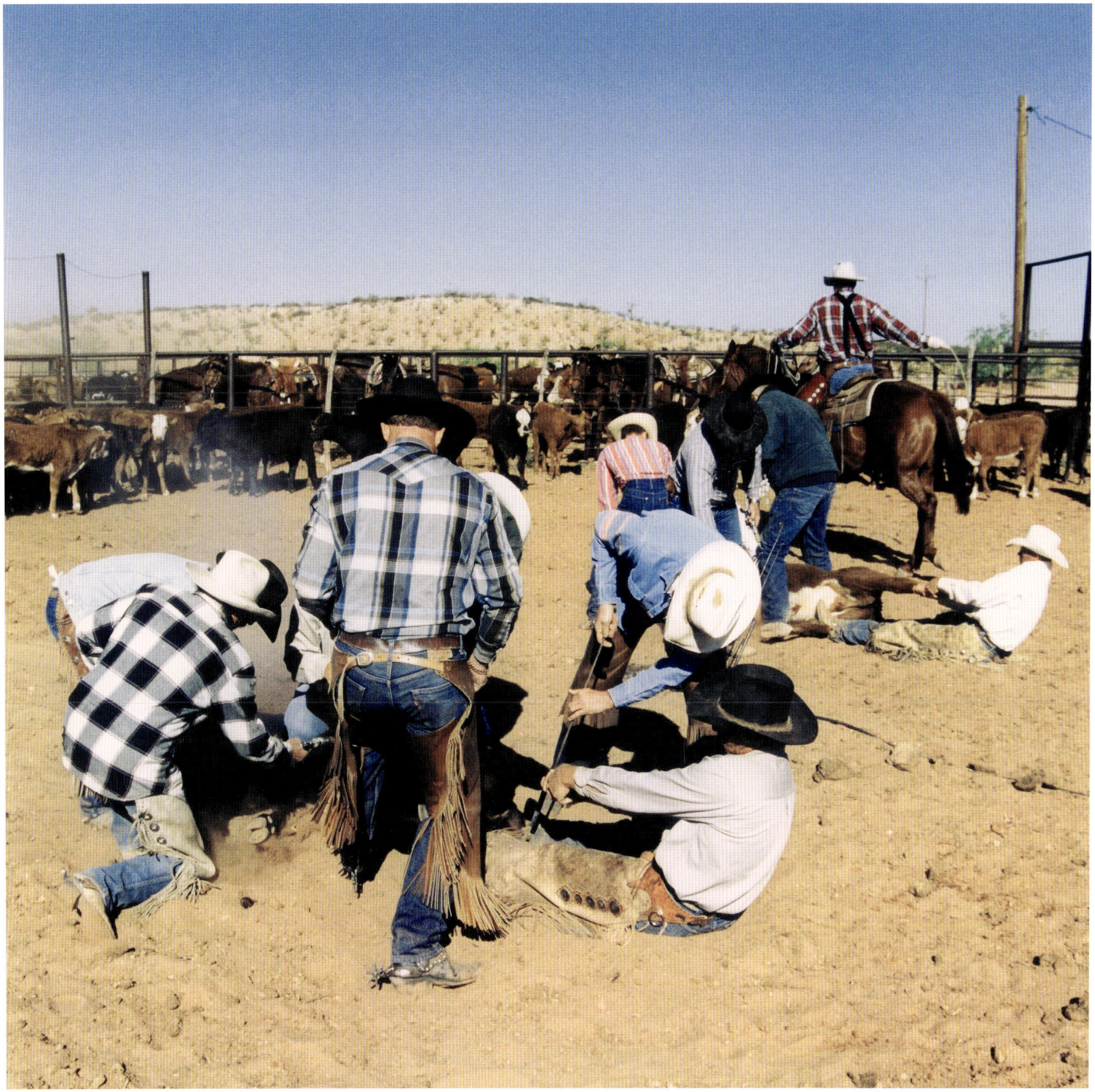

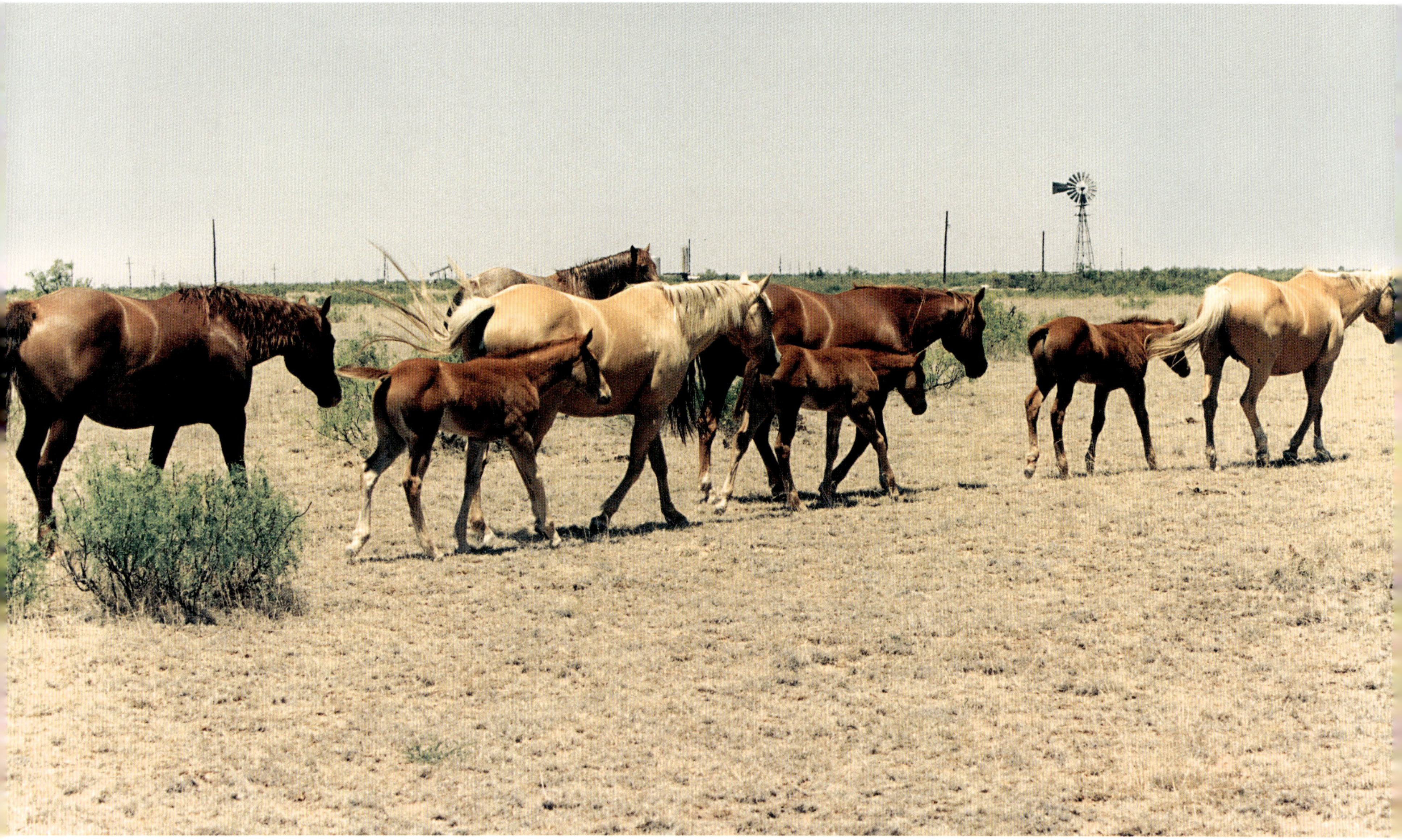

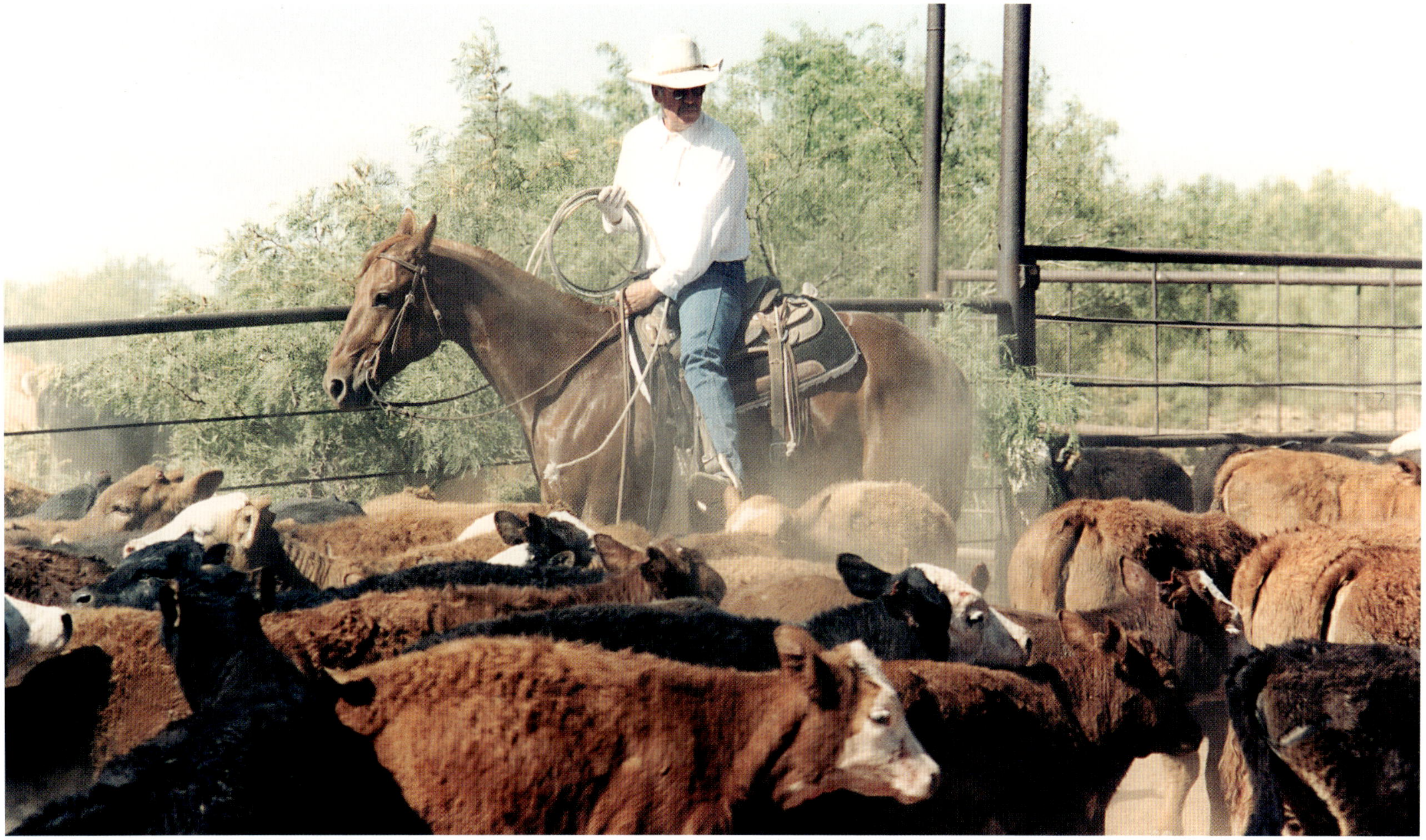

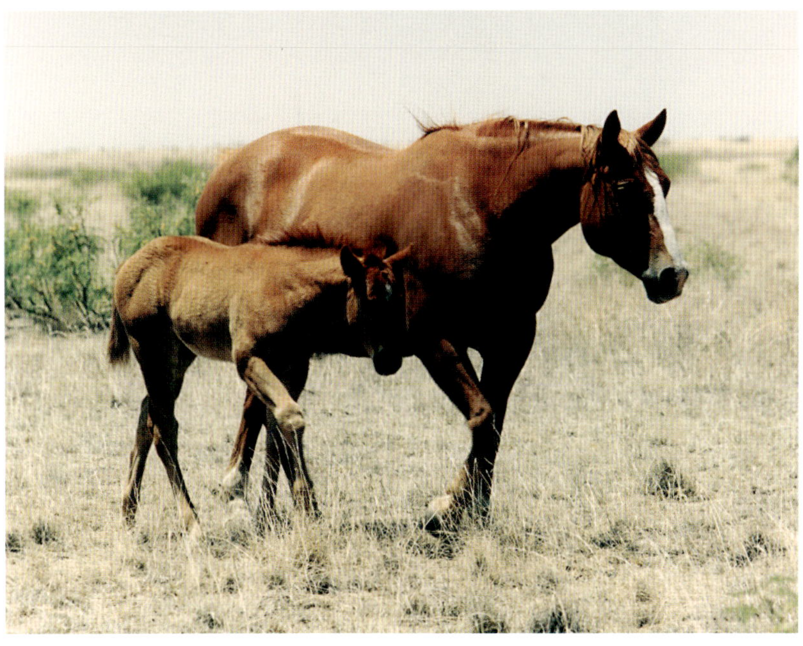

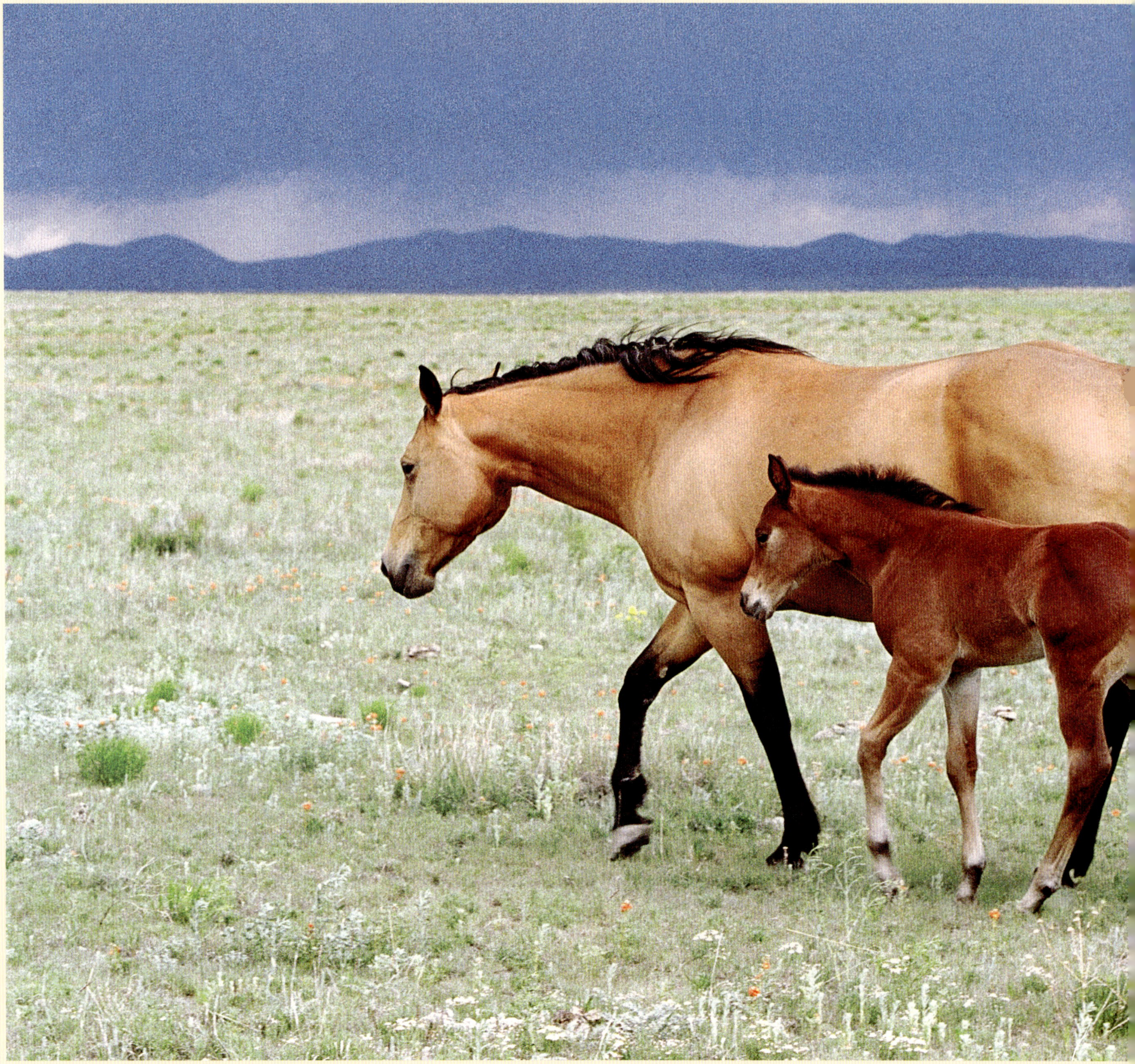

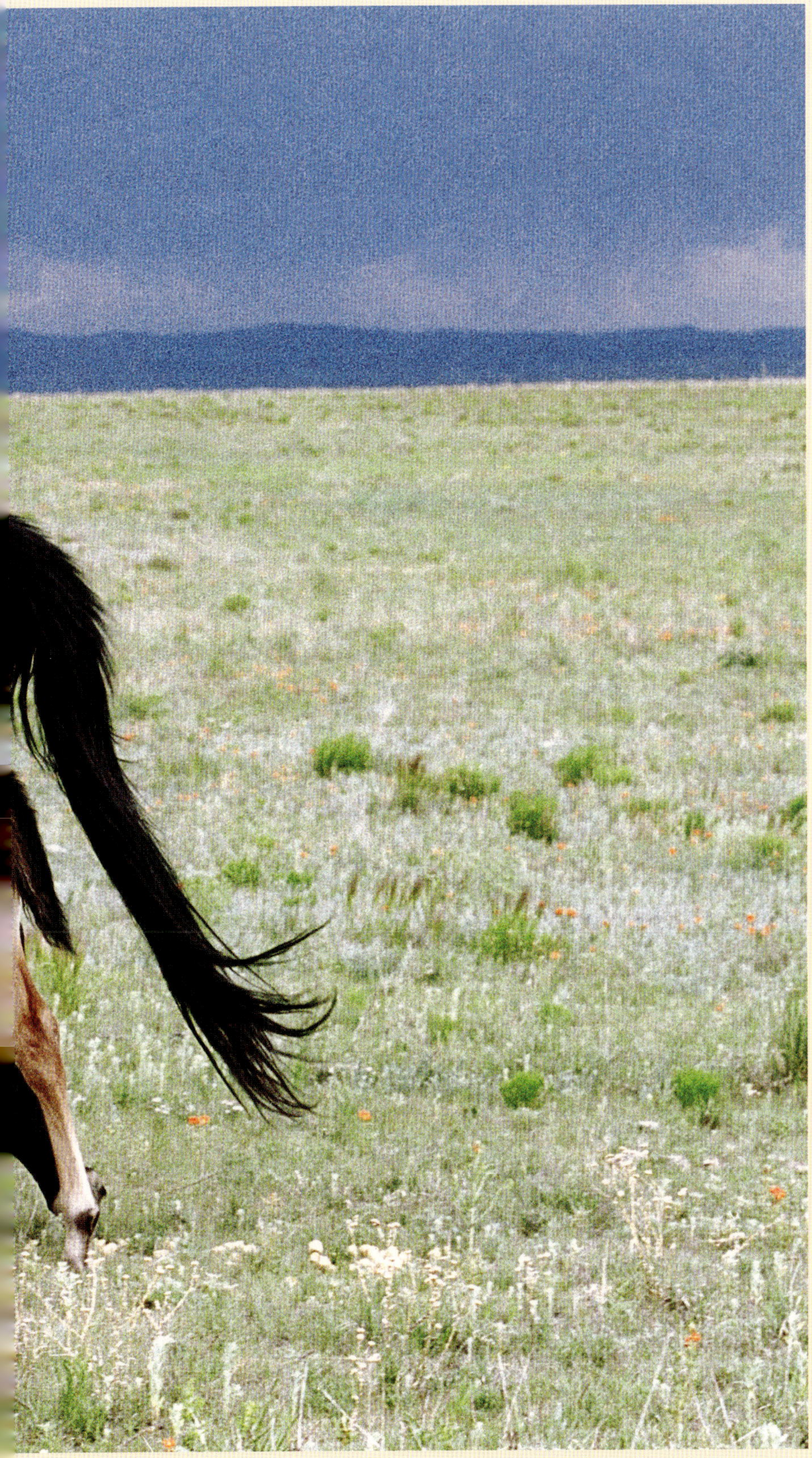

CS Cattle Company

Cimarron, New Mexico
2000

A Family Affair

The Davis family of New Mexico continues to run a ranch that was started by the great-grandfather of today's managers.

When Frank Springer topped a rise and looked down on the little village of Cimarron, New Mexico, in February 1873, two things immediately struck him—the snow-covered peaks to the west and the rolling short grass prairies to the east. It was a beautiful sight to a young man from Iowa, who had come west the year before as a lawyer for the Maxwell Land Grant. Living in Las Vegas, New Mexico, he had ridden north to Cimarron to take care of some company business, and immediately fell in love with the country. As a matter of fact, Springer was so struck with the area that that same afternoon, he bought 1,000 acres 6 miles east of Cimarron, on some of the prettiest pasture land he had ever seen.

That first thousand acres became the basis for the 180,000-acre CS Cattle Company, which in 2000 earned the Best Remuda Award. It was Frank's brother, Charles, who gave his initials to the ranch's name after he joined Frank in 1878 and became manager of the ranch.

Management of the CS Cattle Company has remained in the family to the present time. Charles Springer was succeeded by his nephew Hank, Frank's youngest son, and Hank was succeeded by his older brother Edward. The title then went to Les Davis, a grandson of the ranch's founder who took over in 1946. Les ran the ranch until he died in May 2000. He was succeeded by his wife, Linda, who has lived and worked on the CS for 50 years. She is also the daughter of Albert Mitchell, who served four terms as president of the American Quarter Horse Association and who had run both his own Tequesquite Ranch and the famed Bell Ranch, both located in eastern New Mexico. Given all that family history in the ranching business, it's not surprising that all of Les and Linda's six sons and daughters have returned to the ranch where they serve as vice presidents in charge of specific divisions of the company.

The history of the CS remuda goes back to 1890, when Frank Springer bought at auction two young English-bred Thoroughbred stallions—Uhlan II by stakes winner Uhlan, and Ute Chief by Maximilian—along with 12 fillies, also Thoroughbreds. These horses became the basis for the ranch's herd, but the blood was strengthened in 1912 when Hank Springer bought from Quarter Horse foundation breeder Billy Anson a horse they called "Little Joe." Not to be confused with the Little Joe of South Texas fame, the CS's Little Joe was by Harmon Baker by Peter McCue and is known in AQHA records as Old Joe 2.

Through the years, the ranch continued to add to and upgrade its horse herd, and in the 1920s, manager Ed Springer took advantage of the U.S. Army's remount program. In an effort to obtain cavalry horses, the Army provided mostly Thoroughbred stallions free of charge to ranchers. Then, if the ranchers decided to sell the resulting foals, the Army got first choice. Among the Thoroughbred stallions with which the CS was provided was Chimney Sweep, the sire of Brush Mount.

Brush Mount was one of the horses that Springer declined to sell to the Army. He used him on some of the ranch's mares, but the stallion ended up in the possession of Hank Wiescamp of Colorado and, then, the nearby Moore Ranch of Raton, New

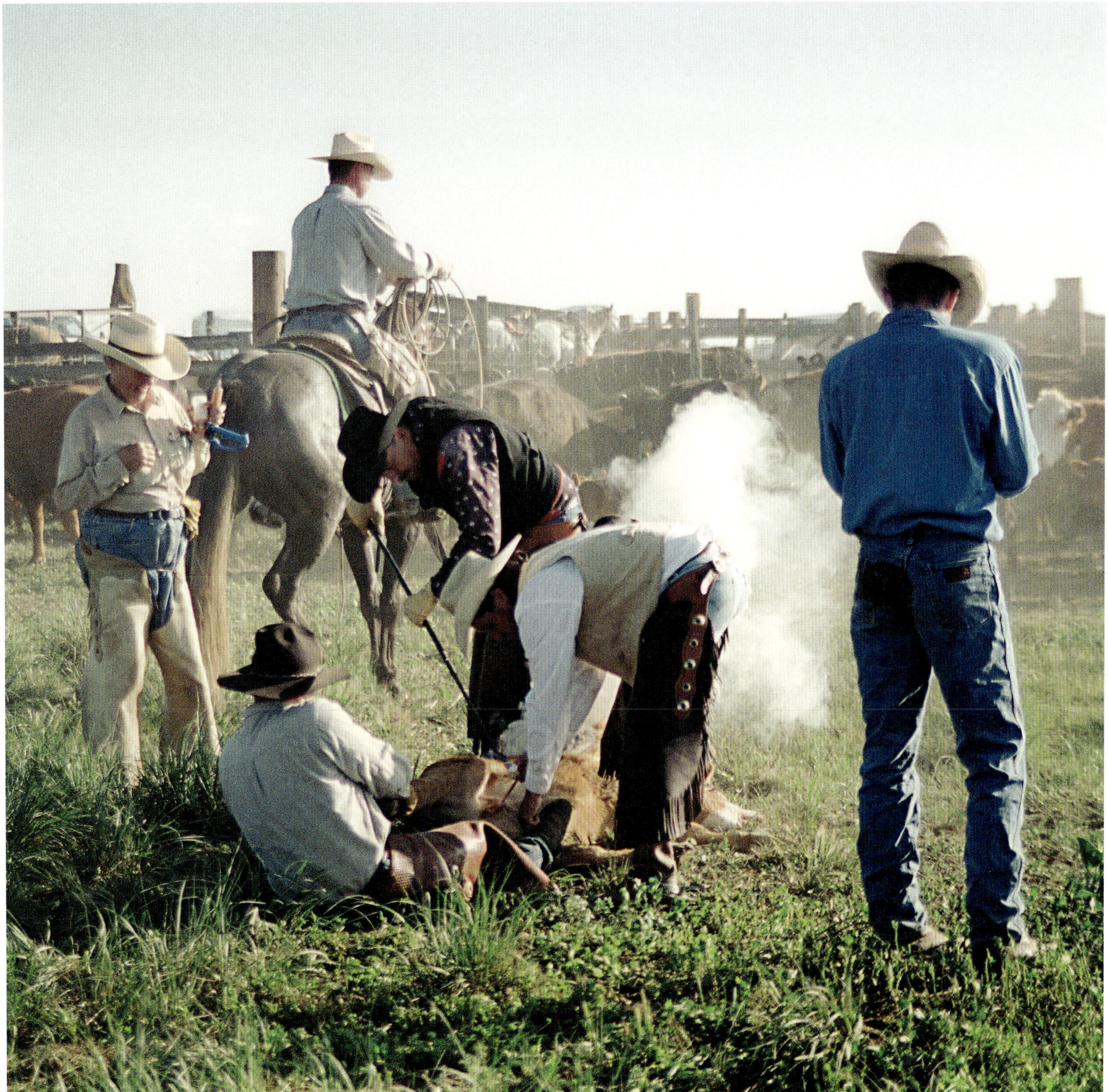

Mexico. Brush Mount became one of the early foundation sires of AQHA. Among his get was Gold Mount, the sire of Maddon's Bright Eyes, one of the top racing Quarter Horses of the 1950s and the world champion racehorse in 1951.

When Les Davis took over the ranch in 1946, he realized what his uncle, Ed Springer, had known all along: Raising cattle was not the only way to make the ranch pay. There was also money in horses. In addition to selling to area ranchers, Les began to market the horses on both coasts as mounts for polo players and as hunter/jumper prospects. Then, with his good friend, Albert Mitchell, serving as AQHA president from 1946-48, he saw the advantages of having registered horses. At that time, the Association was inspecting horses for Quarter Horse conformation before they could be accepted into the registry. In 1948, Les sent word to Amarillo that he had horses ready for inspection. Forty-three of 60 CS mares were accepted and given registration numbers. Thus began a breeding program that has continued through the years. In 1998, the CS Cattle Company was recognized by AQHA as being a Legacy Breeder, one that has registered horses for 50 consecutive years.

Today, the CS runs about 10 broodmares, down, due to the drought, from the 25 head they had a short time ago. They do, however, still have two stallions on the ranch. Mr Remind is by Letters Reminder by Doc's J Jay out of a Mr Snooper mare, and LH Red Hot Affair is a Colonel Freckles- and Sonny Dee Bar-bred stallion.

In addition, they are breeding up to four mares a year to Roosters Shorty by Gallo Del Cielo out of a Shorty Lena mare. Roosters Shorty is owned by Sally Schwartz of nearby Raton, New Mexico.

Cattle numbers, too, are down due to the drought. When the grass is good, there are approximately 2,200 mother cows, and another 2,000 or so stocker cattle on the ranch. And with those kinds of numbers, the ranch managers feel that horses are necessary. "We use our horses all year long," says Kim Barmann, daughter of Les and Linda Davis and vice president in charge of the Clayton Place, one of the ranches that makes up the CS Cattle Company. "We follow a time-controlled grazing management system, which requires us to gather and move our cattle from pasture to pasture throughout the year.

"In the spring, we work our calves by heeling and dragging them to a branding fire," she continues, "just like they did it on this ranch a hundred years ago. And we do it that way because we think it's the best."

Most of the horses in the saddle horse remuda are geldings. However, mares that are prospects for the broodmare band are started along with the colts, and ridden until the owners determine whether or not a particular filly has the characteristics they want continued in their horse herd.

Besides Kim, other Davis family members with specific responsibilities on the ranch include Warren, the oldest son and manager of the Crow Creek division; Julia Davis Stafford, an attorney who handles all the ranch legal work and manages the cattle on the headquarters division; Randy, who is in charge of the hunting operations; Kirk, who handles all the ranch's machinery and oversees the farming operation; and Bruce, who manages the mountain ranch country and handles all the cattle buying and selling. Although each sibling has his or her own specific responsibilities, when it's time to work cattle at just about any place on the ranch, almost everyone is there, horseback.

The CS Cattle Company has been a family-run ranch from the very beginning, when Frank Springer brought his brother, Charles, out to manage things. It's still family run, and it's still in the same family. Frank would be proud.

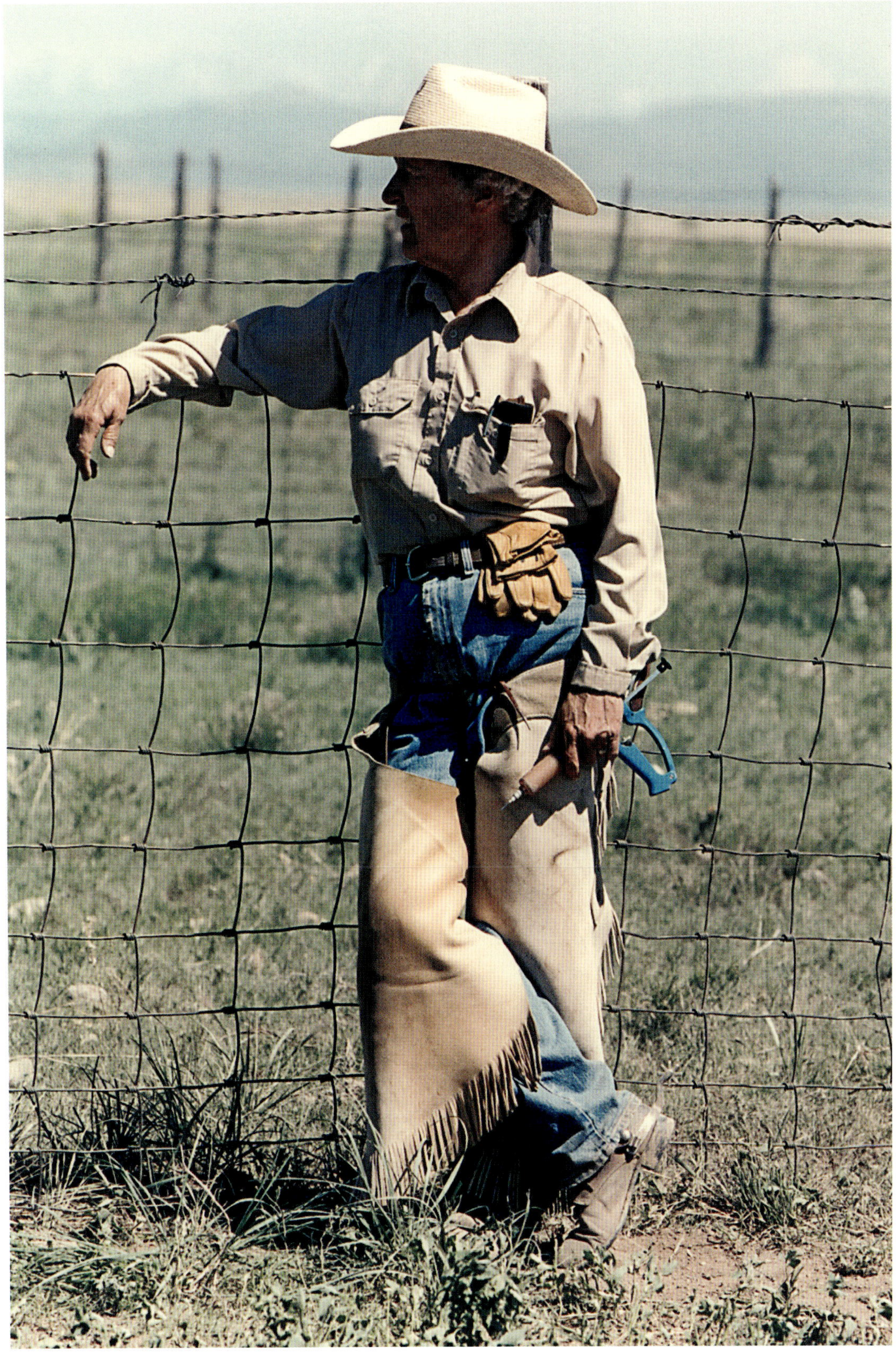

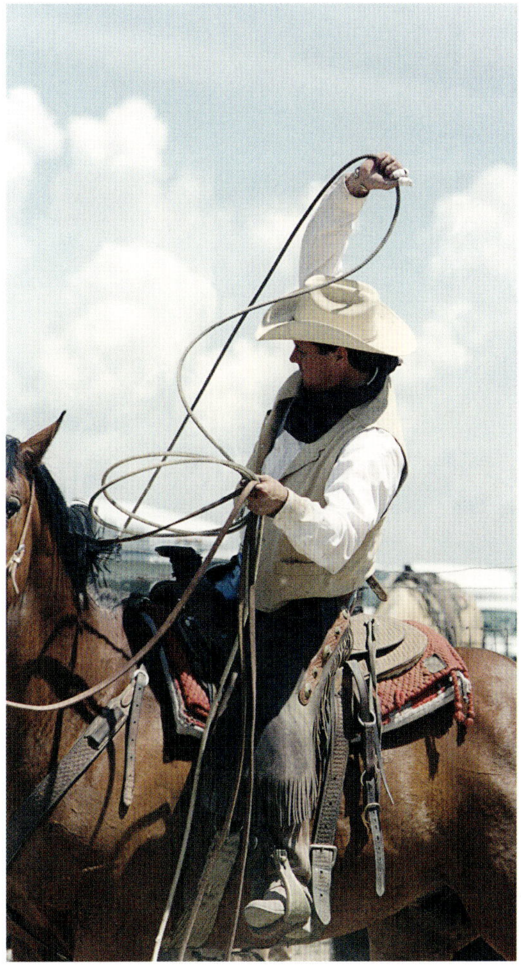

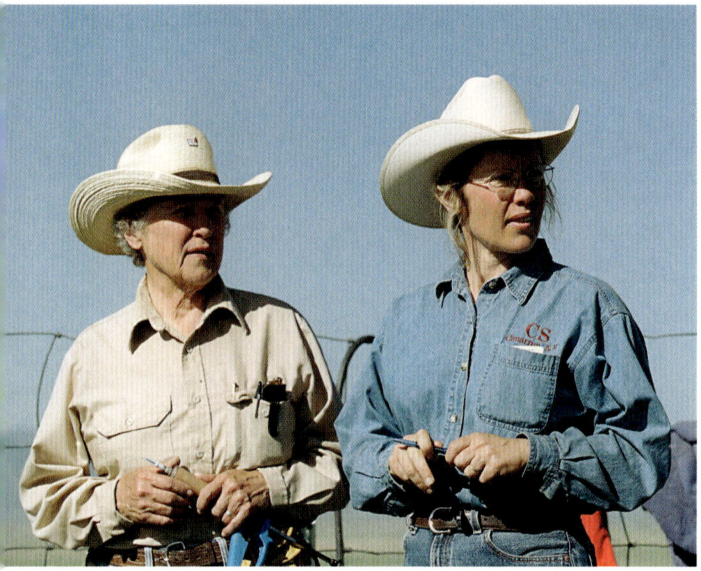

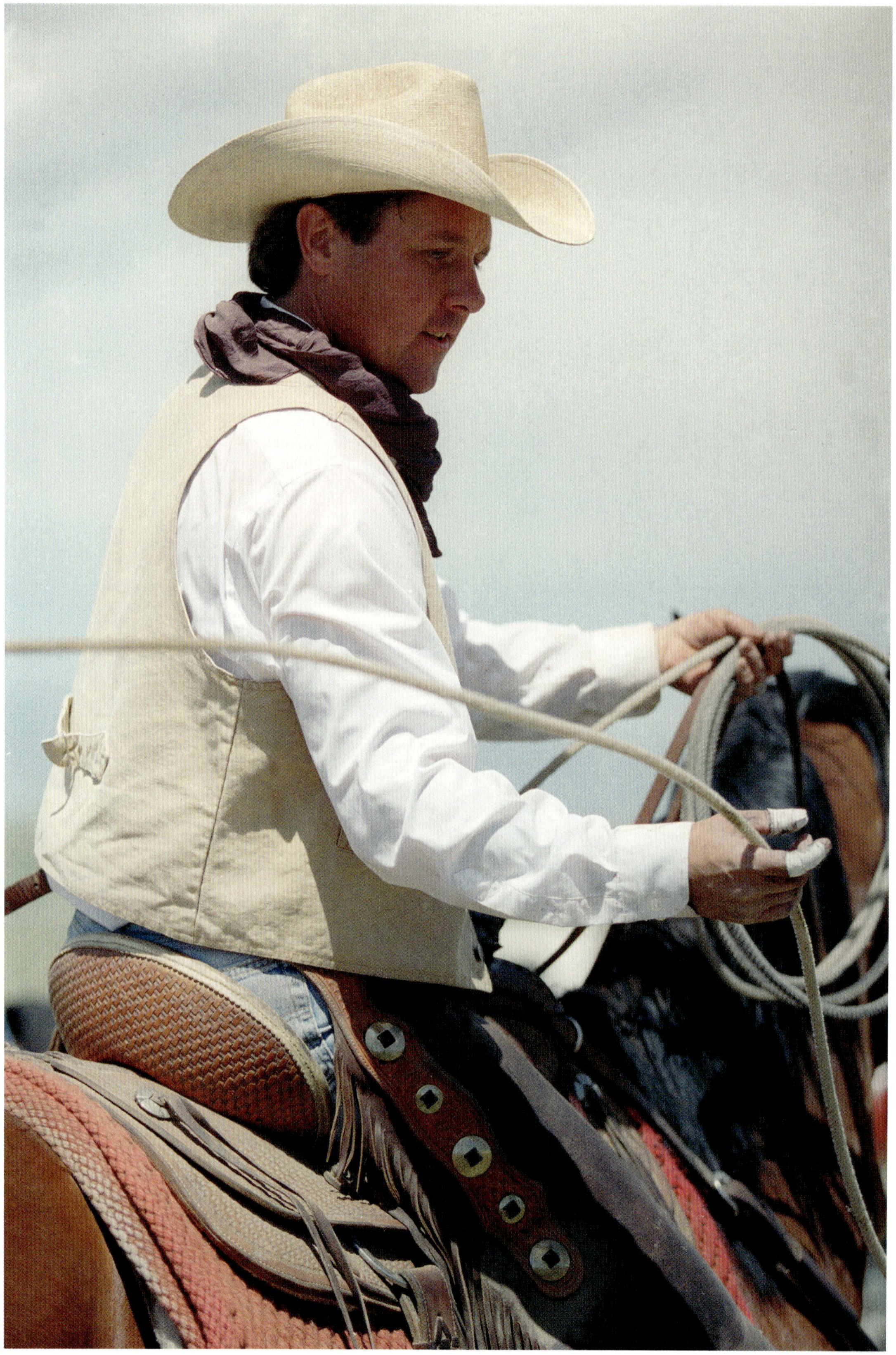

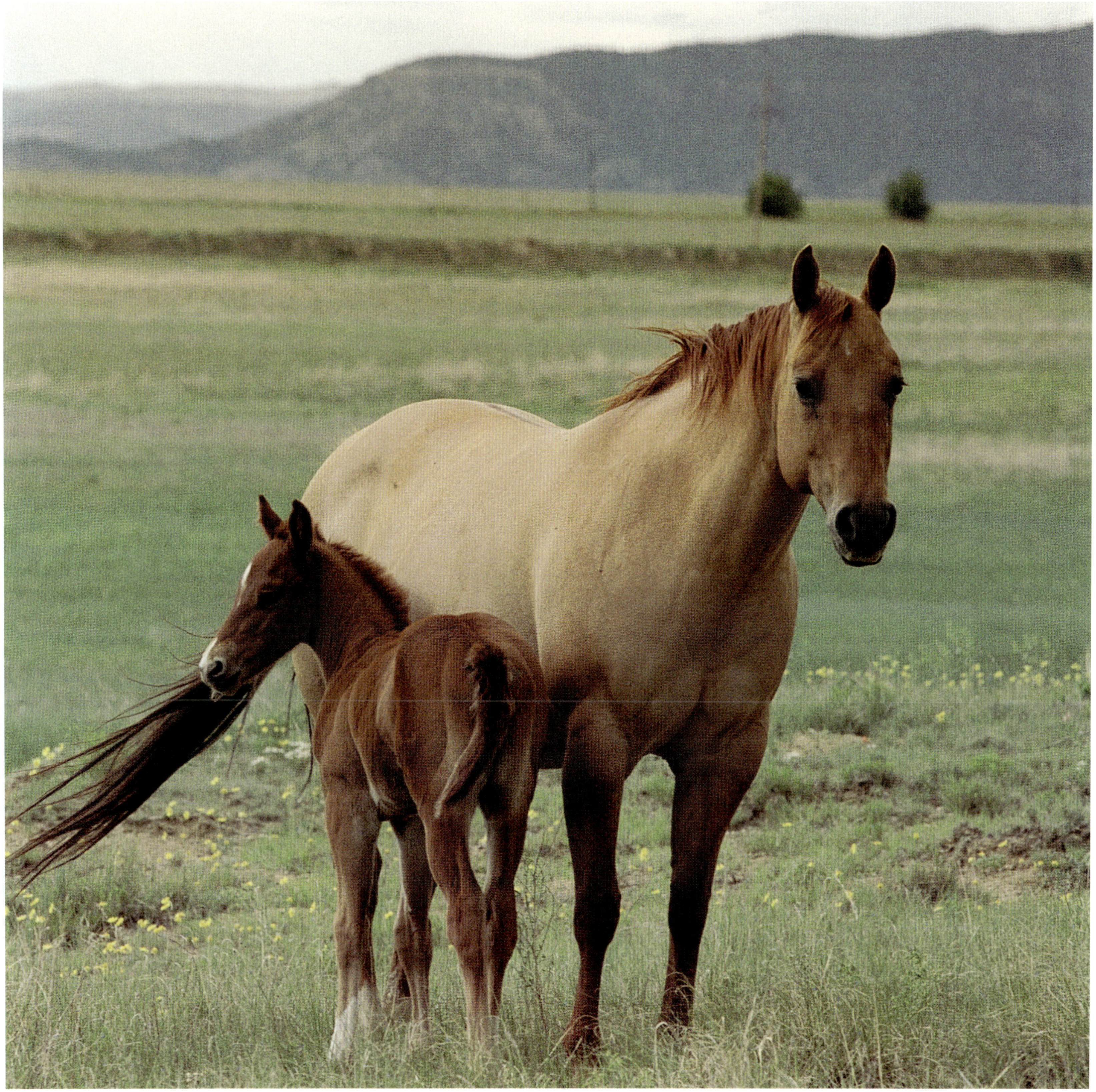

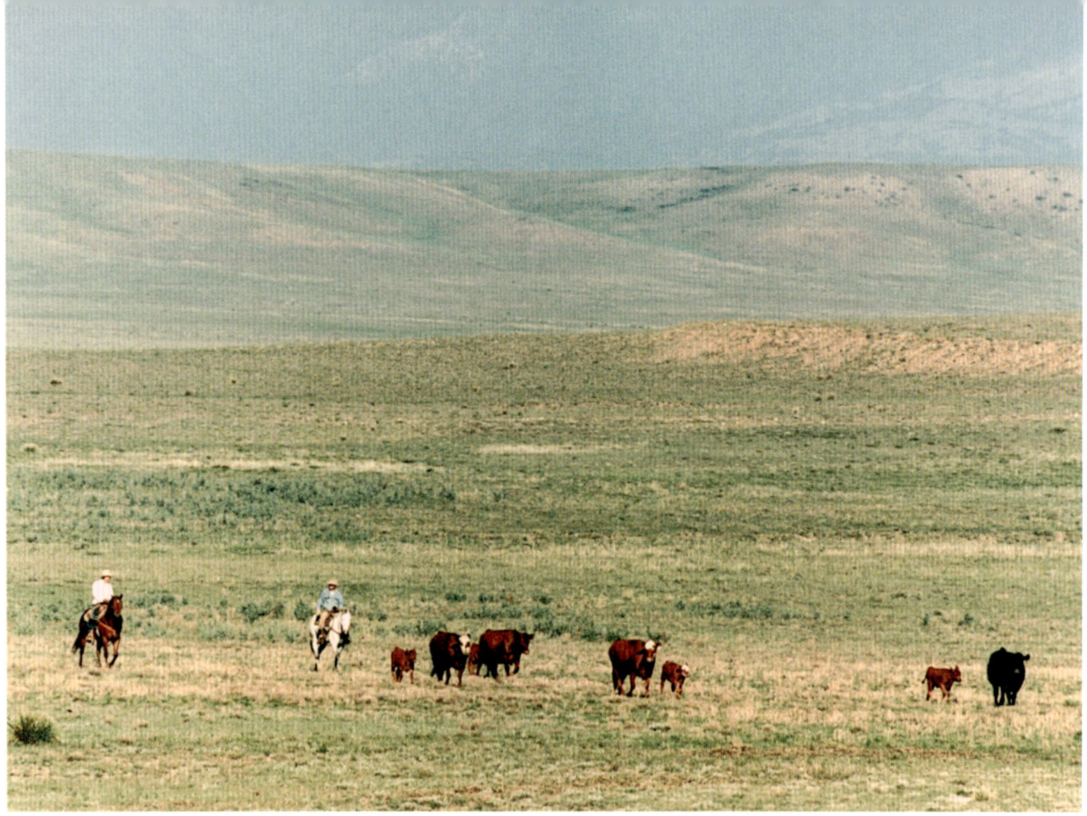

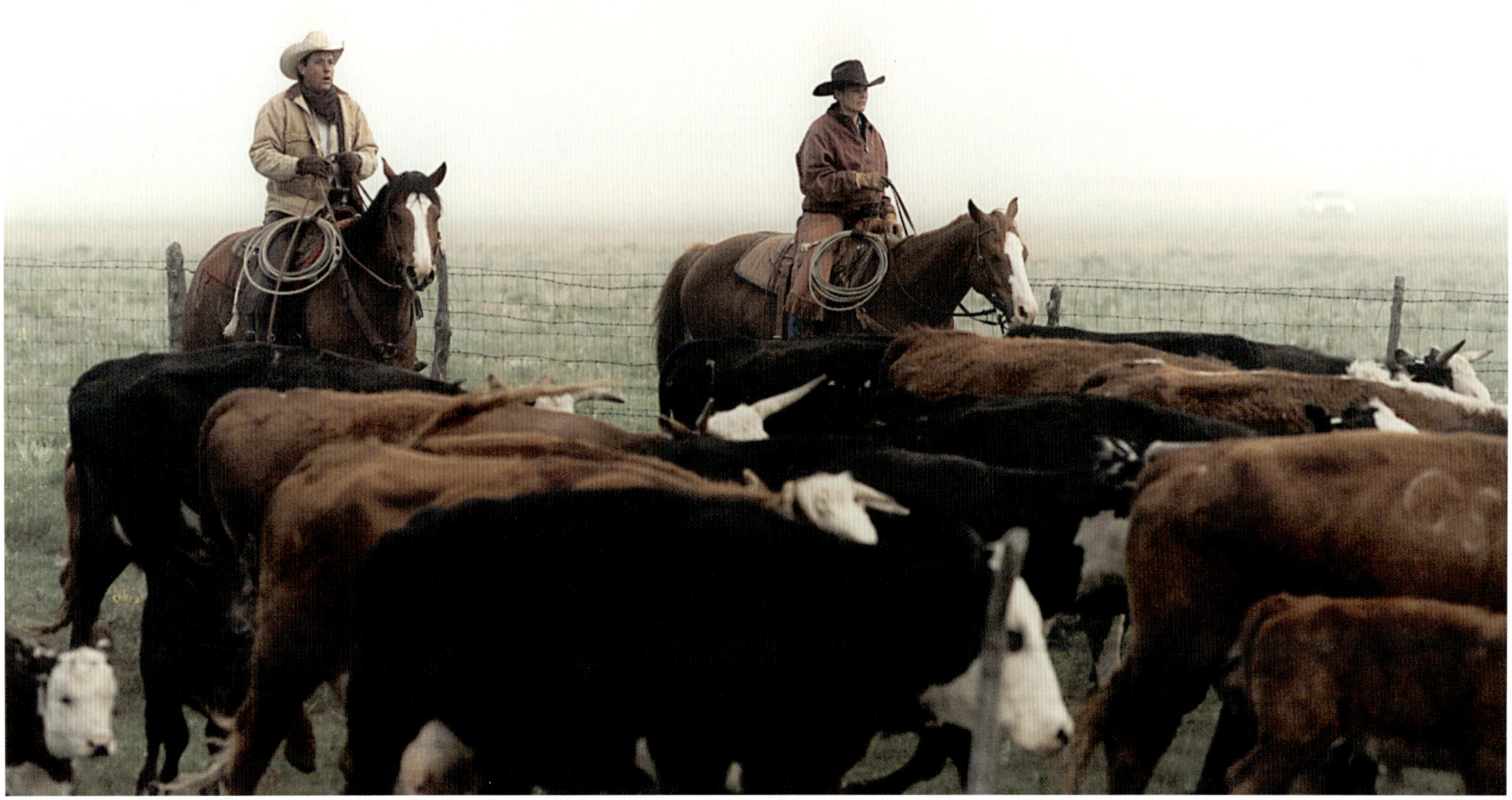

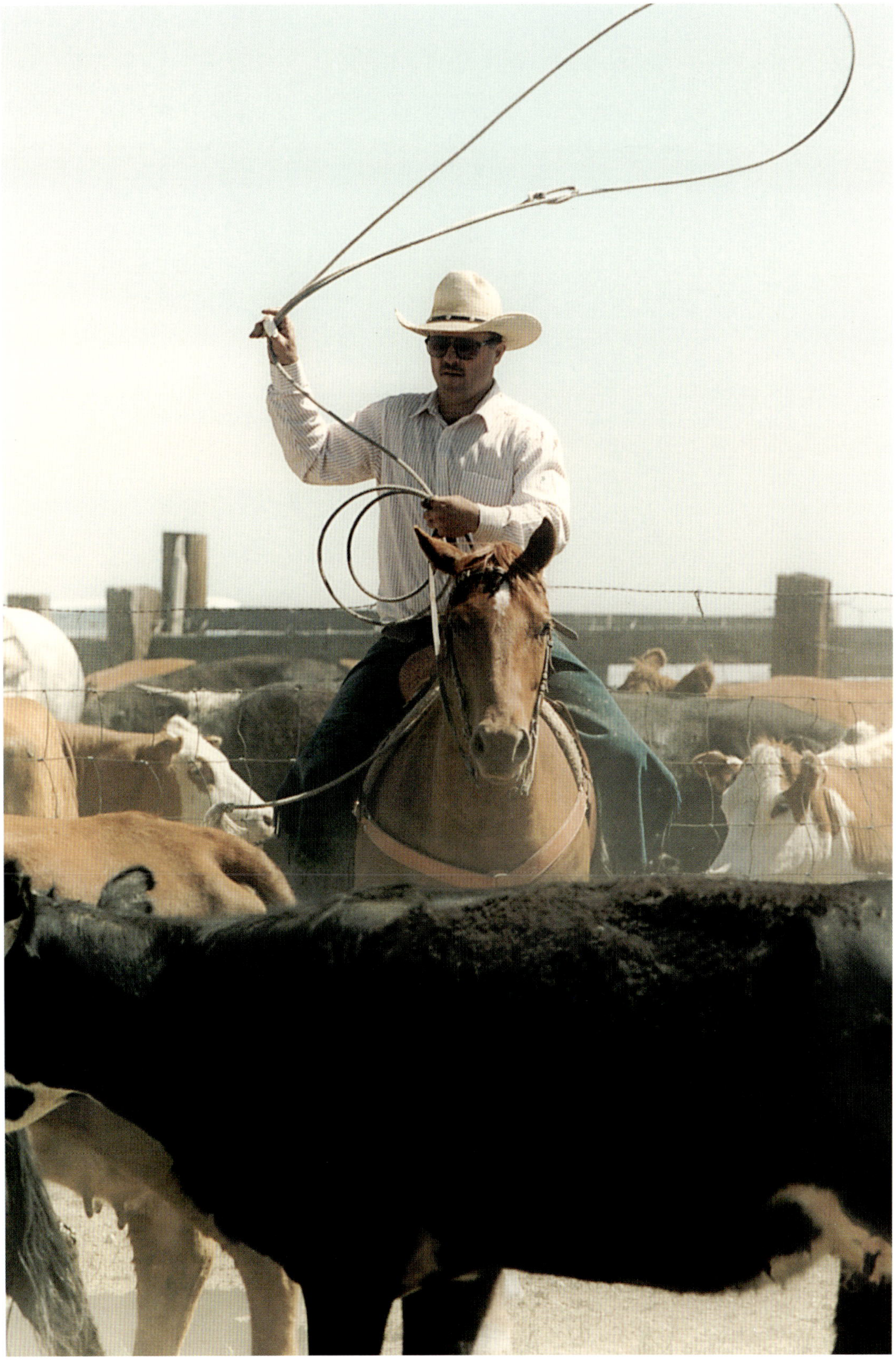

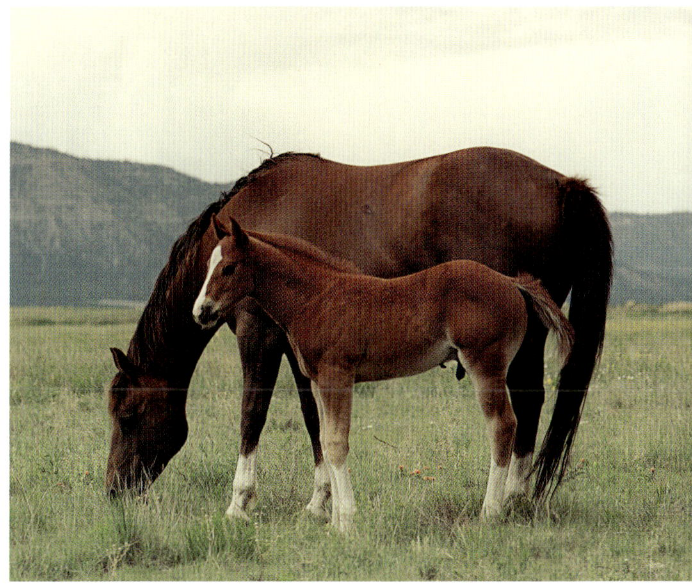

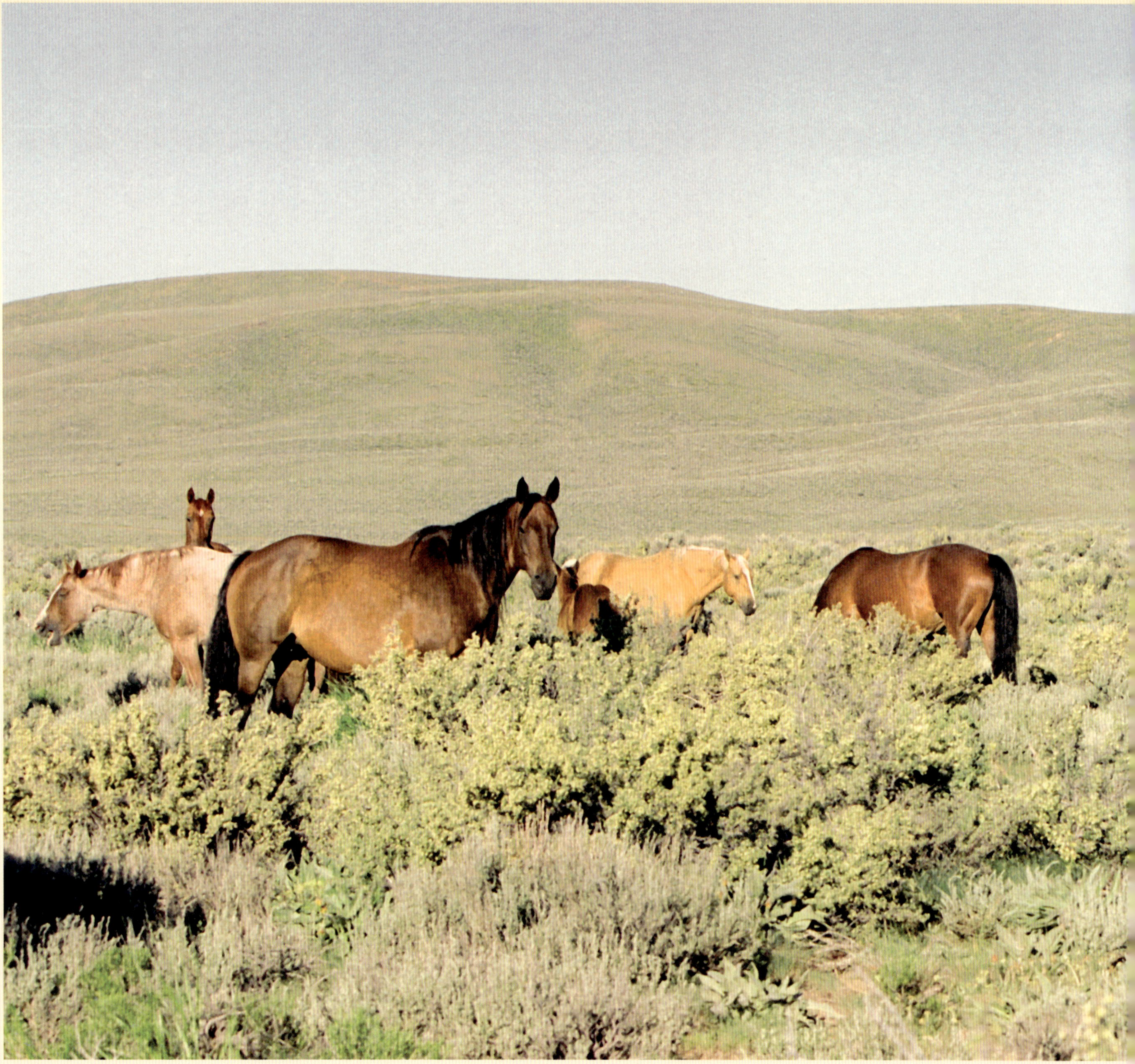

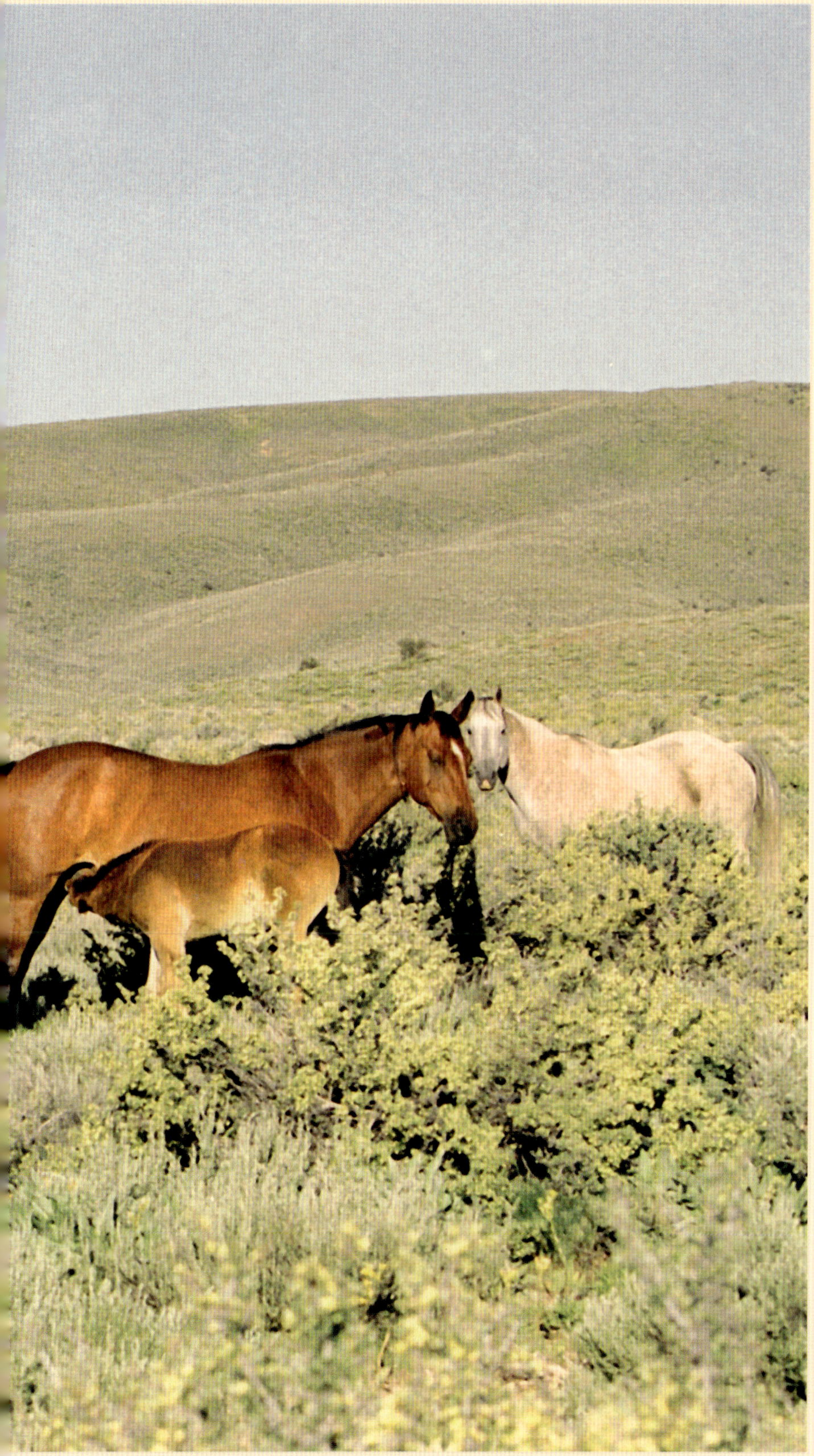

Van Norman Ranches

Tuscarora, Nevada

2001

They Ride Good Horses

The Van Normans ranch in the Great Basin area of the Northwest. They are horseback every day. And their remuda is one of the best.

When I arrived in Elko, Nevada, the evening of June 10, I called Bill Van Norman to see what time he wanted me at his house the next morning.

"Oh, about 6 o'clock, Jim."

"Bill," I asked, "what time does it get daylight here?"

"About 5."

Relieved, I realized that at 6 a.m. it would be light enough to take pictures. Back home in Amarillo, two time zones away, it's barely light enough to see at 6 in the morning.

Bill added, "By the way, Jim, did you bring a jacket? It snowed two inches here this morning."

Not only was I two time zones away from the 90-degree heat in Amarillo, I was 3,500 feet higher and several hundred miles farther north. Elko is in northern Nevada, and although the small community of Tuscarora, near which Van Norman Ranches is located, is only slightly north of Elko, it is even higher, up around 6,400 feet.

But the snow didn't last, and shortly after 6 the following morning we were in Bill's pickup truck, climbing a mountain. We were looking for a group of yearling colts turned out for the summer in a mountain pasture. Bill wanted to show them to me. He was proud of them and wanted me to know why Van Norman Ranches was a winner of the AQHA/Bayer Best Remuda Award. The Van Normans ride good horses.

It was Bill's father, Charlie, who put the ranches together. He and his wife, Della, bought a small homestead in Independence Valley, near Tuscarora, in 1945, and through the years they added to it. In later years, their sons, Bill and Robin, and their respective families, took over operation of the ranches. When Charlie died in 1996, and Della in 2000, the ranches were firmly established and continue to operate today.

Van Norman Ranches is a cow outfit. They run about 1,400 cows, calve in the spring, usually in April, and carry those calves throughout the winter. They are then sold as yearlings the next fall.

But winter comes early in northern Nevada and lasts a long time. Most years find the Van Normans feeding their cattle by mid-December, and they don't quit until the middle of April, when the snow melts. There have been years when they had to start feeding as early as the first of November, and the demands are high, from one and a half to two tons of hay per cow. Winter temperatures can get as cold as 40 below zero, but that's rare. Most winter nights are in the 10 degree range, and daytime temperatures are usually in the 20s and 30s.

Cow outfits in the Great Basin—that high desert, sagebrush sea that touches the states of Utah, Idaho and California, but lies mostly in northern Nevada and eastern Oregon—are typically big and cover some rough country. Horses are a requirement, and the Van Norman Ranches is no exception.

Bill and Robin's dad, Charlie, grew up in northern California, where Charlie's father was superintendent of the Gerlach Livestock Company. Good horses were a way of life on the Gerlach, and when Charlie came to Nevada, he brought that tradition with him.

Charlie and Della bought the geldings they used as they were establishing their ranches, but when Bill returned from the Army,

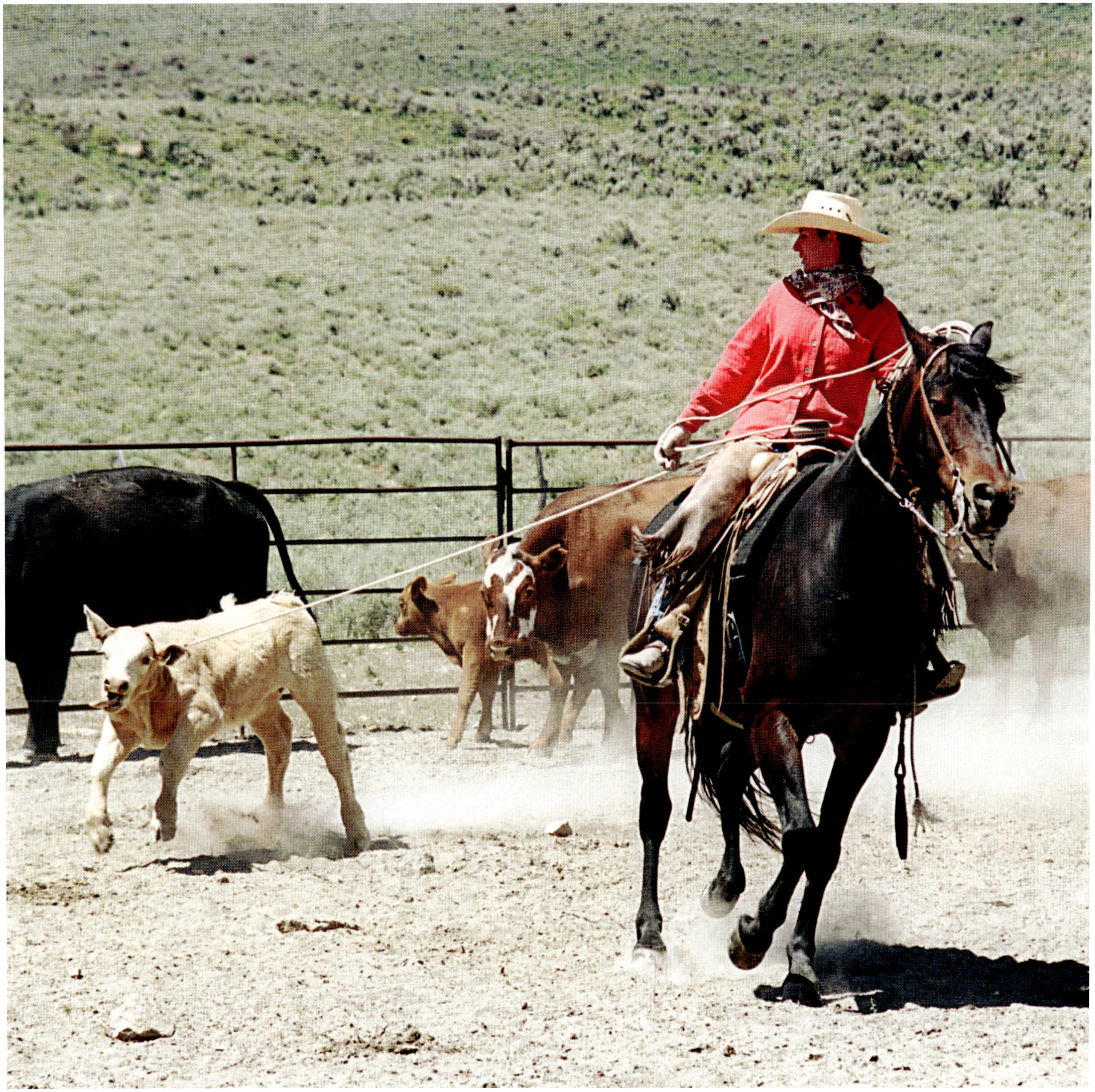

he introduced a new line of thinking to the family. He felt they should start raising their own. They knew the kind they wanted to ride, so why not start breeding them? In 1969, Bill leased a band of mares and a buckskin stallion named Johnny Carlo from Melvin Jones of Carlin, Nevada, and they had their start.

Johnny Carlo went back to Nick Shoemaker on the top side, and Joe Bailey on the bottom. He was foundation bred, and sired cow horses that had good bone, size and disposition. As Charlie and Bill continued to upgrade their horse herd, those were the characteristics they were looking for.

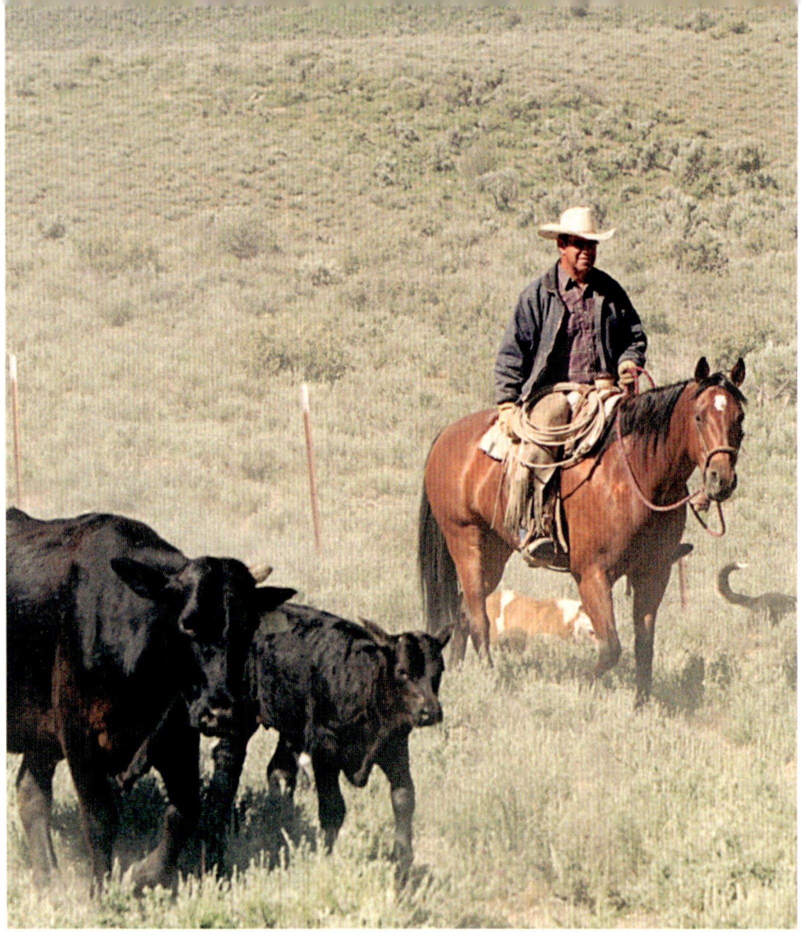

And the breeding of their horses today reflects those characteristics. Most of their 50 broodmares are descendents of Doc Bar, Doc O'Lena, Doc's Lynx, Mr Gun Smoke, Poco Leo Bars, Bert, Bueno Chex and Colonel Freckles. Stallions the ranch is using include Colonel Zippo Pine by Colonel Freckles; Madonnas Blue Bee by Flying X6 out of a Jackie Bee mare; Little Elmer Fudd by Little Lena Doc; and Showstoppin Boon by Peptoboonsmal.

All the mares are pasture bred, with the stallions turned out in May. They raise all the horses used on the ranch, and ride mostly geldings, although they try to ride all the fillies that go back into the broodmare band, and will occasionally ride a mare for ranch work. Those not used are sold, either by private treaty or in their annual sale.

As mentioned before, winters are harsh, but the horses are accustomed to it. Bill says, "We don't feed our mares much hay or grain. There's normally plenty of grass in the hay meadows in the summer, and during the winter, if the grass is down there, they'll paw that snow off.

"Besides," he continues, "they winter with the cows. When we feed the cows, the mares push in there and get some hay. But two months before they foal, they start getting alfalfa.

"These mares were born here, and they know how to survive. You go to California or Texas and buy one, and bring her up here, she's going to die if you don't take care of her. It takes them a couple of years to get accustomed to the weather."

Bill and his brother, Robin, are partners in the operation, and according to Bill, it's an ideal situation. They farm 10,000 acres, most of which is hay that is fed to the cattle during the winter. Farming is what Robin enjoys, so he takes care of that end of it, while Bill is in charge of the livestock. Each thinks he has the best deal.

Bill's whole family is involved in the ranching, including his wife, Geri, their daughter, Tilly, sons Troy and Ty, and Ty's wife, Ronda. It's a family operation, and when there's work to be done, everyone is horseback.

In the Great Basin, cowboys—or buckaroos, as they are called there—take many of their traditions from the early day California vaqueros. That includes their equipment and how they ride their horses.

"We will start riding our colts in the spring of their 2-year-old year," Bill says, "but we will only ride them about 10 times and then turn them out for six months. We'll start using them when they are long 2s. They'll still get some breaks, but from then on they are pretty much a part of the string.

"We'll ride them in a snaffle bit until they are 4 or 5 years old," he continues, "and then put them in a hackamore (bosal) for a year or two. Then they'll be in the two-rein for a couple of years, and then into the bridle.

Bill explains that with the two-rein, the horse is ridden with the hackamore, but under the hackamore is a full bridle with a

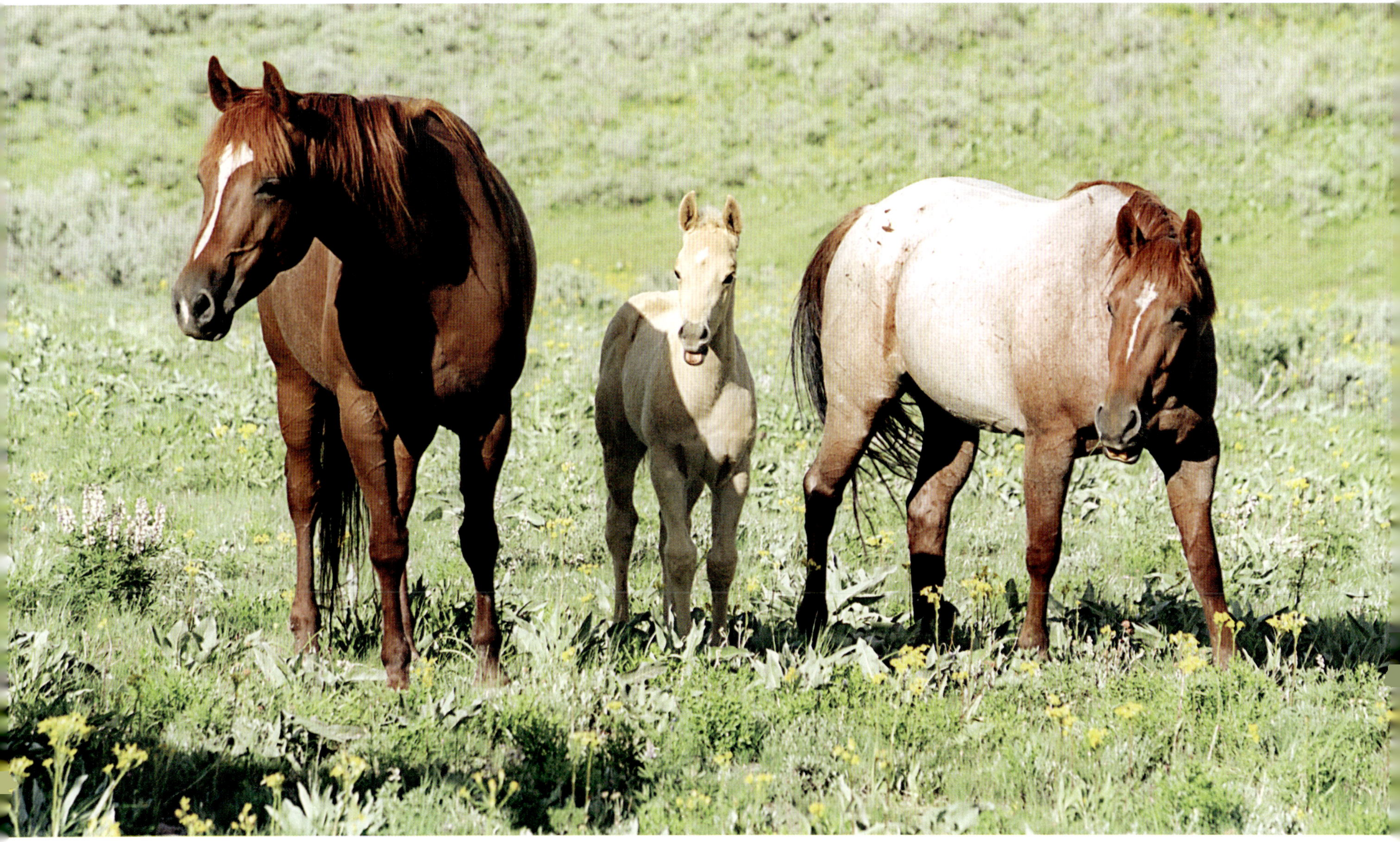

curb bit, usually a spade or half-breed. The horse might carry the bridle with no reins attached to the bit for a few days, as he learns to hold it in his mouth. Then the reins are added, and for the next year or so he is ridden with both the bridle and the hackamore. As the horse gains experience, the reins to the hackamore are used less, and those to the bridle, more.

"We believe their teeth aren't mature enough for a bridle until they are at least 6 or 7," Bill says. "Of course, there are a lot of horses that are not done that way, but if you've got a nice horse and you plan on showing him in the bridle, it's definitely the best way to go."

It's midmorning as we drive down off the mountain. We have already shed our jackets in the rapidly warming morning, and Bill figures that Ty, Troy, Tilly and the others have the cattle gathered and are at the branding pen, which is the plan for that day. Bill's mind hasn't been far from there all morning. Even though our mount so far has been the pickup, he has been wearing his chinks (chaps) and spurs since before 6, and his horse is patiently standing in the trailer waiting on him.

Bill jumps his horse out, pulls up the cinch and steps aboard. As he shakes out his rope, he motions for me to follow him into the branding pen. He wants to show me another reason why Van Norman Ranches was judged to have the best remuda.

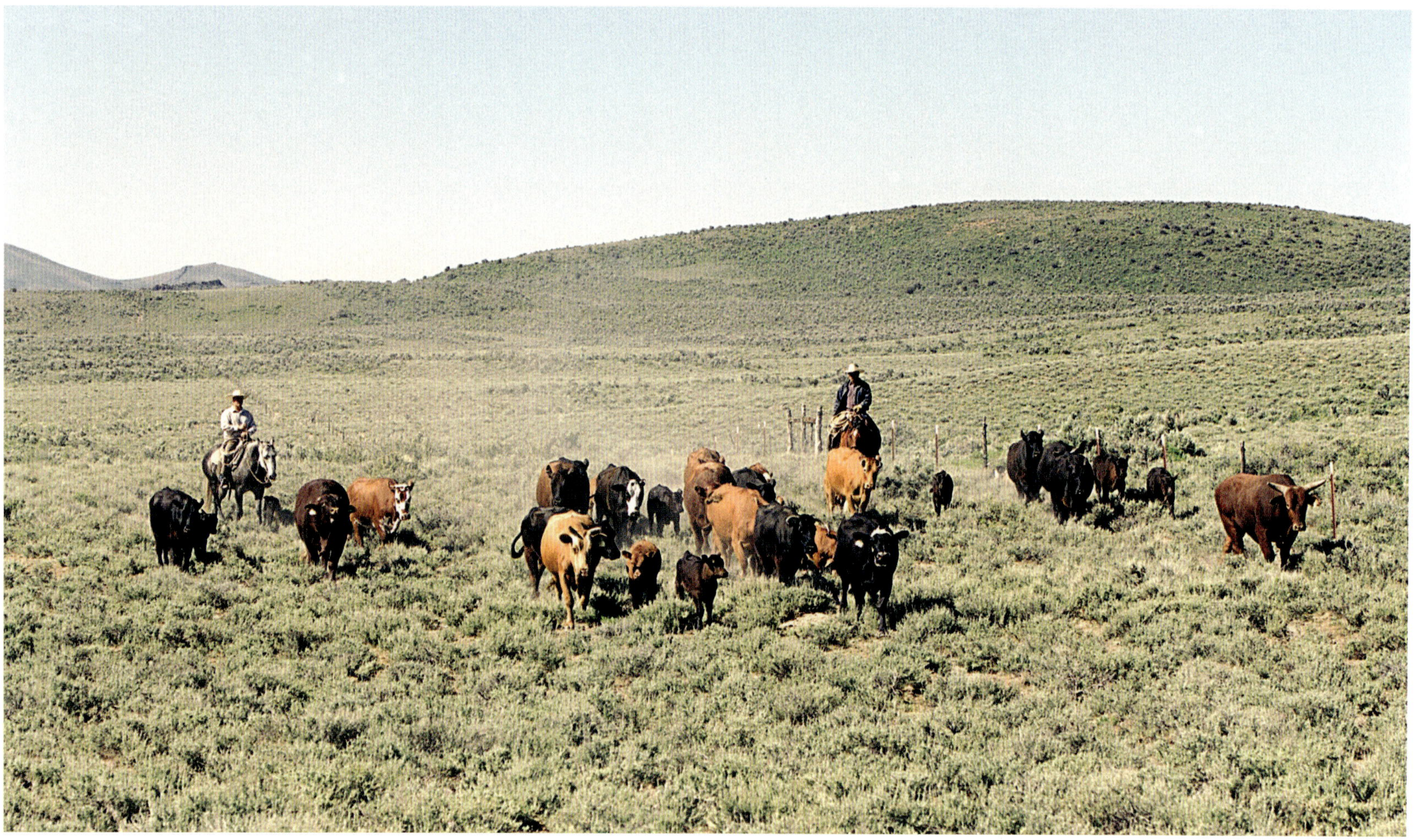

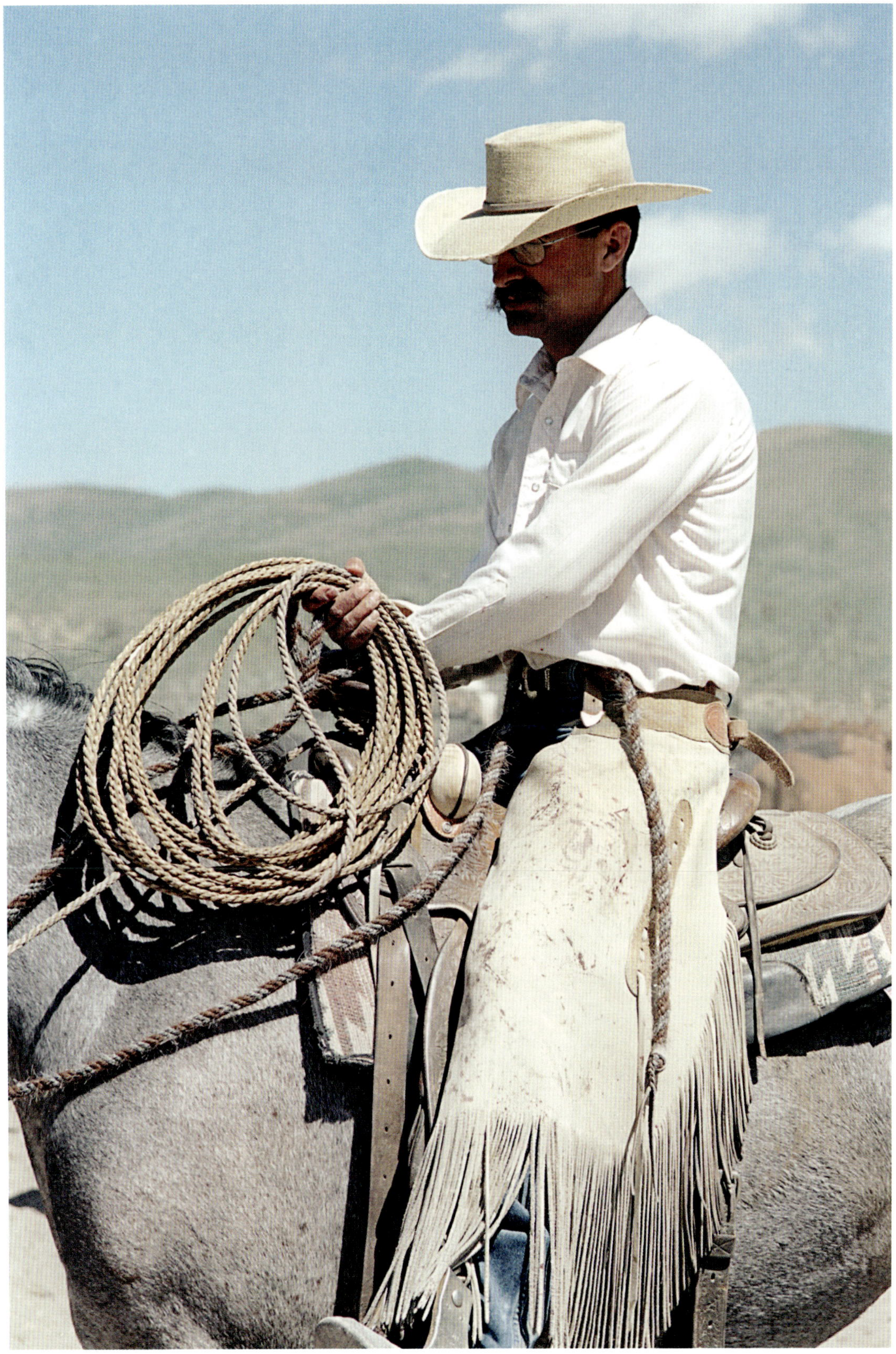

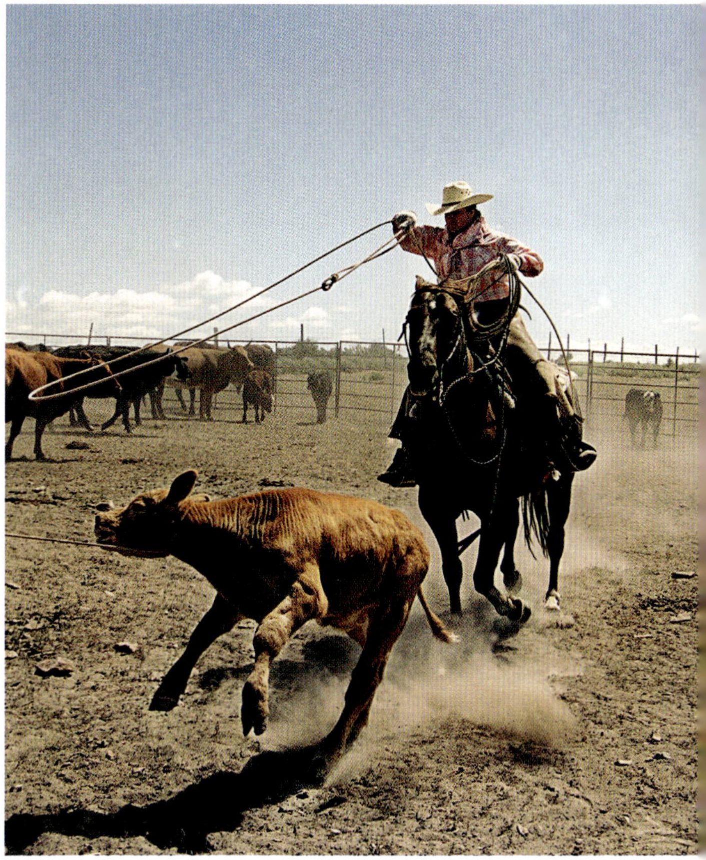

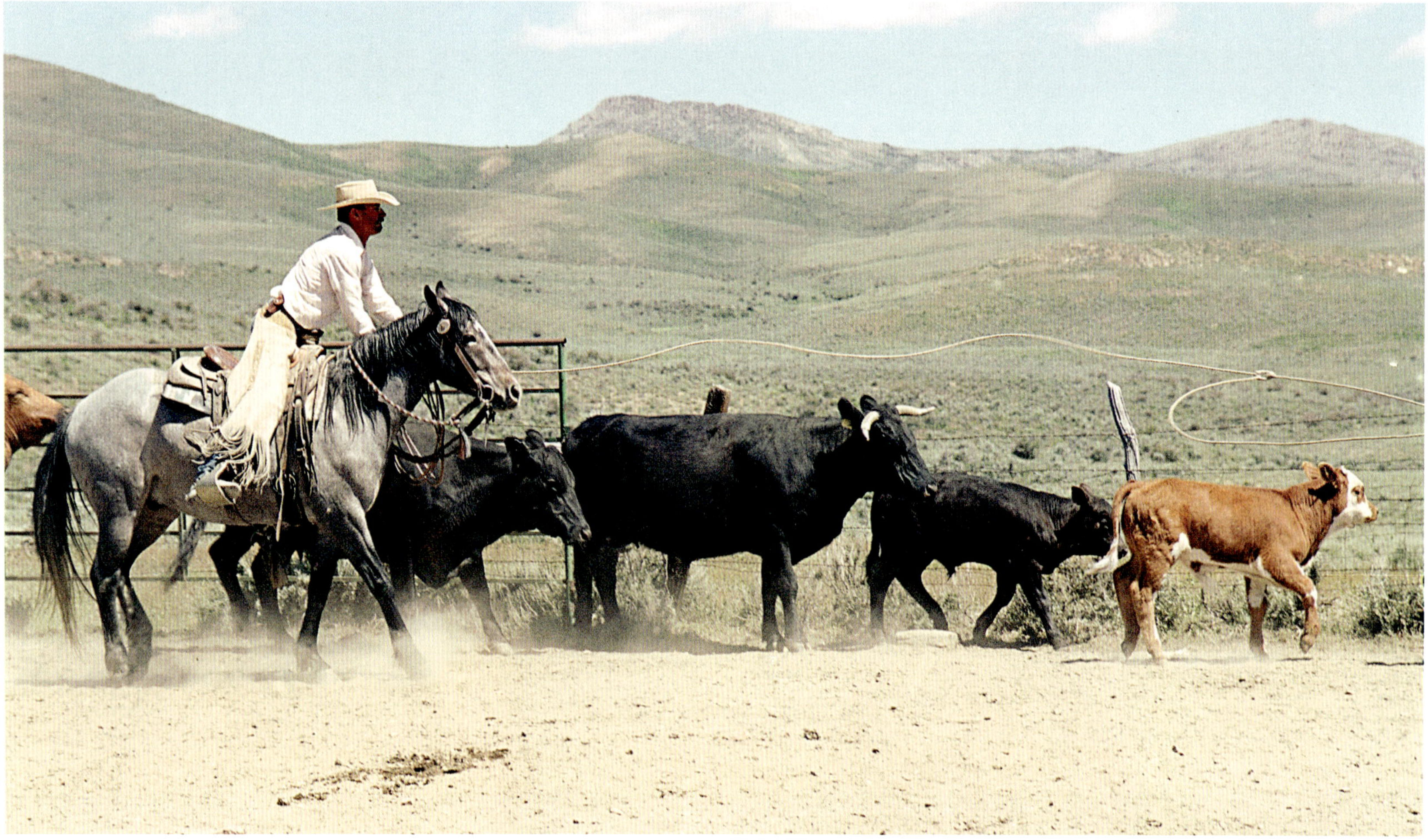

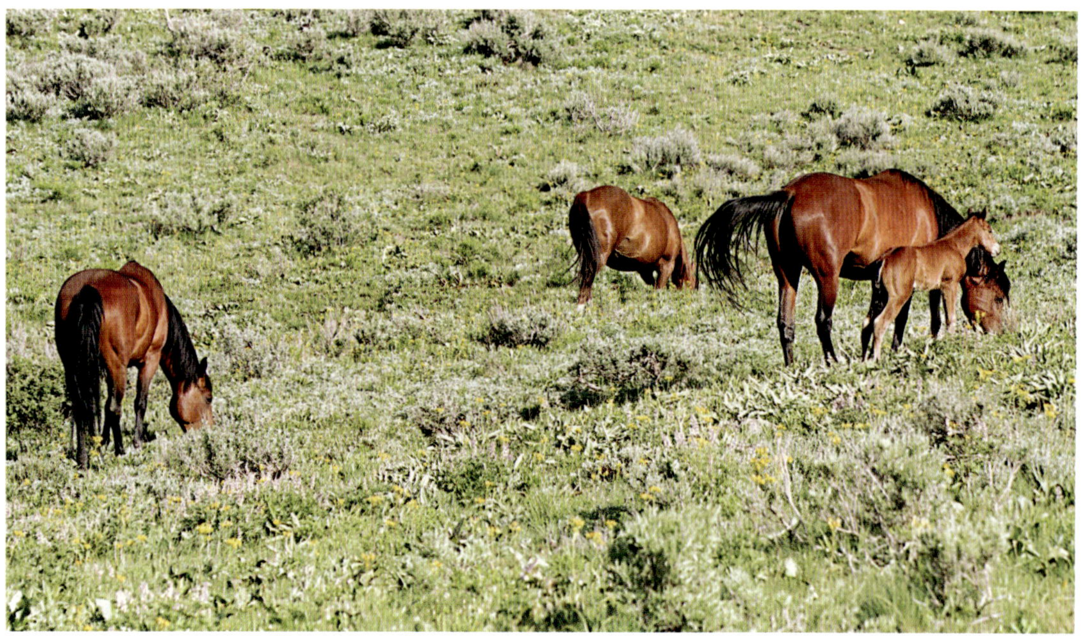

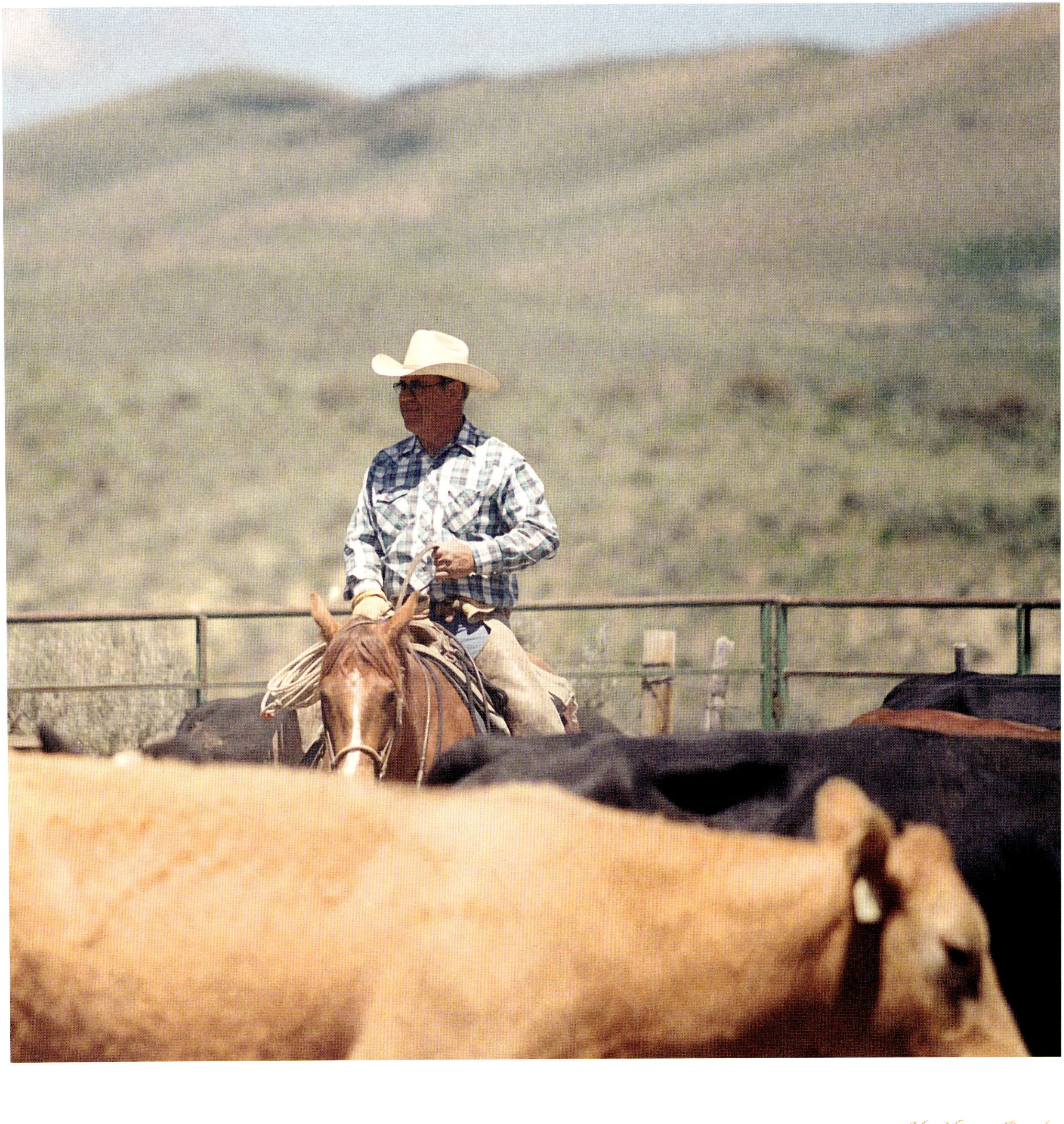

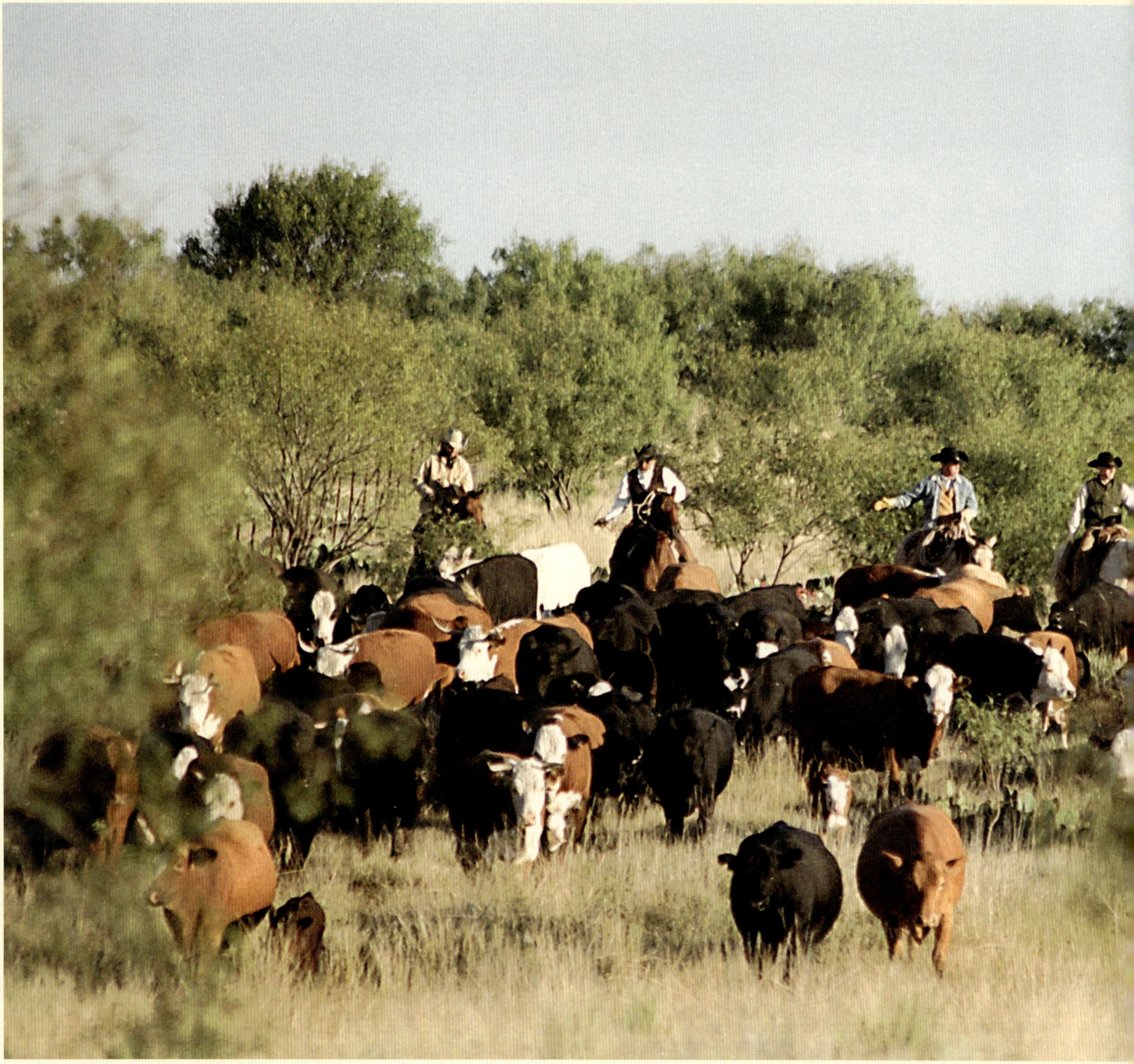

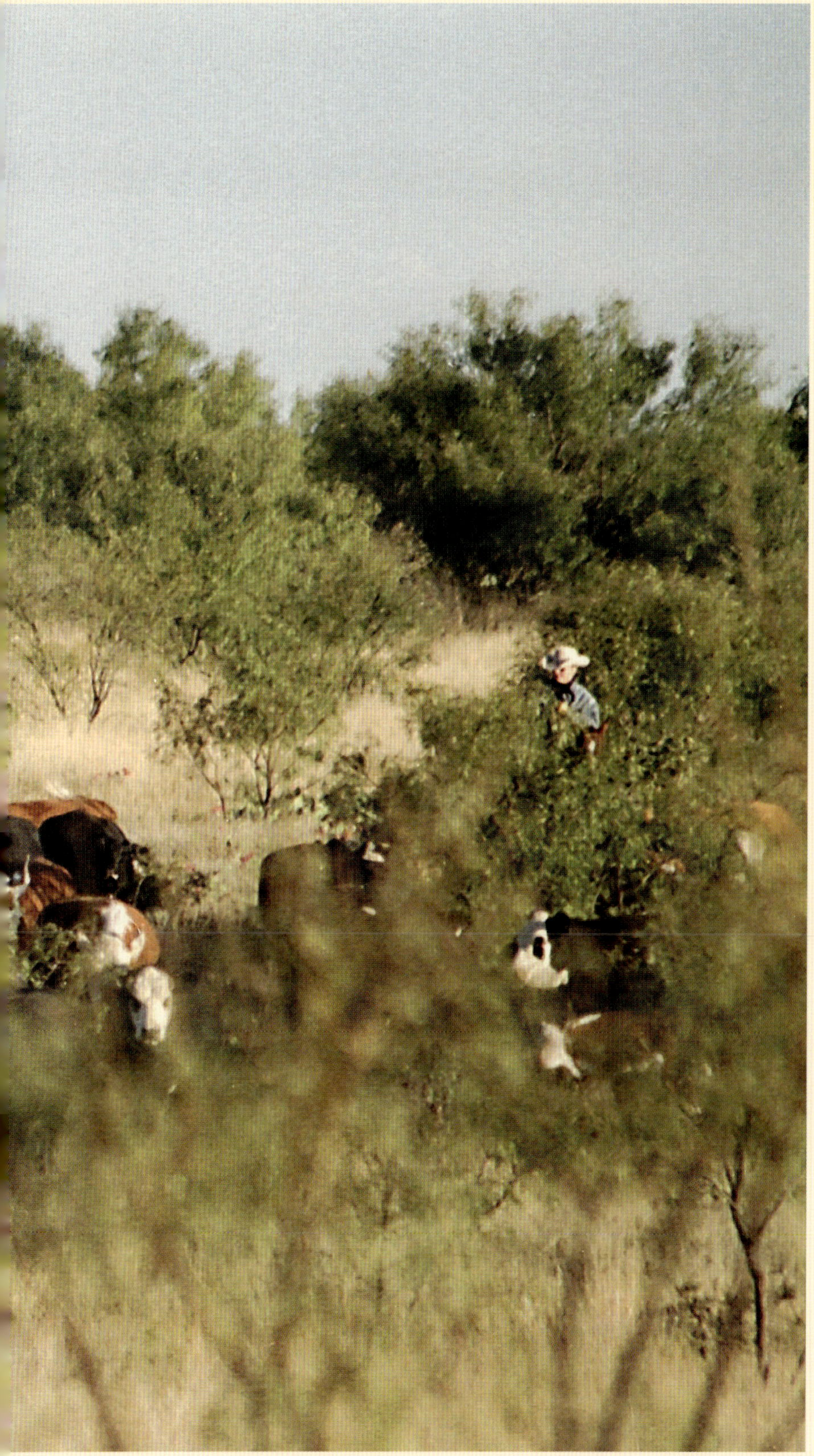

W.H. GREEN CATTLE COMPANY

Albany, Texas
2002

The Horses of the Green

Henry Green, in 1885, said that when his father stepped down from the train, he was the maddest man that Henry had ever seen. And the thing about it, he was mad at Henry.

Thomas Green was a Confederate veteran who, in 1885, lived and ranched in central Texas, near Hillsboro. He also was running some horses—mares, colts and geldings—on the open range in Shackelford County, about 150 miles northwest of Hillsboro. The land on which Green's horses were running actually belonged to the Texas Emigration and Land Company through an old agreement with the state dating back to the 1840s. It was open range and pretty well anyone who wanted to was grazing it. Caring for the horses was an old cowboy who lived in a dugout not too far from Albany, the Shackelford County seat.

But now the old caretaker sent word to Green that the TE and L Company was selling off the land and it was being fenced. The open range days were quickly coming to a close. Green sent his son, Henry, who had just graduated from Trinity College, to round up the horses and sell them, and return with the money. Henry got the horses gathered, and sold them, but he liked the area around Albany so much that he spent the money on leasing some land. He then wrote his father and told him what he had done.

A few days later, the old caretaker rode into Henry's camp with a sheet of paper in his hand. It was a telegram from Henry's father, telling Henry to meet the train in Albany the next day. His father would be on it.

"He finally convinced his father to ride out with him and look at the land," Billy Green said of his grandfather, Henry, "and the farther they went, the more his father calmed down. Finally, he offered to help his son stock the land with cattle. The land my grandfather leased, and then bought, is the land our family is ranching on today."

Billy is William Henry Green III, named for his grandfather and his dad. The ranch is the W.H. Green Cattle Company, named the 2002 winner of the Best Remuda Award. The Green remuda is one of the best.

The Horses

"I guess maybe the fact that my great-grandfather had all those horses here has stayed in the blood of my family," Billy says, sitting across the dinner table from his uncle, Bob Green, who nods in agreement. Bob, like Billy's father, inherited a part of the ranch that Henry established, but it's Billy who has carried on with the horse breeding. Billy's father died in 1994 and left Billy his part of the ranch, on which he runs about a thousand mother cows and another 2,000 stockers. He also has 15 broodmares and says, "I've always liked horses, and I've always wanted to raise them."

It was probably those family genetics that caused Billy to drive over to Seymour to a ranch cutting one day in 1983. He had never been to a cutting, so he thought he would go see what it was all about.

"There was a man there by the name of Glen Bruton, who worked for the Four Sixes Ranch. He won the cutting on a 4-year-old yellow stud, but he was going to have to sell him. The Four Sixes didn't want one of their cowboys riding a stud. I bought him."

The stallion's name was Golden Three Bars. He was by Gold Finger Bar out of Lemac Jan by Dan's Sugar Bars, and Billy quickly decided that he was the best horse he had ever ridden.

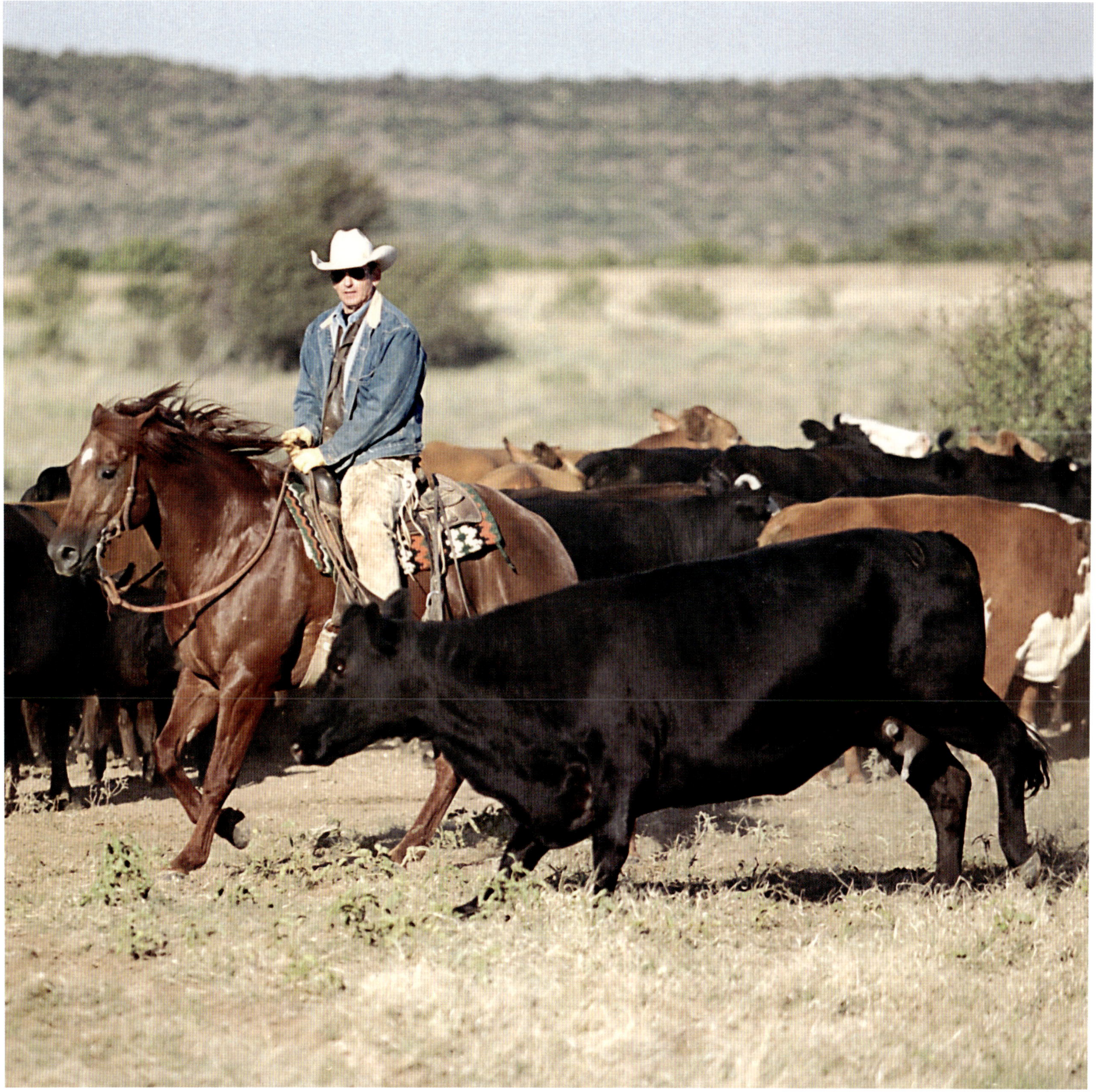

"A friend of mine, Jim Trammel, told me one time that if you could find one good producing mare, that would be the best thing you could have in the horse business. So I thought, 'Well, this is the best horse I have ever had, so I'm going to see if I can find his mother. She should sure be a good-producing mare.'"

Billy found Lemac Jan and bought her. And he was right about her being a good producing mare. With six of her daughters in his broodmare band, she put Billy in the horse business in a big way. She produced some great ranch horses, and some of her foals made him a competitor in the National Cutting Horse Association.

"When I brought Lemac Jan home," he says, "I was still interested in cutting, and Doc Bar was the hot bloodline at the time. Carol Rose had Genuine Doc, and I decided to spend the money to breed to him."

Billy named the resulting foal Genuine Jan. When the filly was 2, he broke her. When she was 3, he began to use her a little in the pasture, riding through some cattle and just generally giving her some experience. He realized early on that she had some talent.

"We were calving heifers," Billy says. "What we do is pull all the heifers up in a corner and cut out the heavy springers (cows about to calve), and put them in a trap to watch. I was riding this filly, and on her own, she started rating those cows. When the cow stopped, she would stop, and then she would turn back with the cow. I had never ridden one that would work a cow like that, especially a horse that young. Well, I had always wanted a cutting horse, and I thought I just might have one."

Not knowing a whole lot about cutting horses, Billy enlisted a little help from friends Tom Merriman and Lonnie Morris, both of whom rode Genuine Jan and helped Billy learn to ride cutting horses. Then, he sent her to trainer David Holdsford.

Holdsford agreed that the mare was a good one, and entered her in the NCHA Classic/Challenge in 1992. Holdsford split seventh and eighth in the open competition, but Billy won the amateur division on her.

"In the meantime, we had been breeding Lemac Jan to some local stallions, but when we saw what she produced out of Genuine Doc, we knew we needed to be breeding her to some better horses. So the next year we went back to Genuine Doc."

In succeeding years, Billy also bred her to High Brow Hickory and Sonitalena, and the foals of both did well in the cutting arena. Then he took her to Haidas Little Pep. From that breeding he got Haidas Jan.

"She was another one of those that I could tell by riding her in the pasture was going to be a good one," Billy says. "We rode her on the ranch until she was about 4, and then I took her down to Greg Welch's place. Greg didn't start showing her until she was 5, but she did really well. She's the best one we've ever raised."

Welch was a leading cutting horse trainer. He died in 2005, but under his direction, Haidas Jan earned more than $150,000 in NCHA competition. In 1999, she won the NCHA Bonanza Classic, and in 2000, she was reserve champion at the Bonanza and at the NCHA Super Stakes.

"She's retired now," Billy says. "Greg was showing Snorty Lena at the same time he was showing Haidas Jan, and he won over a hundred thousand on him, so we bred Haidas Jan to him in 2001. Her first colt was Jans J Uno."

Both Jans J Uno and the mare's 2004 foal, Jans Scoot by Smart Lil Scoot, were in cutting training in early 2006

Lemac Jan's breeding to High Brow Hickory produced Hickorys Lowenbrau, a 1992 stallion that Billy is using in his breeding program today and still showing some. Randy Rollins broke the stallion, and then Welch showed him. As a 6-year-old, Billy showed him in the amateur division at the 1998 NCHA Classic/Challenge.

"Greg was doing so well with him I just couldn't stand it," Billy says, "so I showed him in that same cutting I had won before. We were leading up until the last horse, but he went in there and beat us. We got reserve."

Hickorys Lowenbrau so far has NCHA earnings of more than $70,000. All of Lemac Jan's produce have, up to now, cumulatively earned more than $250,000. Her grandsons and granddaughters have already earned more than $63,000.

"It was never my intention to raise those kinds of horses," Billy says. "We are breeding for ranch horses. I don't want a horse that I can't ride on this ranch. Part of our ranch is rocks and hills, and you've got to have a good-boned, stout horse that will carry you. I just wouldn't feel good about breeding something that wouldn't

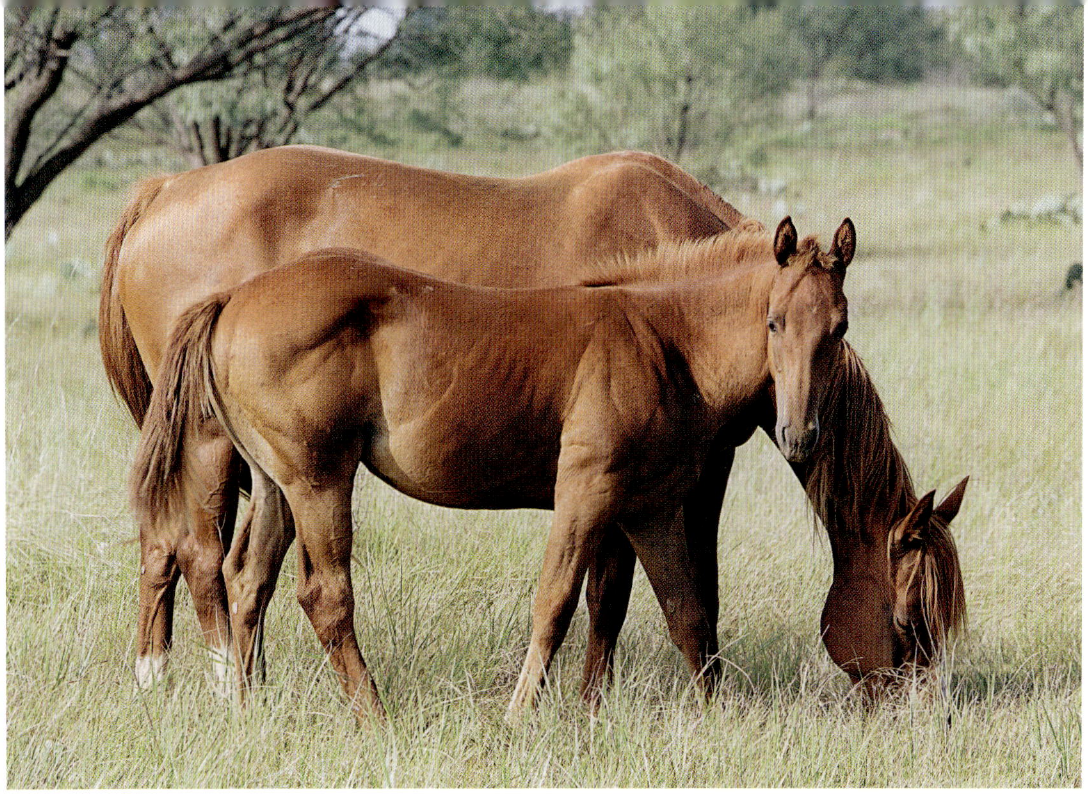

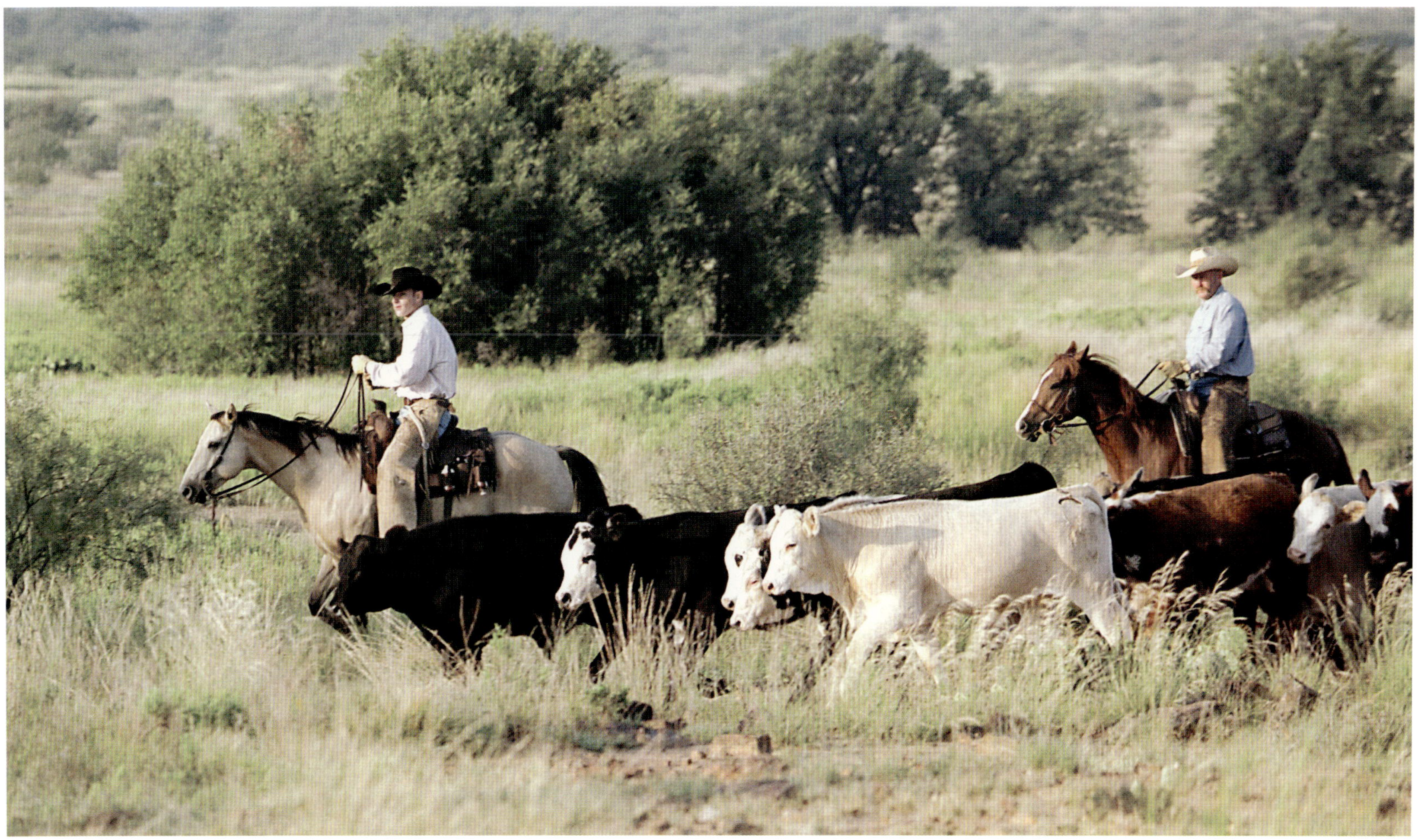

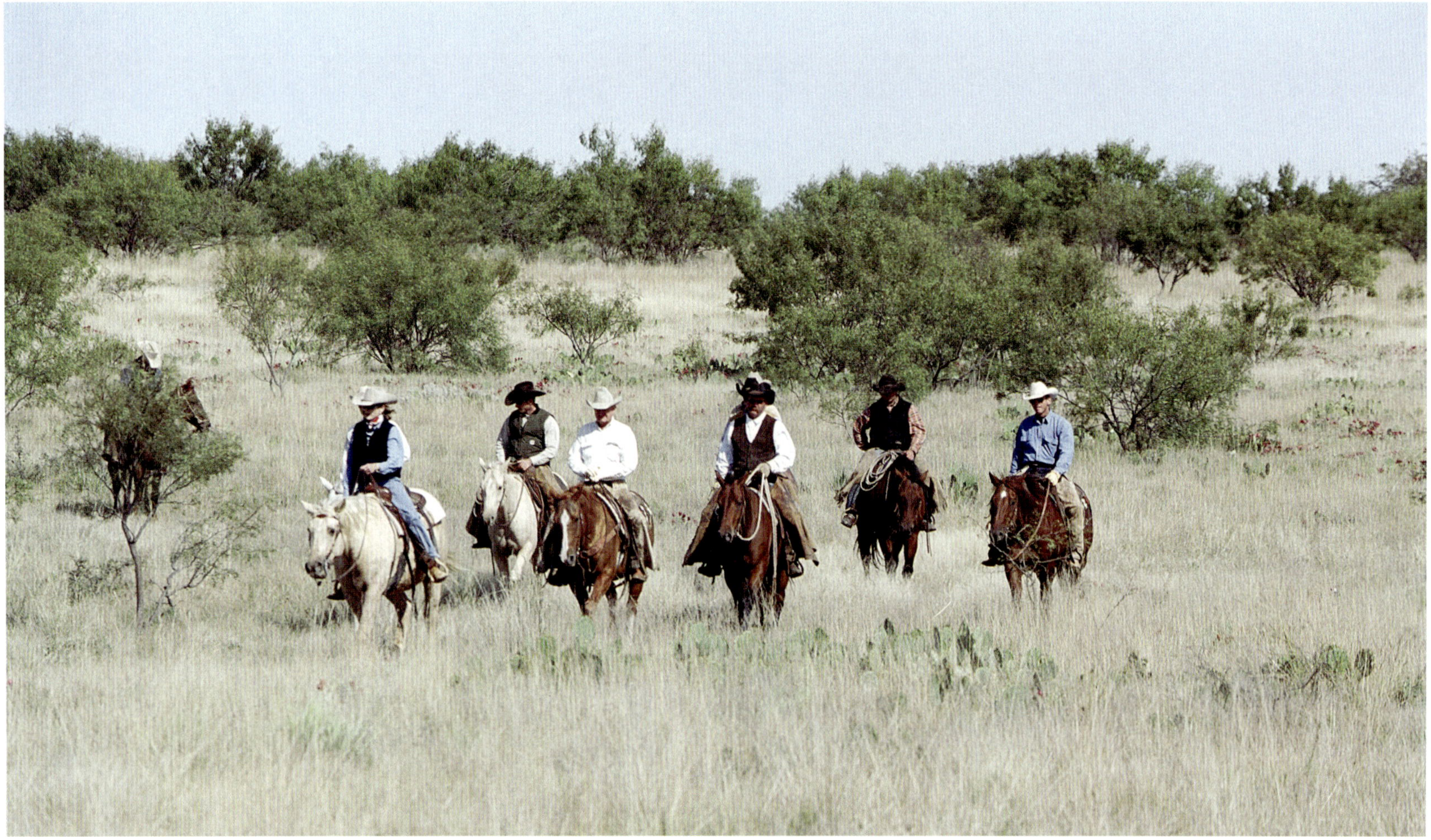

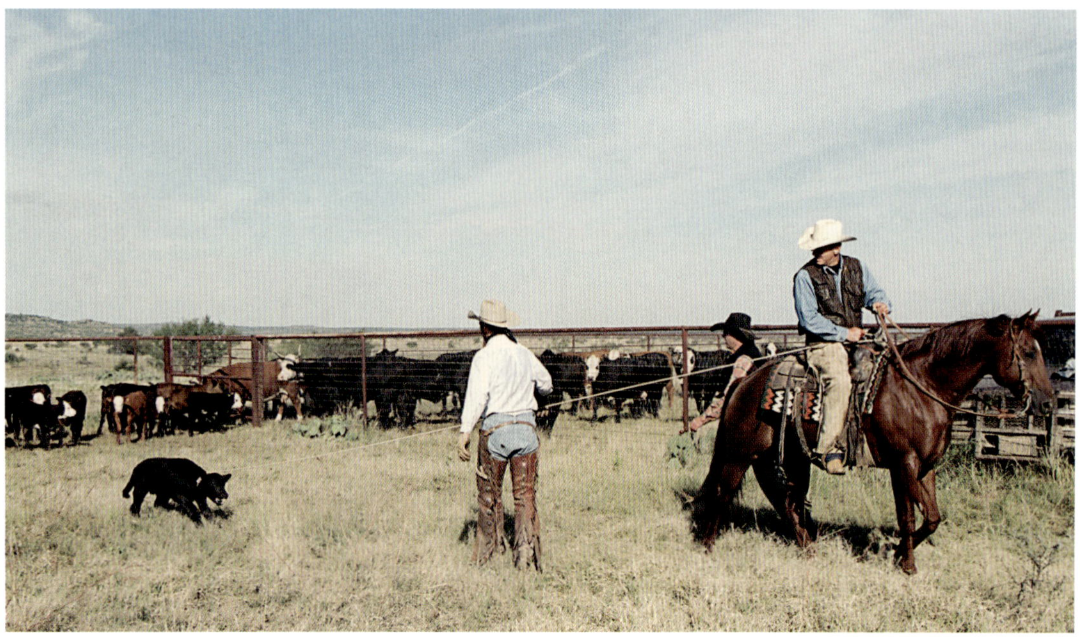

do that. But I've always felt that if you were going to progress, you need to breed to a stallion that's better than the mare.

"I didn't have the money to go to the best stallions, but that mare (Lemac Jan) made up for it. A mare like that will make anyone look good. And it's been a big plus and a lot of fun to have these cutting horses. It's added a lot of spice to life, and I love it. To feel that horse move underneath you is one of the greatest feelings in the world."

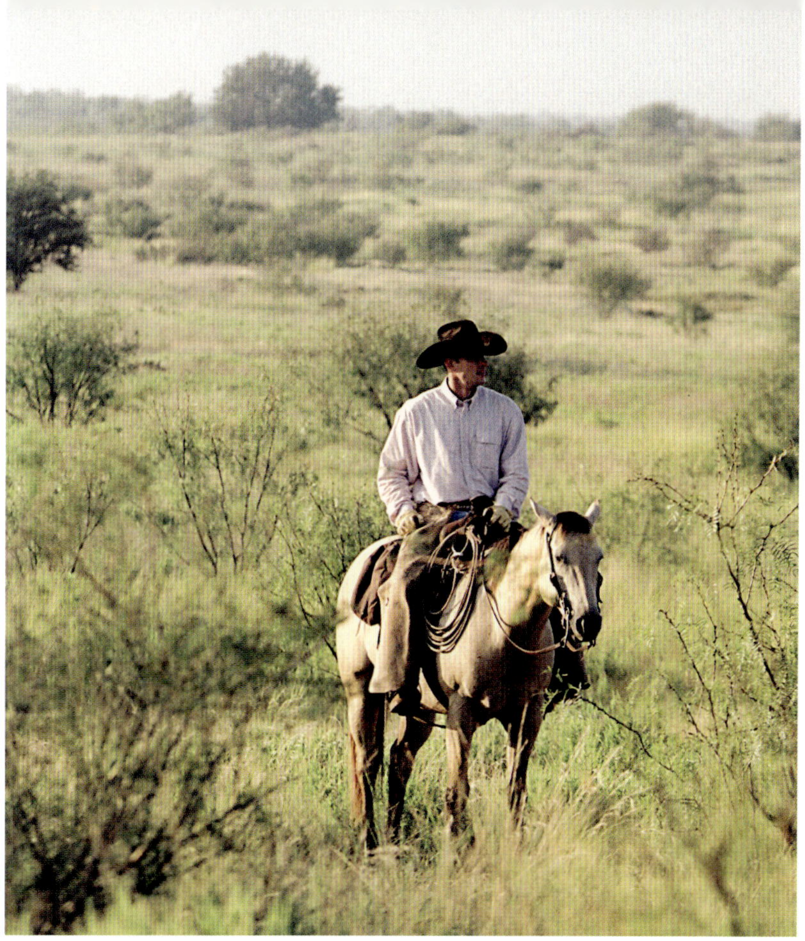

The Ranch

Almost a dozen cowboys move through the dark as they catch horses and begin to saddle. Away from the light spilling from the saddle house door, only the occasional glow of a cigarette bobbing up and down can be seen. Breakfast was an hour before, and sunup is almost an hour and a half away.

Billy had originally planned for everyone to haul their horses to the backside of the pasture, but changed his mind. "I think we'll just trot back there," he says. "It'll be good for the horses, and it'll be light by the time we get there." Riding out with Billy is his wife, Liz, and his son, Henry, the fourth in line to carry the name. Both Liz and Henry work very closely with Billy on the ranch. The three of them, along with everyone else on the cowboy crew, is riding a Green-bred horse. Several of the horses can be traced back to Lemac Jan.

The Greens run approximately 1,000 cows. In addition, in the good years, when there's plenty of moisture, they also buy and pasture about 2,000 yearling calves, called stockers, which they run throughout the winter. Their cows are bred to calve in the fall, and they wean and ship the calves in the late spring and early summer.

"The number of stockers we run really depends upon the condition of the pastures," Billy says. "That's something I learned from my dad and my uncle Bob. If you don't stock all your country with cows, when you have a drought, you just don't buy any yearlings. That way you may not have to sell your cows.

"Normally, those stockers can do well in the winter in this country. If it rains, there is something green here all winter. They can kind of hold their own through the winter, and when spring gets here, they will really gain."

This morning they are gathering cows and calves in a 2,500-acre pasture. The calves will be branded and vaccinated, and the bull calves castrated, and then turned back out with their mothers. Also, dry cows, those that didn't calve this year, are cut out to be sold, which gives Billy a chance to work his horse. He's riding Hickorys Lowenbrau, one of his successful cutting horses.

But he's also a good ranch horse.

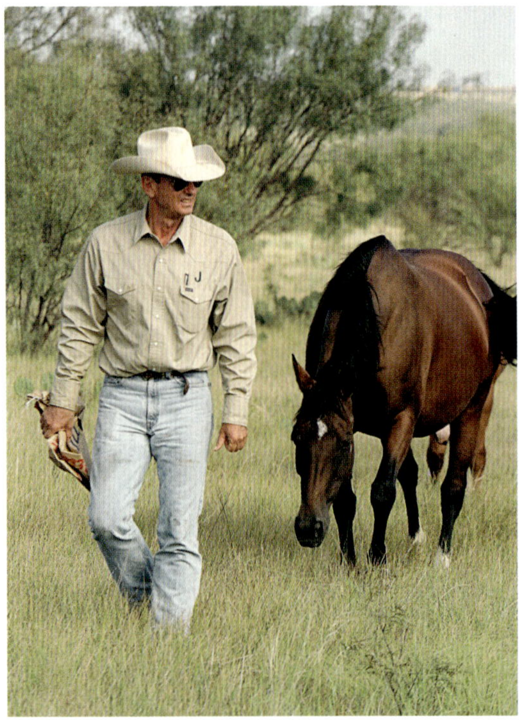

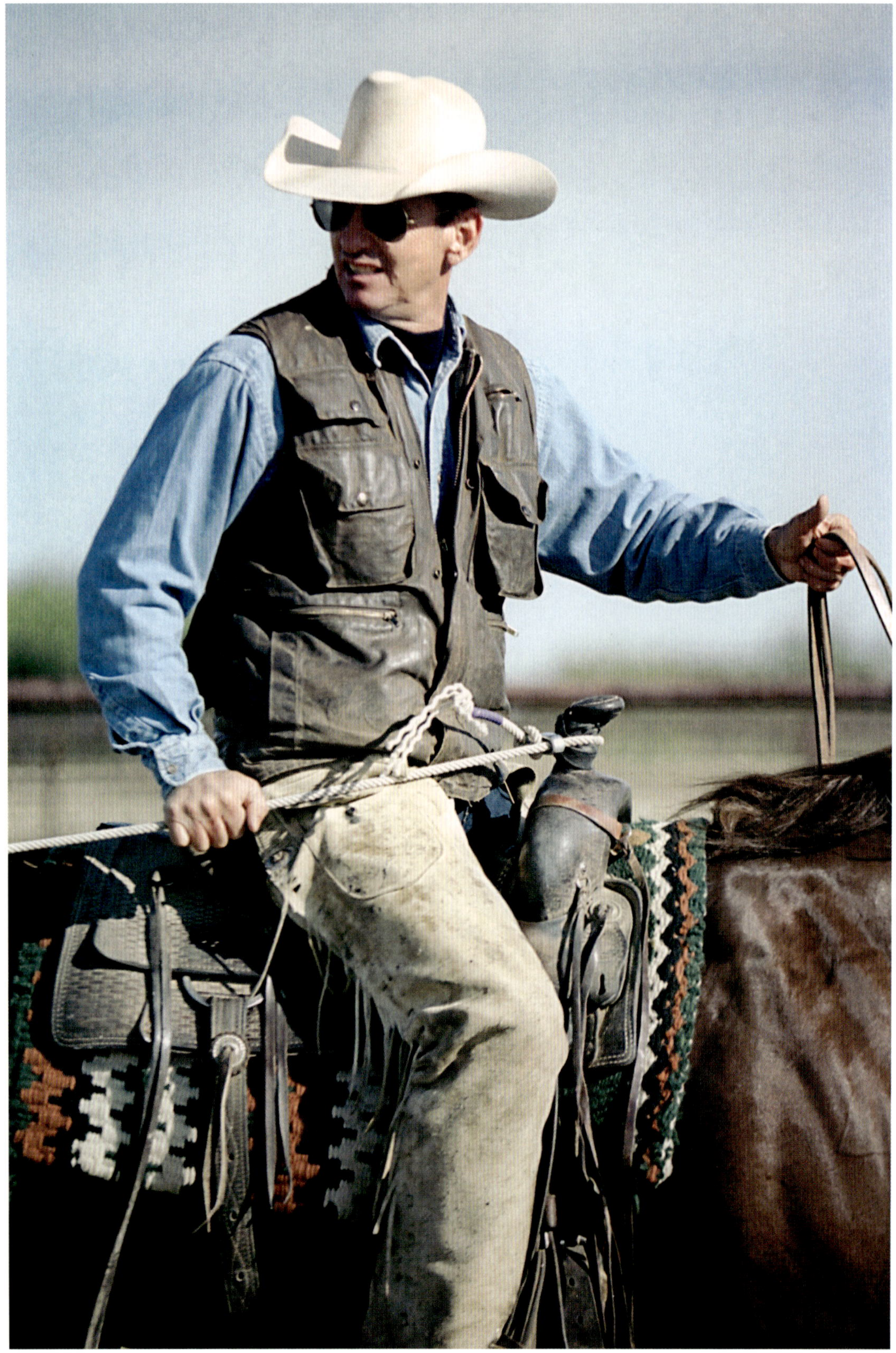

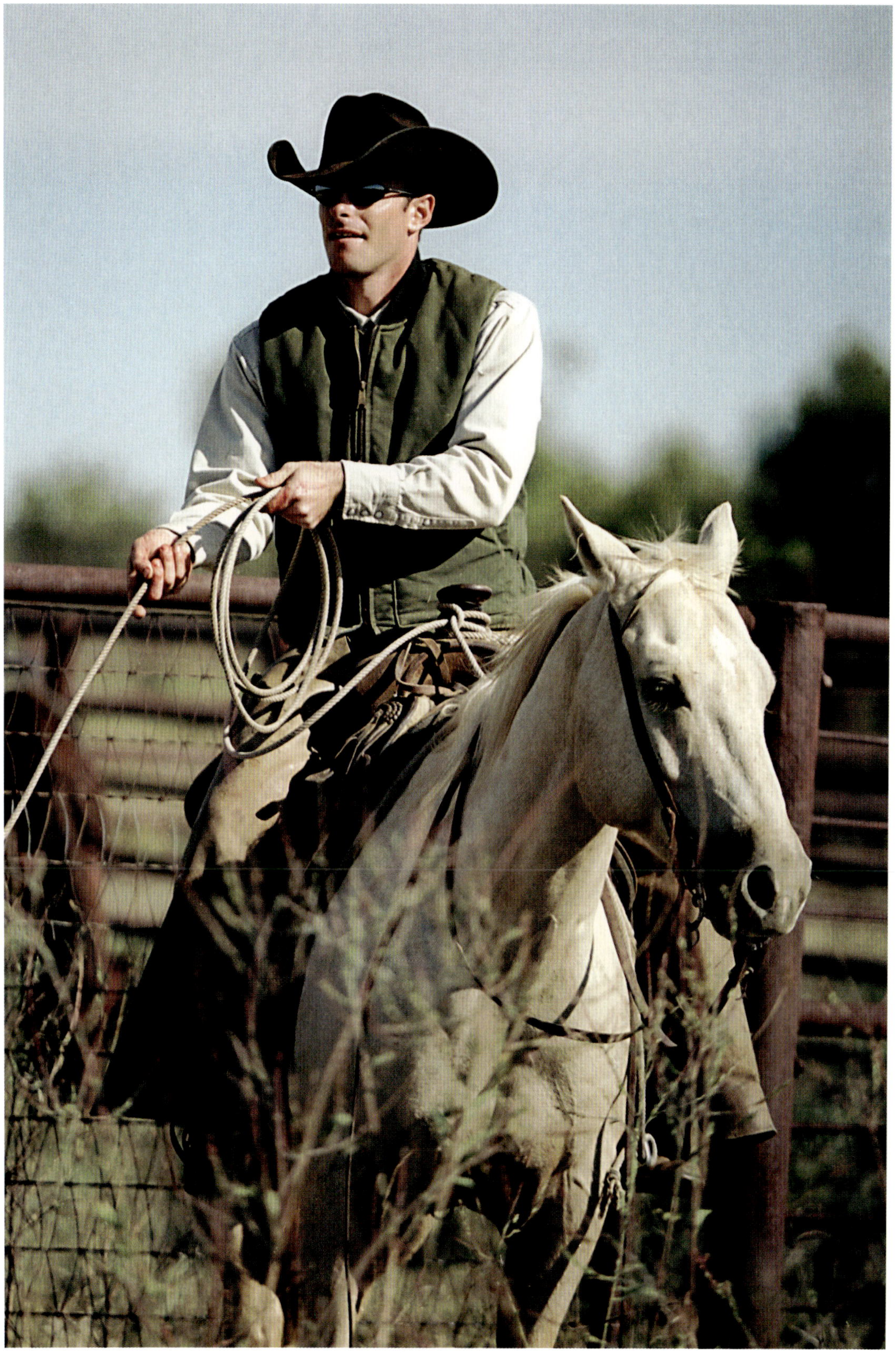

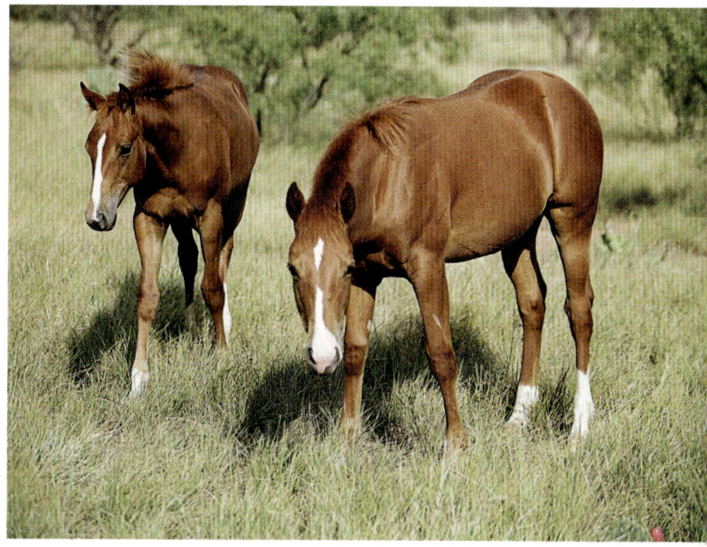

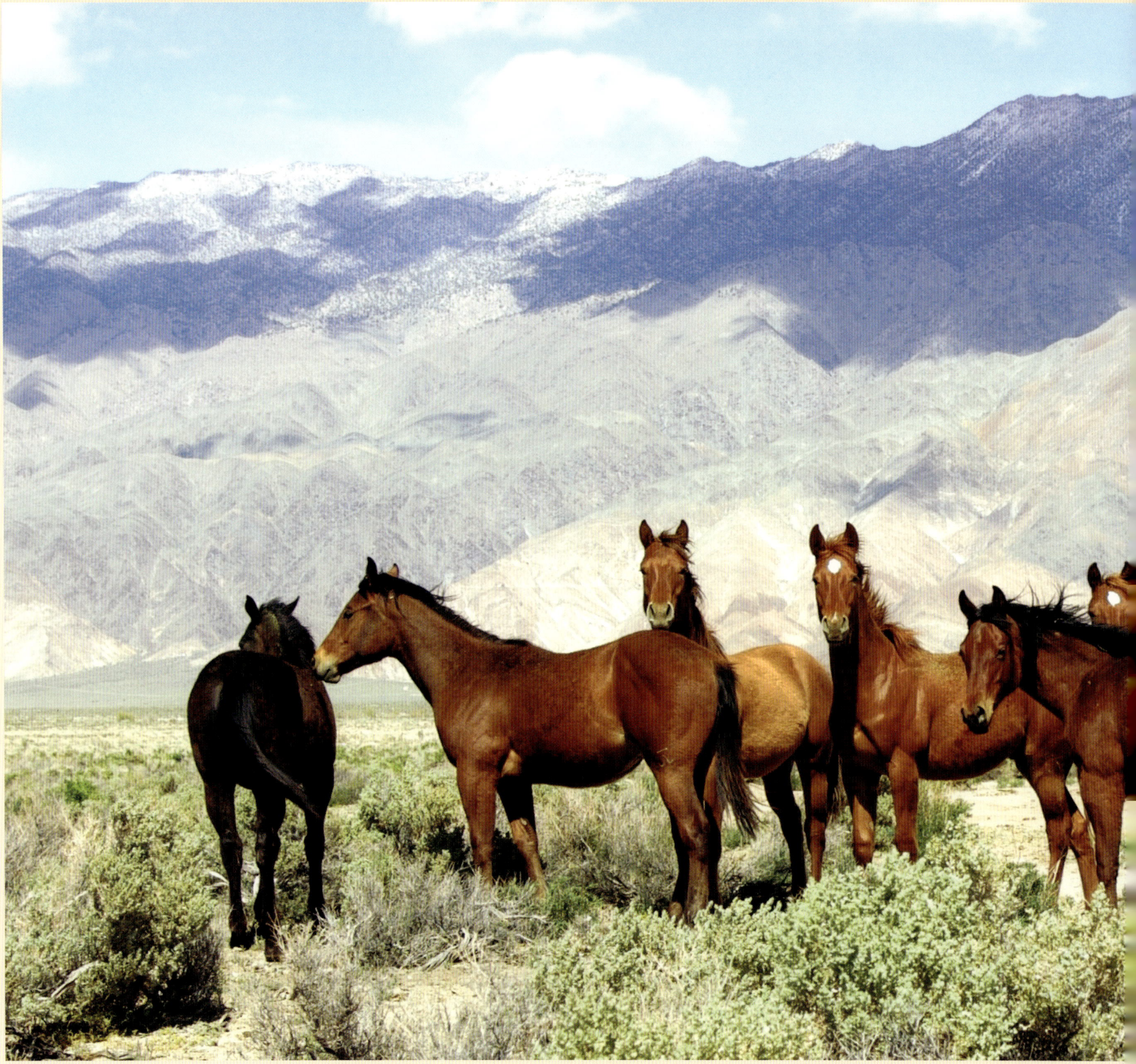

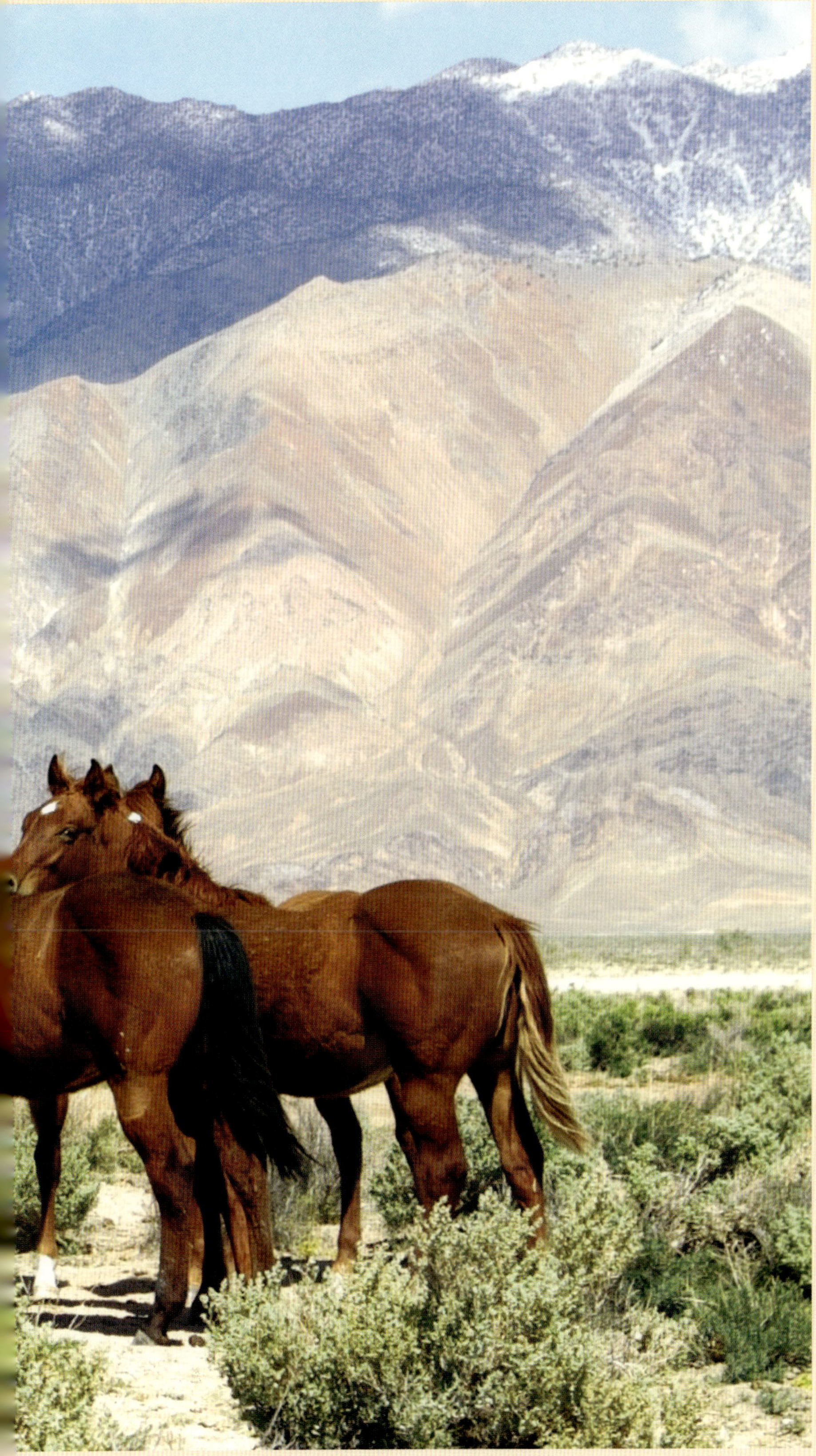

LACEY LIVESTOCK

Paso Robles, California

2003

Horses in the Desert

Almost everyone's vision of a desert is that of a dry, sandy area, almost devoid of vegetation and with unrelenting heat. That's not necessarily so—especially in the winter and early spring.

John Lacey pulls his heavy, canvas coat closer around his neck, and makes an effort to tug his felt hat farther down on his head. He is thankful for the gloves that cover his hands as he works the stiff plunger of a syringe. Almost unconsciously, he turns away from the strong north wind. It is the middle of April in the Mojave Desert, and John is cold.

"Spring comes late up here," he says, and pointing to the clouds that obscure all but the base of the mountains, adds, "It can snow in April."

John Lacey is on his ranch in the Owens Valley of California, a couple hundred miles almost straight north of Los Angeles. His land is in the upper reaches of the Mojave Desert, near the communities of Olancha, Lone Pine and Independence. Owens Valley is bordered on the east by the Inyo Mountains, the only barrier between it and Death Valley. On the west are the Sierra Nevadas, containing the largest mountain in California, Mount Whitney. In all but the warmest months, snow covers the peaks of Mount Whitney and several other nearby mountains that are more than 14,000 feet tall. Owens Valley sits at 3,800 feet. John's family has ranched in this valley since his grandfather came here from Missouri in 1870.

Lacey Livestock is pretty well centered in the Owens Valley, but it branches out into a good portion of California. John and his son, Mark, who lives with his family at Independence, lease 60,000 acres in the valley and north of there, in both Inyo and Mono counties, on which they run about 2,000 cows and summer approximately a thousand yearlings. They also have another thousand yearlings on a couple of places in the San Joaquin Valley.

In addition, John and Mark have partnered with David E. Wood and formed Centennial Livestock. "That partnership leases land in Monterey County," John says, "where we run yearlings in the winter and then move them—about 4,000 head—to Mono County, near Bridgeport, for the summer." Centennial also leases about 200,000 acres of the Tejon Ranch in the southern San Joaquin Valley, in both Kern and Los Angeles counties, and runs around 3,400 cows and 3,000 two-season yearlings there.

John explains that a two-season yearling is one that is kept through two grass seasons. The cattle are bought weighing between 280 and 350 pounds in March, and sold about the first of May the following year, weighing around 800 pounds.

"If you are running 3,000 two-season cattle, there are about six weeks there when you have 6,000 head. You want to bring in the light cattle when the grass is still green. By the time the grass dries up from the summer heat, the young cattle are over any sickness and they are going well," he says.

Breeding Horses

That many cattle on that much ranch land can only be handled one way—with horses. The Laceys have always had good horses, going back even to John's grandfather, and for that reason they won the Best Remuda Award in 2003.

"I got interested in the horse business at an early age," John says. "My dad didn't raise many horses; he bought most of his, but we rode good horses. A friend of my dad's raised Quarter

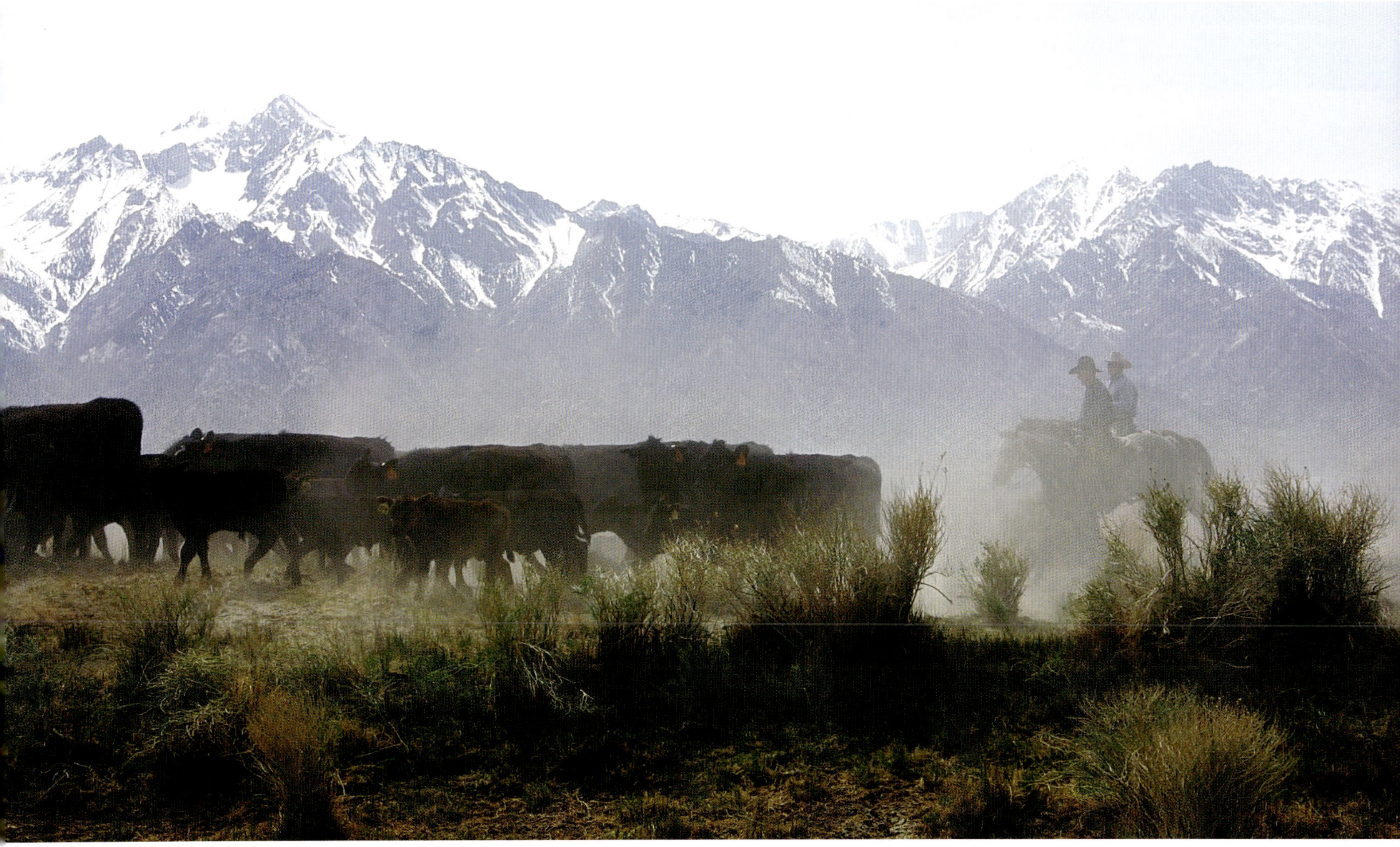

Horses, and I kinda got the incentive to go into the horse business from him. I bought my first mares when I got out of college in 1960 and raised my first foals about 1964. I guess you could say that I poor-boyed it in the beginning, just picking up mares where I could. I didn't have a lot of money to buy them, so I tried to breed them up."

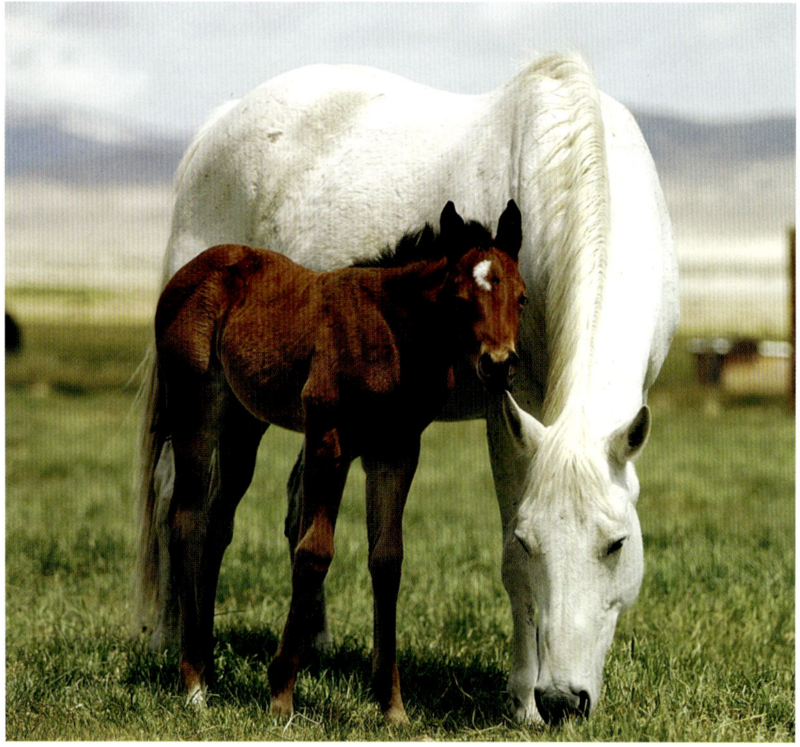

John bought his first stallion in 1976, a buckskin with both Anchor Watch (TB) and Question Mark breeding. He admits that he was probably more interested in the buckskin color in the beginning, rather than the quality, but through luck or otherwise, he ended up with some really nice horses. It was then that he figured out that there were four things he really needed in a horse—athletic ability, good looks, disposition and a lot of cow sense. And that's what he's been breeding for ever since.

His next stallion was a son of Gay Bar King called Long Live The King, which he bred for about seven years. Then he bought a horse called Docs First Fri Dee by Docs Fri Dee Bar. Docs First Fri Dee was a Pacific Coast reining and cutting champion, but the Laceys were only able to use him two or three years before he died. John replaced him, however, with a son of Mr Gunsmoke, which he bred for about three years.

Next, still looking to improve their foals, the Laceys bought a son of Colonel Freckles called Jeftee, who, as a 3-year-old, had made the semi-finals of the NCHA Futurity. They were showing that horse in cutting, and breeding him, when tragedy struck again, and he died after only one breeding season.

Many of the broodmares have been raised on the ranch and reflect the breeding of nearly all those stallions, but John has also bought some mares to add some outside foundation blood. Today he is crossing his mares on a horse called RSI Doctor Pepper, a good cow horse with Mr San Peppy breeding that they bought from the Renderbrook/Spade Ranch in Texas, and Wranglers Starlight, a horse that has qualified for the senior cutting at both the 2004 and 2005 World Shows. Wranglers Starlight is by Grays Starlight out of Wranglers Connie by Doc's Wrangler, making him a full brother to Starlights Wrangler, one of today's top reining stallions.

Using Horses

Unlike most big ranches, which, as a rule, ride only geldings, the Laceys ride some mares. As a matter of fact, on the Owens Valley ranches, they ride almost exclusively mares.

"When I bought my first mares, I decided I liked riding them," John says. "I think they are tough, and we never worry if they are in season or not. We ride them and we run them in a bunch. Our program was founded on the idea of riding the mares and proving the seed stock that we were going to use in our breeding program. We even rode the mares we bought before we decided if we were going to add them to the broodmare band. My philosophy is to have nothing in the broodmare band that hasn't proved itself to be what we want."

On the Tejon Ranch, however, cowboys ride mostly geldings. Most of the geldings that are raised on the ranch in Owens Valley, and aren't sold, go to the Tejon Ranch.

"I've always thought that for your horses to have real athletic ability, you needed to integrate some speed breeding," John says. "We haven't brought speed in through our studs, but we have through some of the foundation mares that we have added over the years. We have raised some pretty good rope horses that have been able to compete at the highest level. I think blending in the speed and cow sense, and then riding them on the ranch and

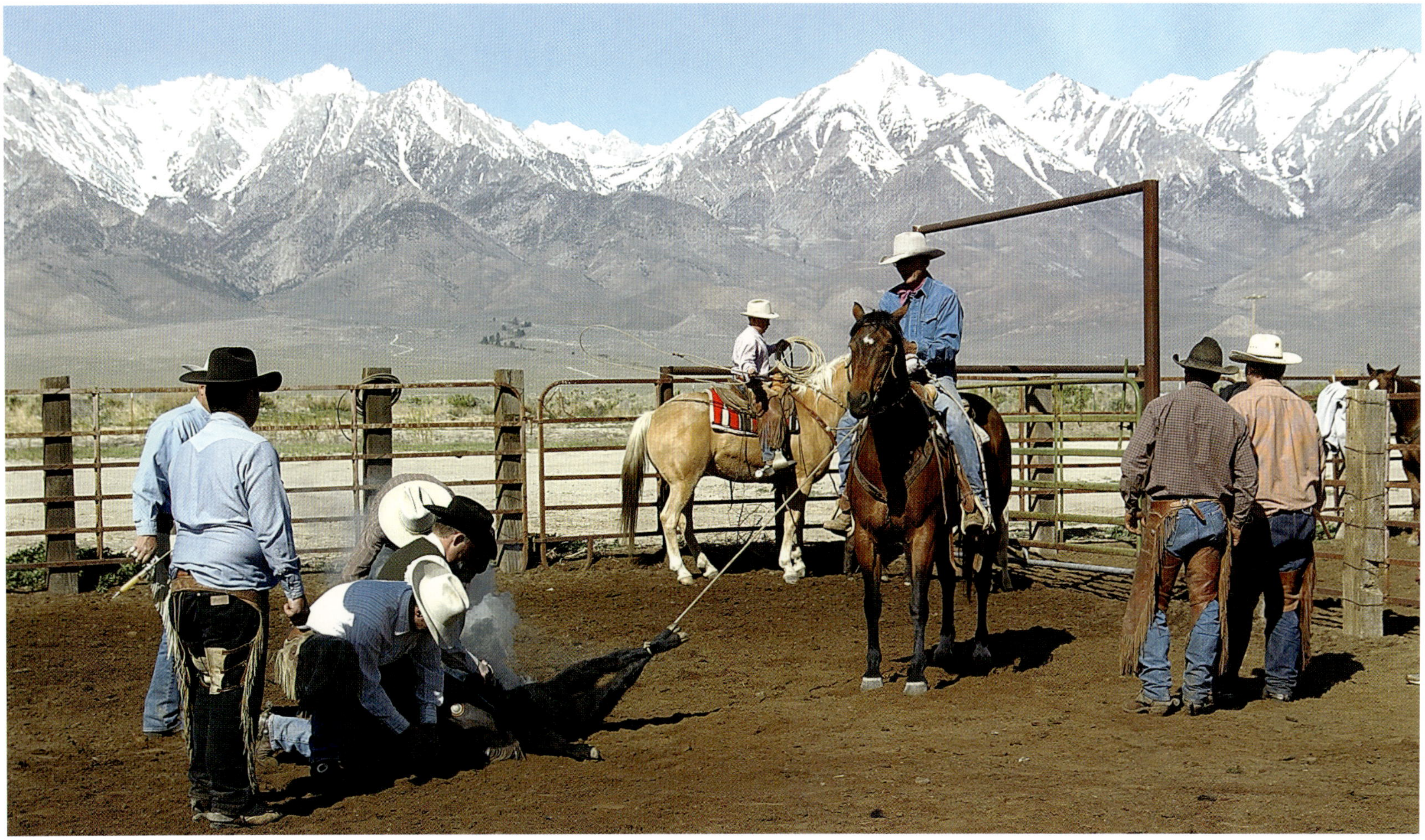

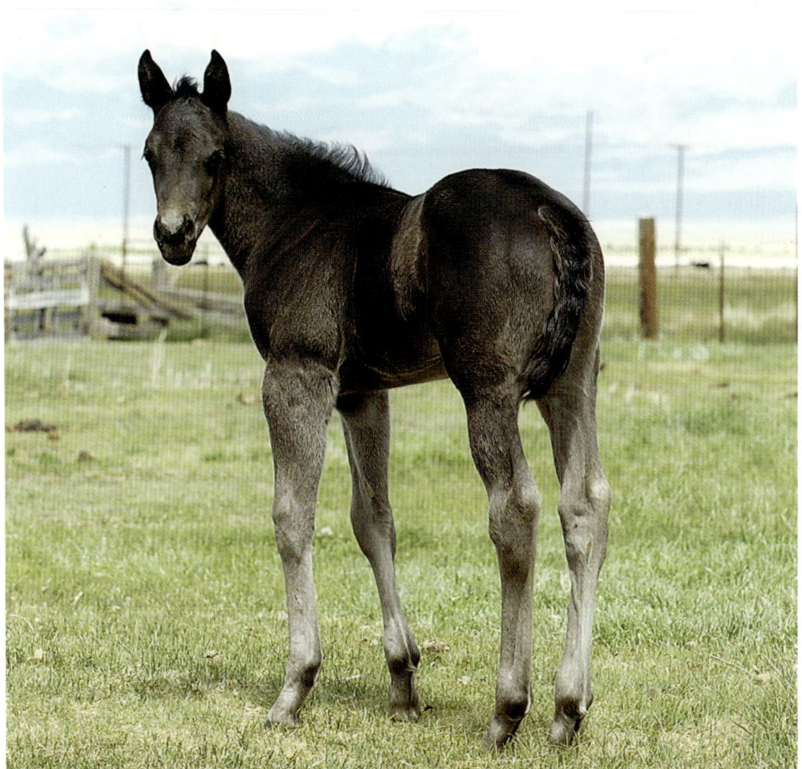

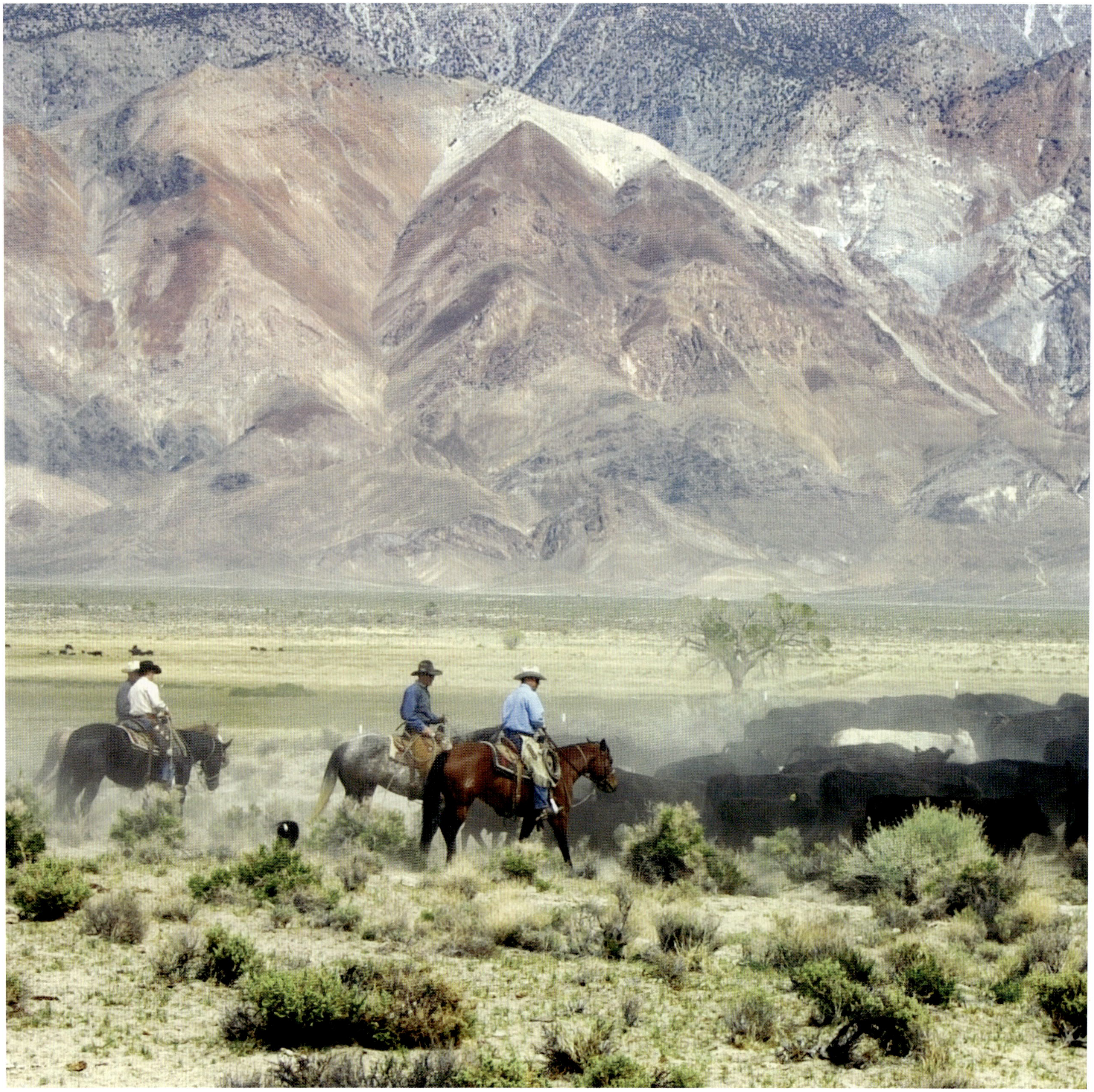

making sure they do everything right before they get into competition is what helps us there."

Livestock Industry

John spends a lot of time on the Owens Valley ranches, and maintains a bunkhouse there for his use, although his son, Mark, lives there and runs those ranches. John lives on the other side of the Sierra Nevada mountains with his wife, Dee, in Paso Robles. There he is closer to the Tejon Ranch. He is a graduate of California Polytechnic State University—better known as Cal Poly—and still wears every day the buckle that he won on the college judging team in 1960. John and Dee also have a daughter, Nicki, and although she is married and living in Hawaii, she is still active in the family partnership on the ranch.

John has been a leader in the livestock industry for many years, and has served two terms as president of the National Cattlemen's Beef Association. He has also been very active in the beef cattle industry in his home state.

But he's always been interested in horses, too, riding first the ones his dad bought and even some they caught wild on the desert. But his goal was to raise his own, and as quick as he was able, he bought those first mares.

"We hope someday in the near future to not have to buy any outside foundation stock," John says. "I want to create our own band of mares that we can say are Lacey-bred. That's what we have been shooting at for 40 years, and we are almost there."

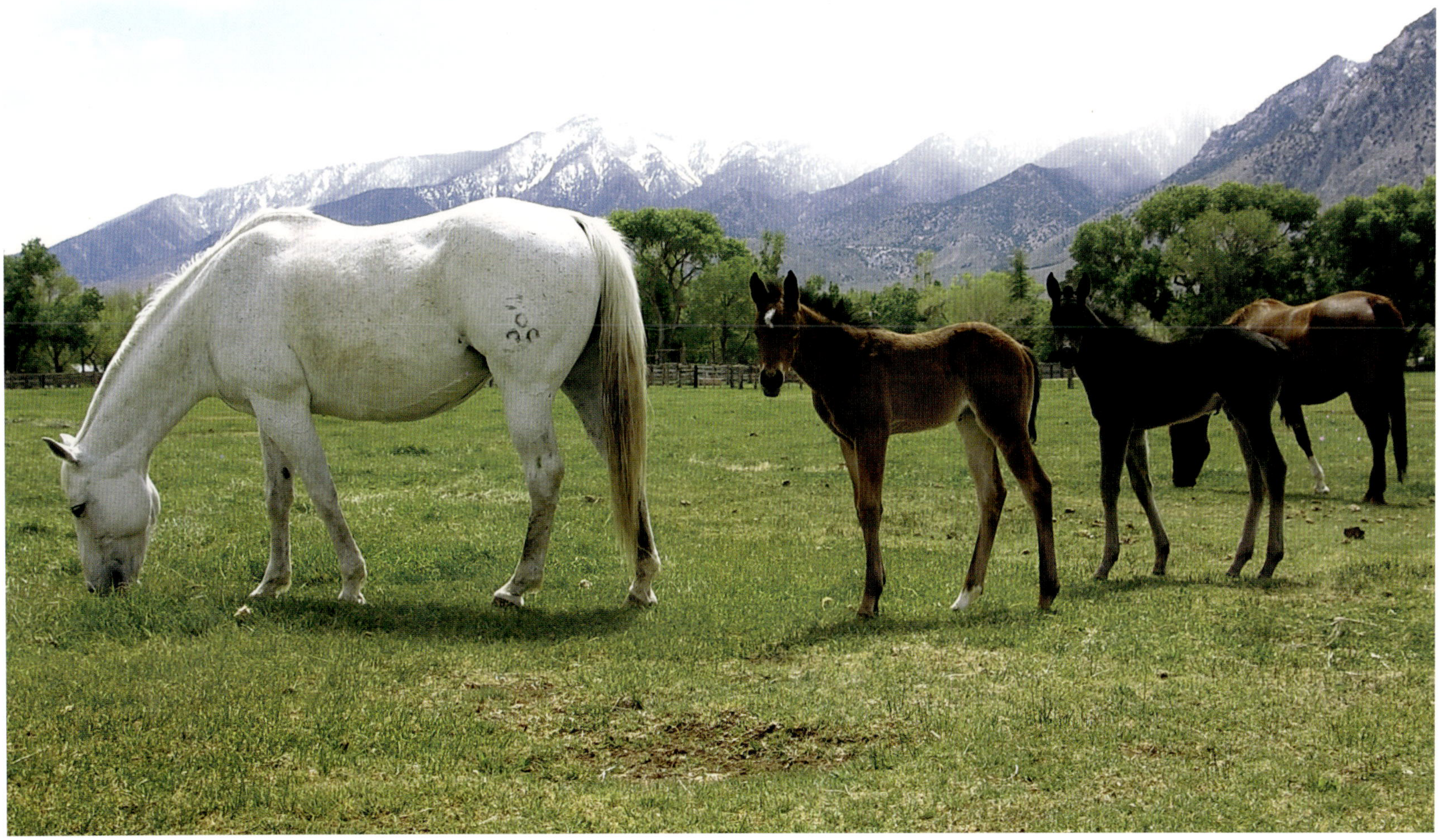

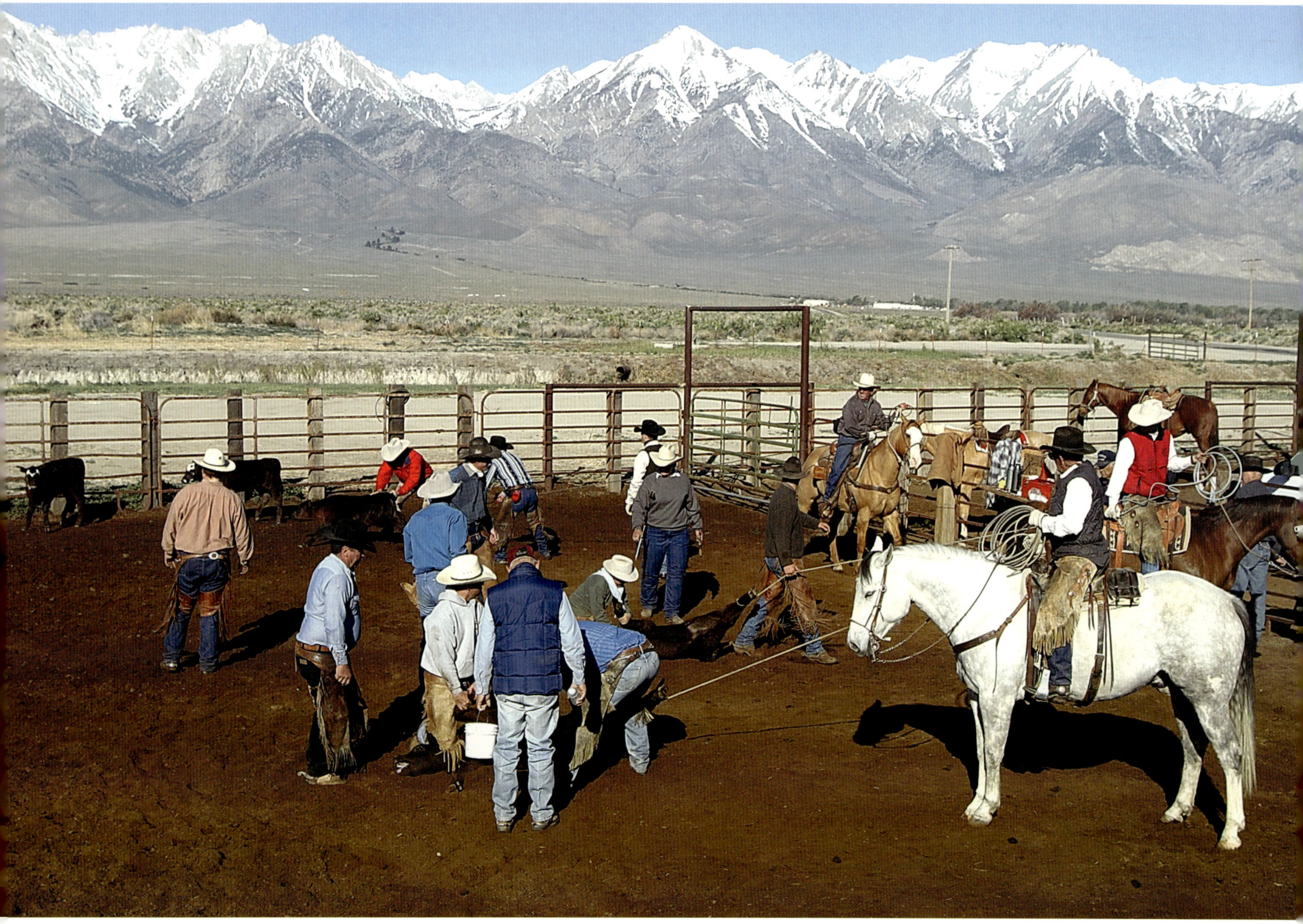

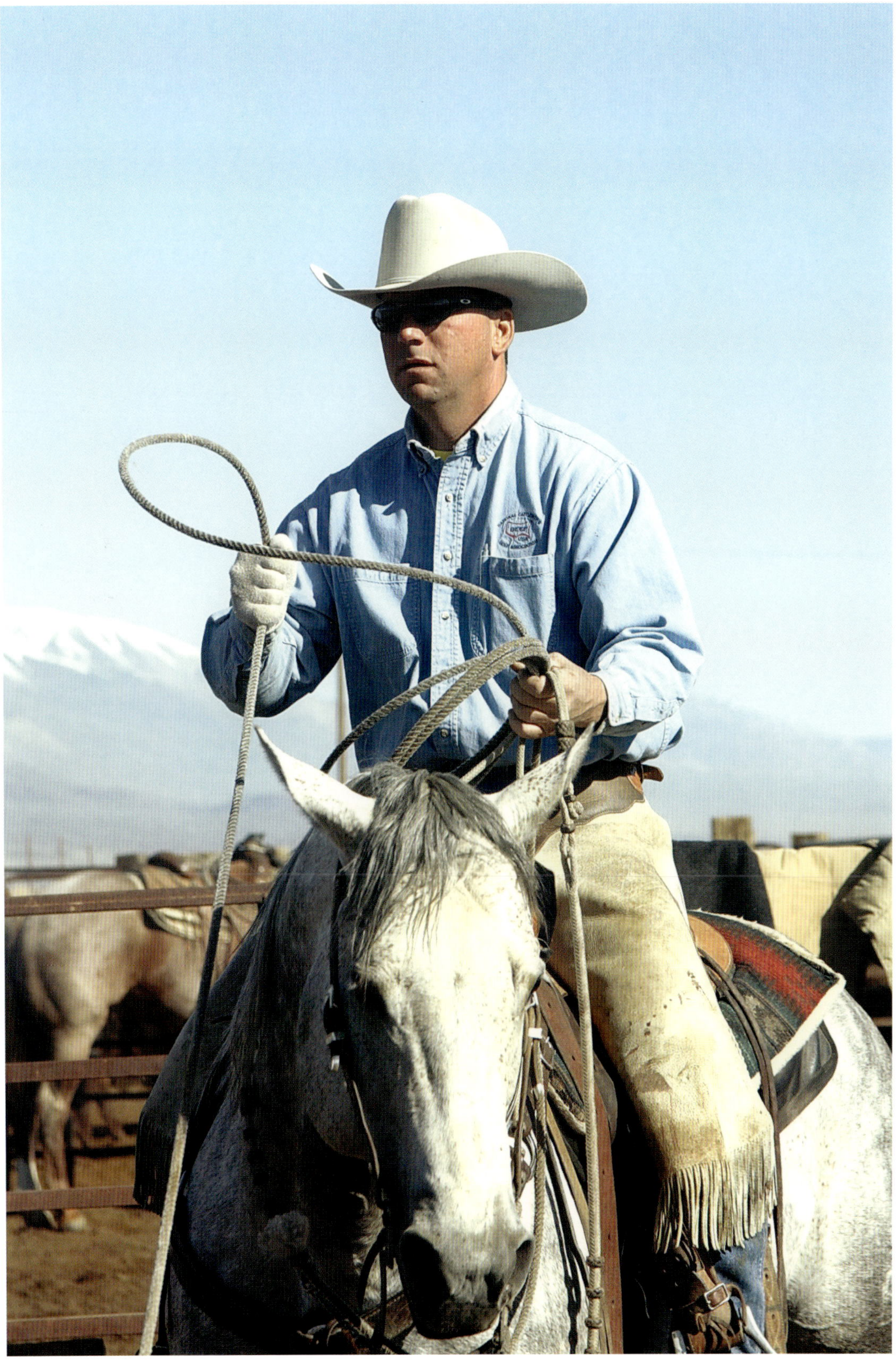

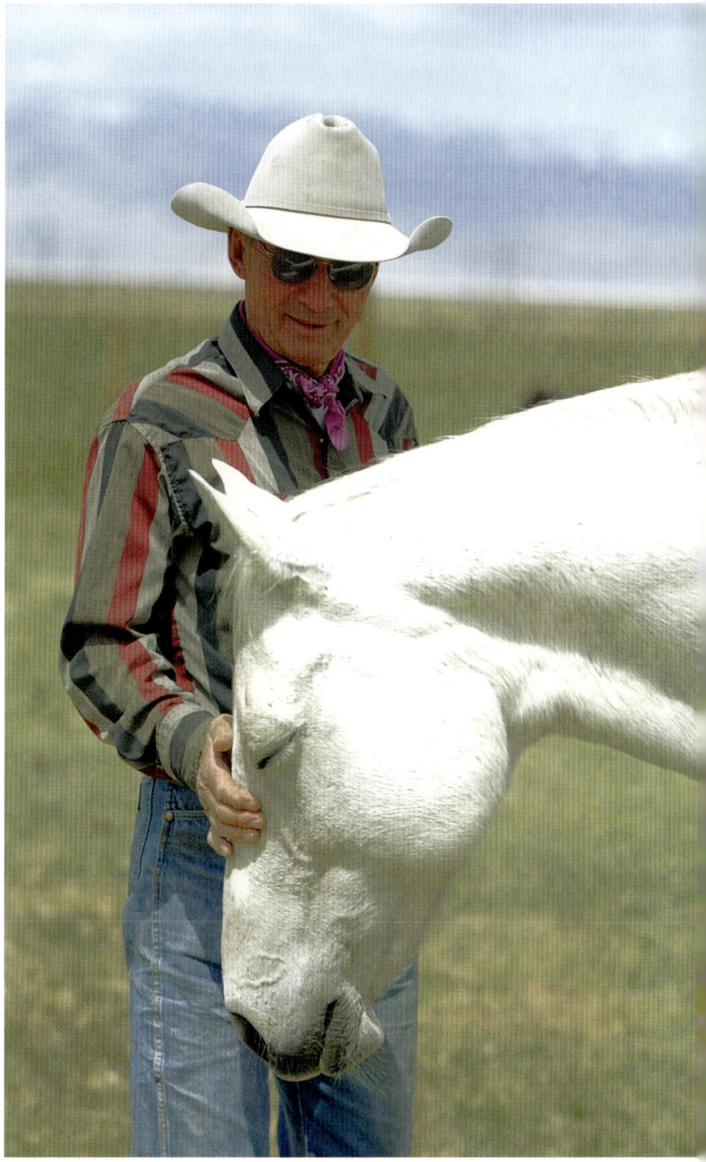

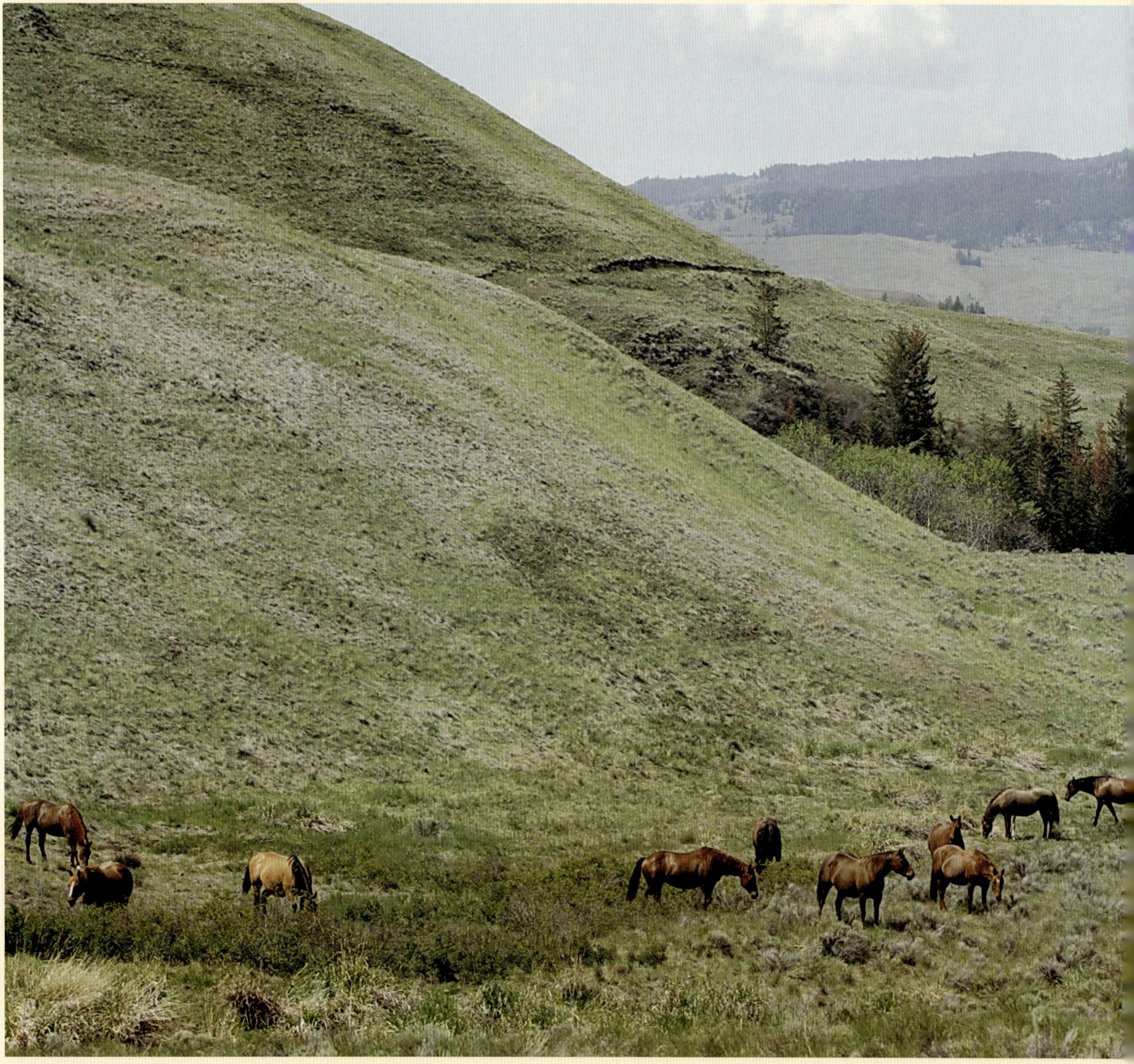

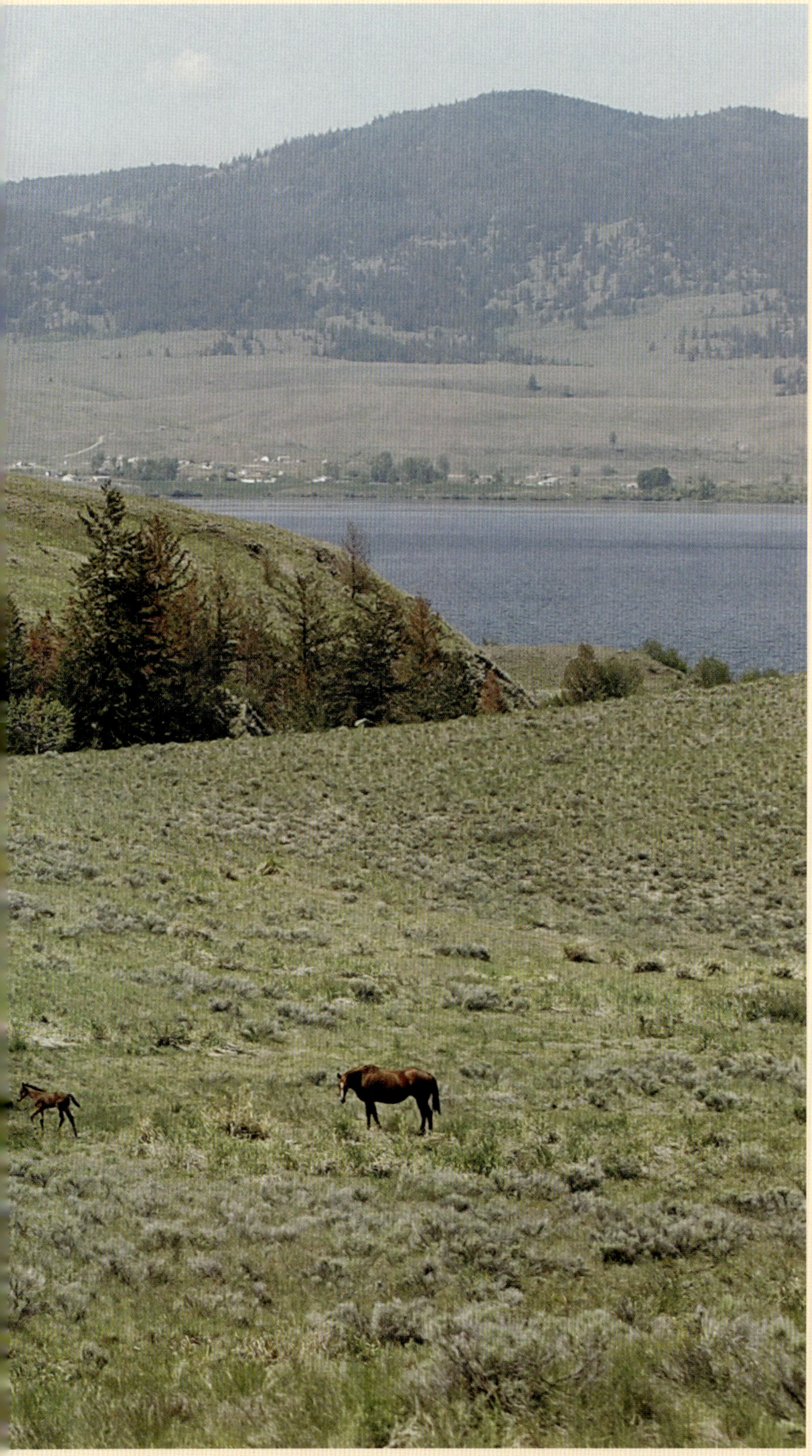

Douglas Lake Cattle Company

Douglas Lake, British Columbia

2004

The Cowboy Way

Douglas Lake Cattle Company is a town of itself. It has a store, post office, school and church. It's also the largest ranch in Canada.

The radio is turned off in the truck, but the digital clock on the dial reads 3:30. It doesn't have to say "a.m." The fact that nothing is visible outside the range of the headlights says it's not afternoon. But we don't need a reminder. We had gotten that when the bedside alarm had gone off an hour before. It was then we questioned exactly why we were getting up at that time of the day.

Or night. Technically, it's still night when we reach the cowboy camp—the Dry Farm Camp—on the south end of the ranch. Dark, anyway. But light spills from the window in the little cookhouse, and there is movement in and around the various cabins. Movement, but little noise. Almost no conversation penetrates the cool morning air, as horses are led into the semi-light of the interior of the camp and saddled.

The aroma of breakfast hits us as we step into the small cookhouse. A young woman stands at a large black stove that monopolizes one wall; huge cast-iron skillets of eggs, potatoes and bacon cover the burners. A long table, surrounded by benches, fills the other half of the room. The cowboys come through the door single file and fill their plates from the skillets. The eggs are seasoned with the various sauces and jellies on the table, and washed down with coffee and juice. But there is no time wasted over second cups. A few minutes later, the boys from the Dry Farm Camp trot their horses out of camp and into the blackness. It's still dark, but the day has begun.

The Ranch

Lying in the rolling grasslands of British Columbia, about 200 miles east of Vancouver and 30 miles east of the little town of Merritt, Douglas Lake Ranch is the largest cattle ranch in Canada. Encompassing 164,000 deeded acres and another 350,000 acres of government grazing land, its borders stretch 60 miles from top to bottom, 35 miles side to side.

The ranch traces its history back to 1872, when John Douglas happened onto a pretty little lake, named it for himself and then homesteaded 320 acres alongside it. Douglas later sold the land to Charles Beak, who, with some partners, founded Douglas Lake Cattle Company in 1886. The ranch expanded as it went through a couple of other owners through the years, until 1959, when Charles "Chunky" Woodward and John West bought the property. West died in 1968, and Woodward became the sole owner.

It was under Chunky Woodward's ownership that the ranch came to prominence in the Quarter Horse industry. However, Woodward died in 1990, and his family sold the property in 1998 to Bernard Ebbers, who sold it to its present owner, businessman Stan Kroenke, in 2003. Through all the owners, the original purpose of the ranch never wavered. Douglas Lake is probably the No. 1 producer of beef cattle in Canada.

"We are what you would call a yearling ranch," says general manager Joe Gardner. "We take calves that are weaned in the fall, and background them on feed all winter. Then we put them back on grass in the spring and sell them as grass yearlings that fall."

The 350,000 acres of government land that the ranch has leased—known as Crown grazing land, which is similar to BLM land in the United States—is used from May to the middle of October. All the cattle are brought back to the ranch-owned land during the

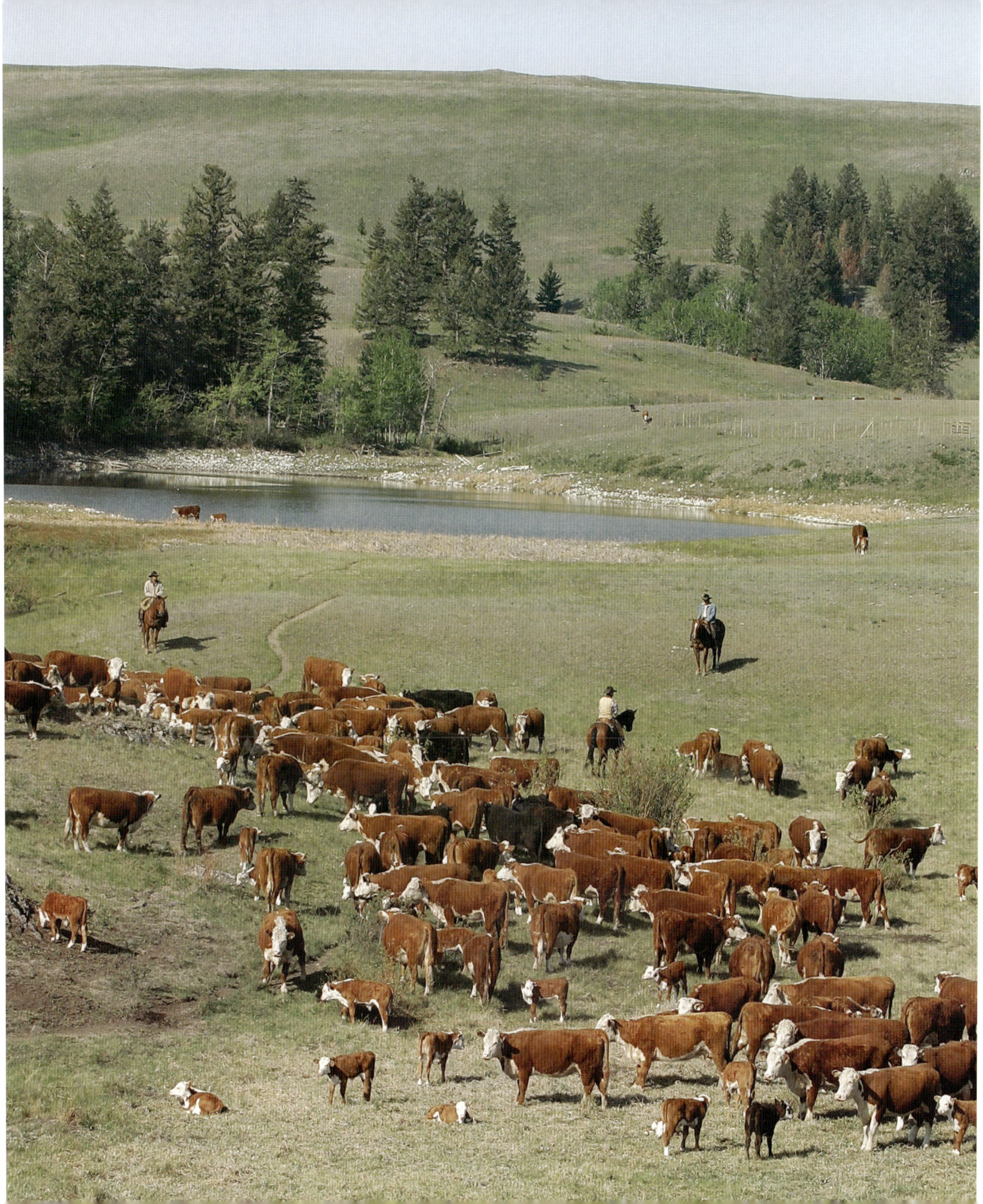

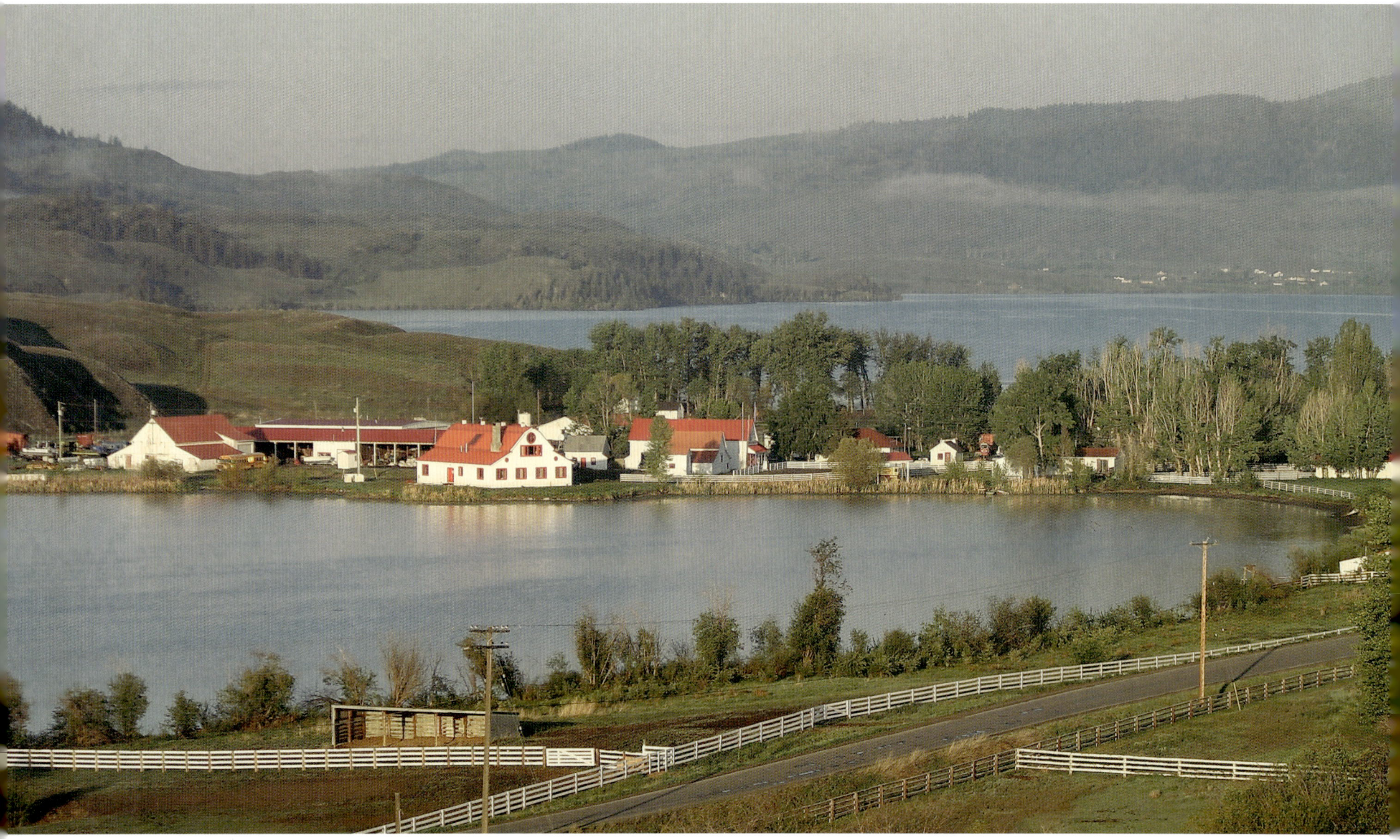

winter months. At any one time, cattle owned by the ranch will include about 7,000 mother cows—both Hereford and black baldies—450 bulls, some 4,500 yearlings on grass and maybe 1,500 yearling heifers that will be used as replacements for the cow herd.

"That's what brings the horse into the picture," says Joe. "Horses are the only way we handle our cattle. This grassland is very delicate. We don't allow any kind of four-wheelers or motorbikes. With our low annual rainfall, any kind of disturbance like that will result in erosion and an influx of weeds, and we can't stand that."

Joe has been manger of the ranch for 25 years. He says that the managers before him had a very strict policy toward trespassers, and he has maintained that policy. "We're not running a popularity contest," he says. "We feel that the single most important asset we have is grassland, so we protect it at all costs."

The predominant grass on the ranch is bluebunch wheatgrass, which is native to the northwestern United States and western Canada. It is resistant to cold winters and dry summers, but is easily destroyed by overgrazing or traffic, a fact Douglas Lake employees are very cognizant of. Even ranch vehicles don't get off the pasture roads.

During the summer, the cattle subsist very well on the native grass, without supplements. However, when winter snow covers all the pastures, everything has to be fed.

Winters can be extreme, but not really, by Canadian standards. The average lows are around 18 degrees Fahrenheit, although 20-30 below zero is not uncommon. Highs in the summer average about 80 degrees, but there are days that are much hotter. The ranch headquarters sits at only 2,600 feet, so temperatures there are more moderate than they are in other areas of the ranch, which go as high as 6,000 feet.

The headquarters is visible for miles. Literally sitting on the shores of Douglas Lake, every building is painted white with a red roof. More a company town than a ranch headquarters, most of the employees live there, and due to the remoteness of the location, the ranch tries to service their needs. A general store handles most daily necessities, and a church offers services every Sunday evening. A ranch-owned school takes care of the educational needs of employees' children.

"The government ran a school out here for years," Joe says, "but the numbers got so low that they had to close it. Then our kids had to travel an hour and a half each way into Merritt to school. This all happened while Chunky owned the ranch. He and I knew it would be hard to keep young families if we didn't have a school, so we started our own. We set up a program that followed the provincial curriculum, got approved for funding for an independent school and hired a teacher."

Actually, there are two teachers, one for the lower grades and one for high school. Several students have graduated from the school, but two have gone all the way through, kindergarten to graduation. Both of them are now in college.

"We only have 10 or 11 children in the school right now," Joe says, "but our kids are getting a wonderful education. They do lots of extracurricular activities—maybe not the same ones they would do if they were in town, but they don't have to waste that three hours a day on the bus. They can spend that time at home, riding their horses or doing other activities they like."

The Horses

Although Douglas Lake Ranch has always had horses, it wasn't until ranch owner Chunky Woodward became interested in cutting horses in the early 1960s that there was a particular breeding program. Woodward began with the purchase, in 1963, of Peppy San, a 4-year-old son of Leo San out of Peppy Belle by Pep Up. Peppy San was bred and owned by G.B. Howell of Texas. The year before, the horse had won second in the first NCHA Futurity under the saddle of trainer Matlock Rose. Woodward brought Peppy San back to Douglas Lake, used him occasionally for ranch work and turned him out with some mares. But in 1966, Rose asked Woodward if he could campaign Peppy San for the NCHA world championship. Woodward agreed, and in 1967, Peppy San was the world champion, earning $20,000, more money than had ever before been won in one year by a cutting horse.

Woodward also bought the mare Stardust Desire, which Rose had made world champion in 1966. Breeding her and some other Douglas Lake mares to Peppy San resulted in the NCHA and AQHA world champions Peppy's Desire, Sonita's Last, Peponita and Royal Santana, plus Canadian champions Peppy Isle,

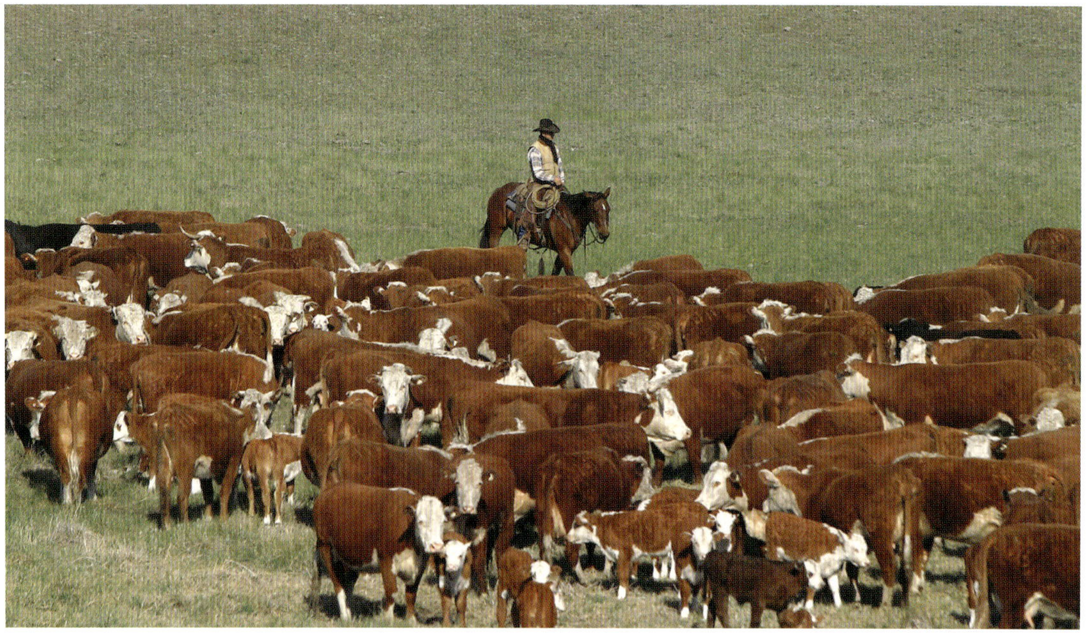

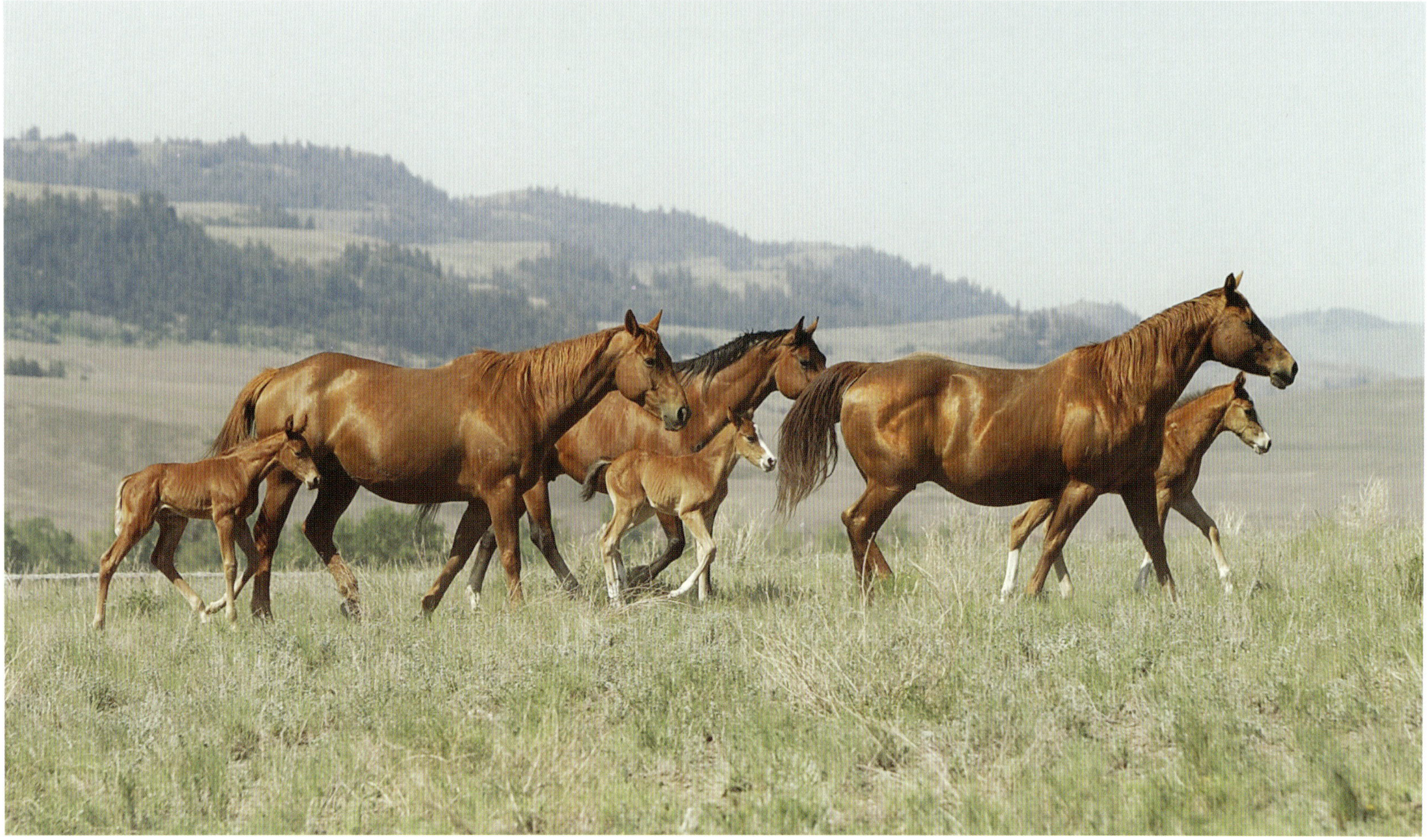

Peppycali, Tip It San, Booger San, Chunky's Monkey, Popular Peppy and Bonita San. Never before had one horse sired as many champions, all bred by one ranch.

Today, Douglas Lake owns approximately 35 broodmares, many of which carry Peppy San breeding. The ranch is currently using two stallions, one of which is Fritzi Badger, a son of Peppy San Badger, who is by Mr San Peppy, a full brother to Peppy San. Another is a son of Mr San Storm by Peppy San. They also have some young stallions that they are about to try. A few of the Peppy San-bred mares were bred to Massasuta and Drifts Chip—both standing at the Four Sixes Ranch in Texas—to get some outside blood, and sons of both of those crosses will soon be added to the ranch's stallion battery. Massasuta is by Dash For Cash, and Drifts Chip is by Double Drift.

There are almost 200 geldings in the saddle horse remuda.

"The purpose of our Quarter Horse program today is to sustain mounts for the cowboy crew," says Stan Jacobs, Douglas Lake cow boss. Stan is a cowboy himself and has been with the ranch about 20 years. "At one time, we had a cutting horse division, but that came to an end with the passing of Mr. Woodward. But the bloodlines and the foundation work that he did remain here today in these horses.

"I have had guys ask us why we have horses bred like that when all we do is cowboy on them," Stan continues. "I say, 'Well, jeez, isn't that terrible. Taking horses that have been bred to work a cow, and actually use them for ranch work. Shame on us.' The guys working in our cowboy crew really appreciate the caliber of horses they get to ride."

Each cowboy on the ranch is allowed 10 horses in his string. These will include some young horses, and some older horses that he will use in the weaning pen or for doctoring in the feedlot during the winter. Everyone will have special horses that they can use to rope a big cow and not get themselves in trouble, and they'll have three or four horses that they'll ride in the branding pen. The horses they use for calving usually have the summer off, so by fall, they are ready to come in and do some weaning and sorting.

All horses being ridden are shod, and from late December until early February, everyone will keep at least two horses "sharp shod" for work on the frozen ground. A horse that is sharp shod will have small, sharp projections on the bottom of its shoes.

Foals are normally weaned by early October. In January, all the yearlings are brought in, halter broke, tied up and played with for about two weeks. The colts are gelded, and everything is branded. In the spring, they are turned back out until the following spring, when the cowboys all come together to get their pick of the now 2-year-olds.

After the selections are made, the cowboys will spend about 10 days working with their new horses, and then they are turned back out. They are brought in again as 3s.

"We'll use those 3-year-olds in the fall when we start to gather the high country," says Stan. "The days are longer, but they are not real demanding, and we don't want to ask a whole lot of those horses just yet. We want to bring them along slowly. Then, in the spring, when the horses are 4, some of the cowboys will have them in the branding pens. That fall, they'll go to the high country again, and that next spring, when they are 5, they are ready to go to work."

Douglas Lake is more than a cow ranch. The company has dealerships for farming and construction equipment and Sooner trailers in area towns, and a recreational business—fishing, hiking and horseback riding—on parts of the ranch. However, it's a cow ranch first and foremost, and that part of the ranch has changed very little over the years.

Cow boss Stan Jacobs grins under his drooping moustache. He is watching almost 500 cows with calves pour through the gate of a wooden corral back onto the rolling grassland. A group of cowboys sit on their horses in front of the cattle, preventing them from running. Turning, Stan says, "We do it the traditional way here, the cowboy way. We do everything horseback.

"Like this herd of cows and calves. We gathered them this morning and branded the calves, 487 of them in a little over four hours. Every one of them was roped and dragged to the branding fire. When we move cattle, it's all done horseback. When we wean, there's a cowboy on a horse sitting in the gate, letting the cows out. When we sort steers, it's done horseback, and when we ship cattle, someone on a horse drives those cattle all the way into the truck.

"The first thing the boys on the cowboy crew do every morning is saddle a horse," Stan continues. "When you hire onto this cowboy crew, you know you're going to be horseback."

Then he grins again. "It's the cowboy way."

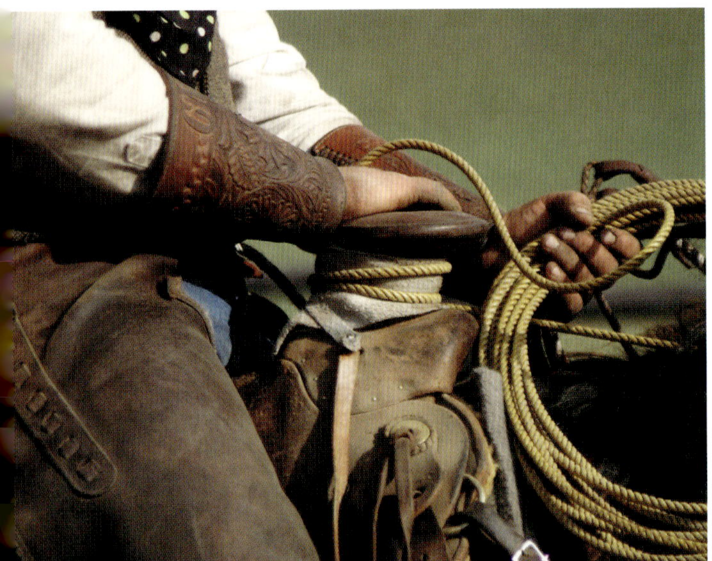

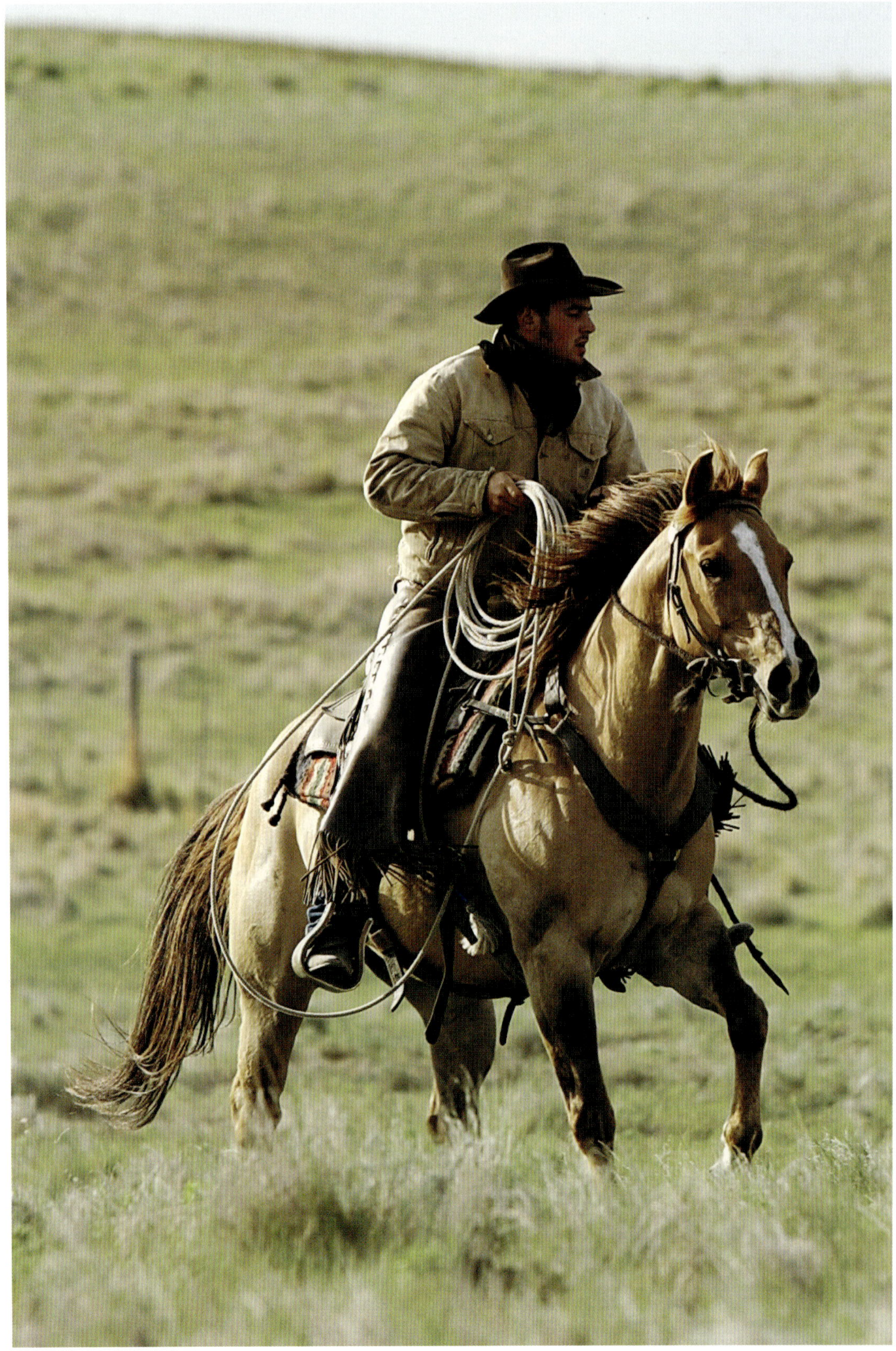

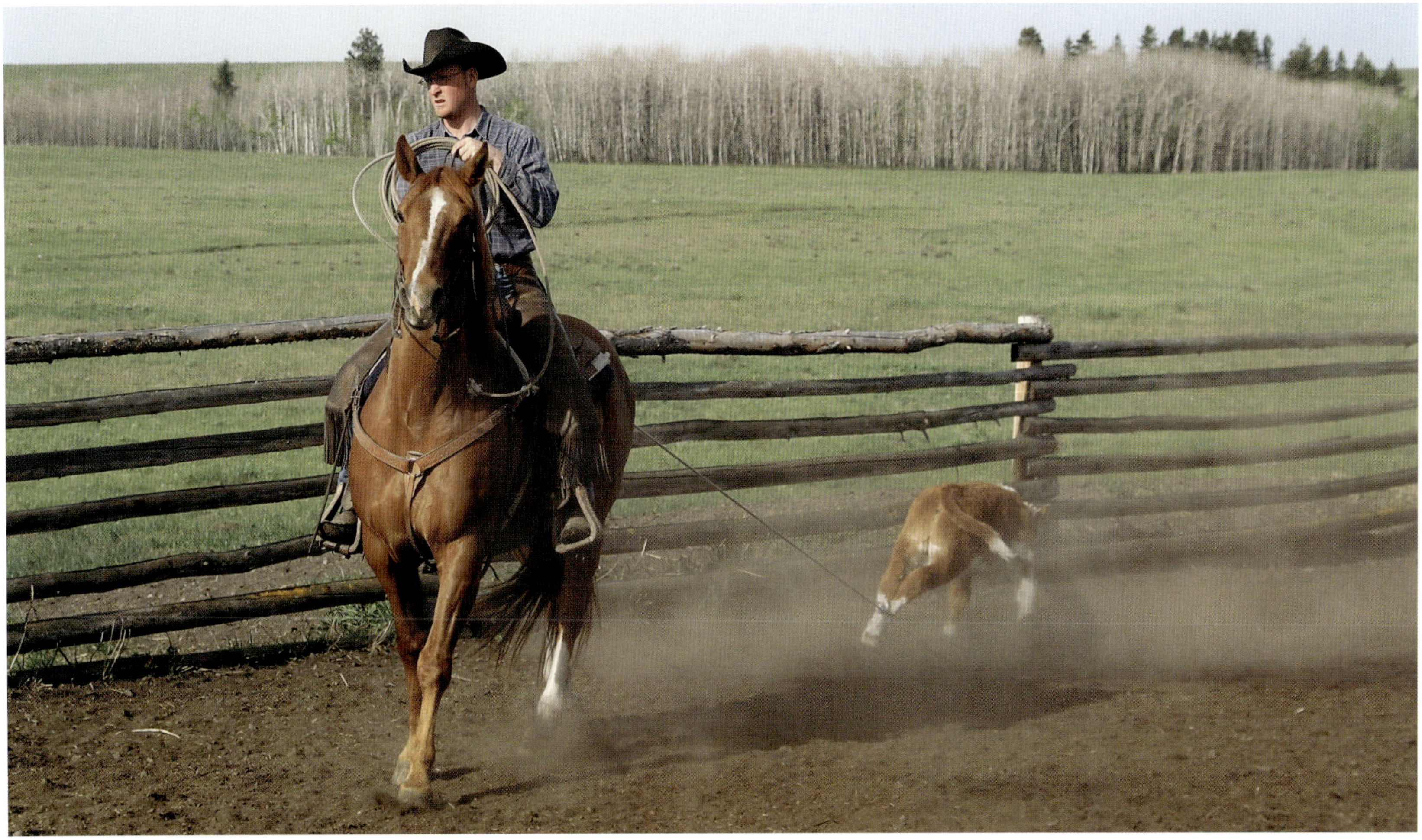

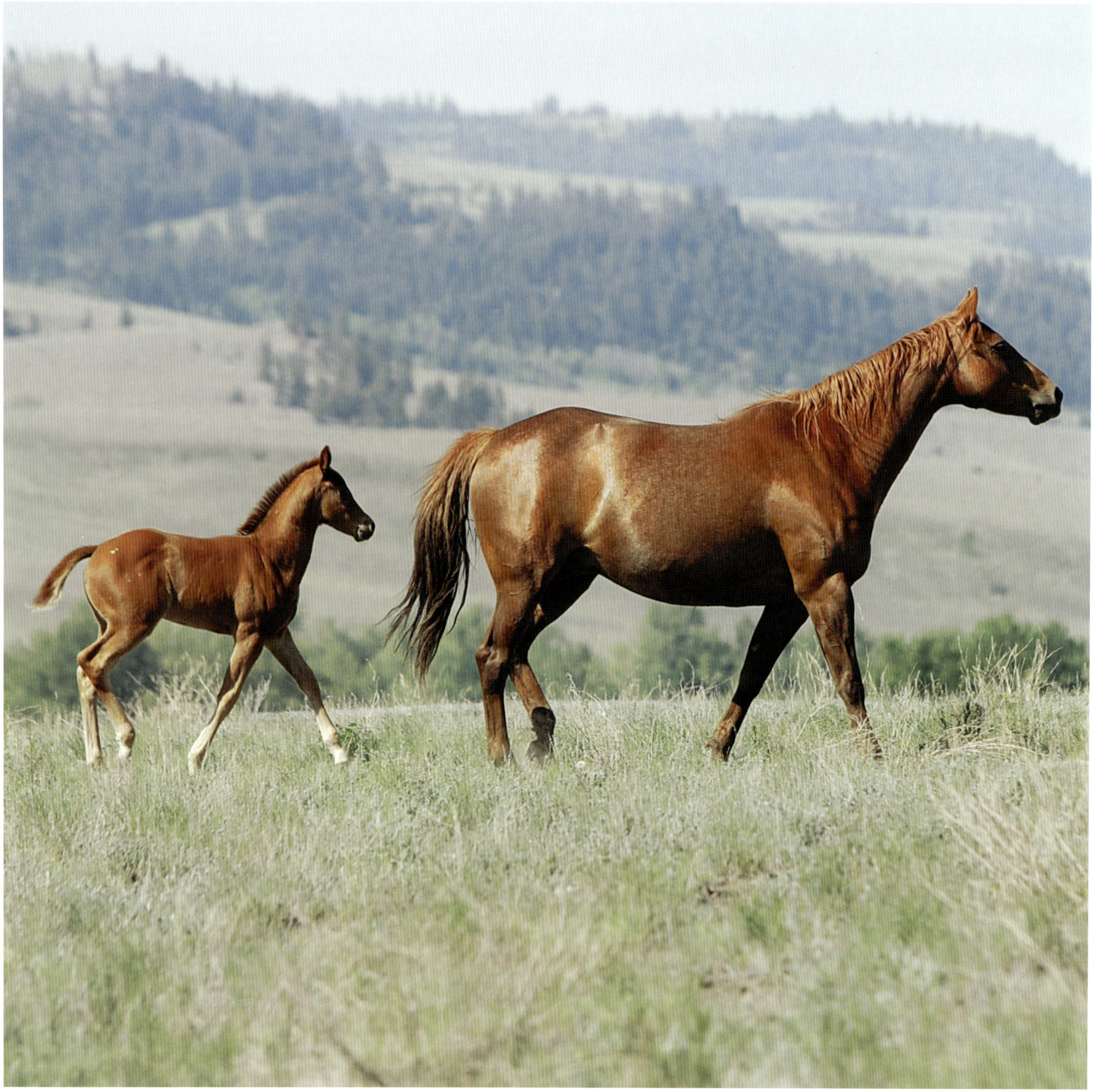

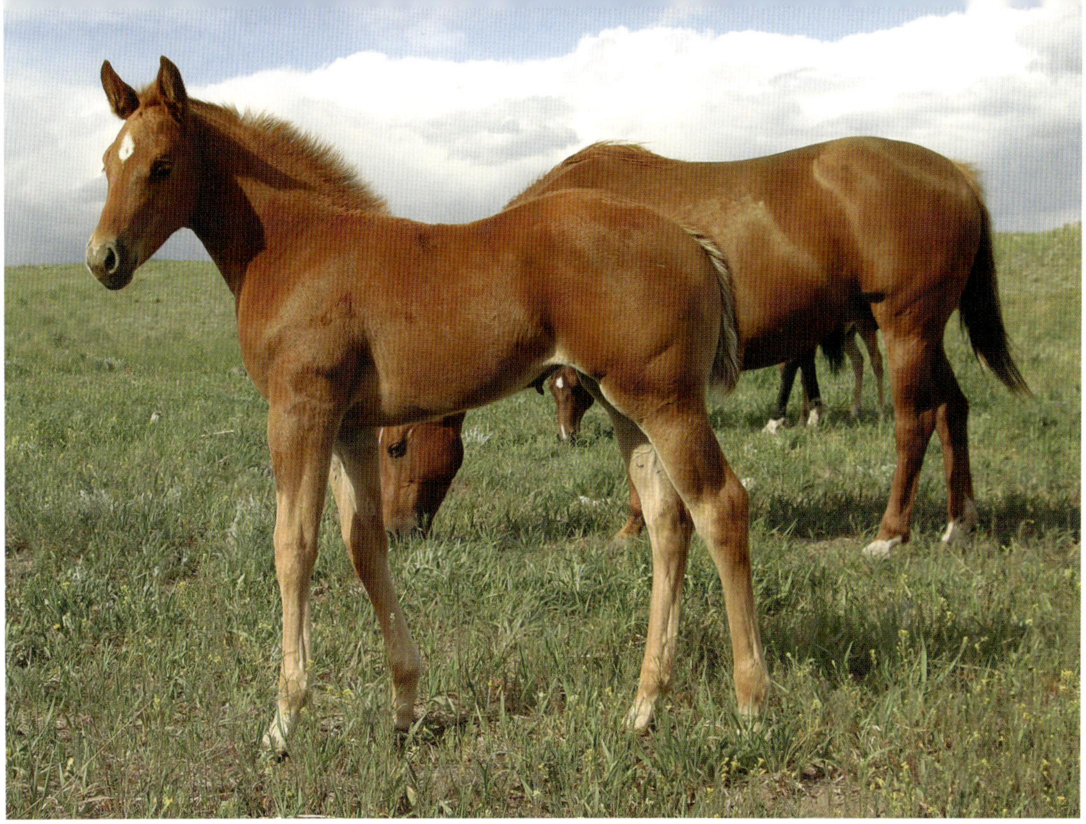

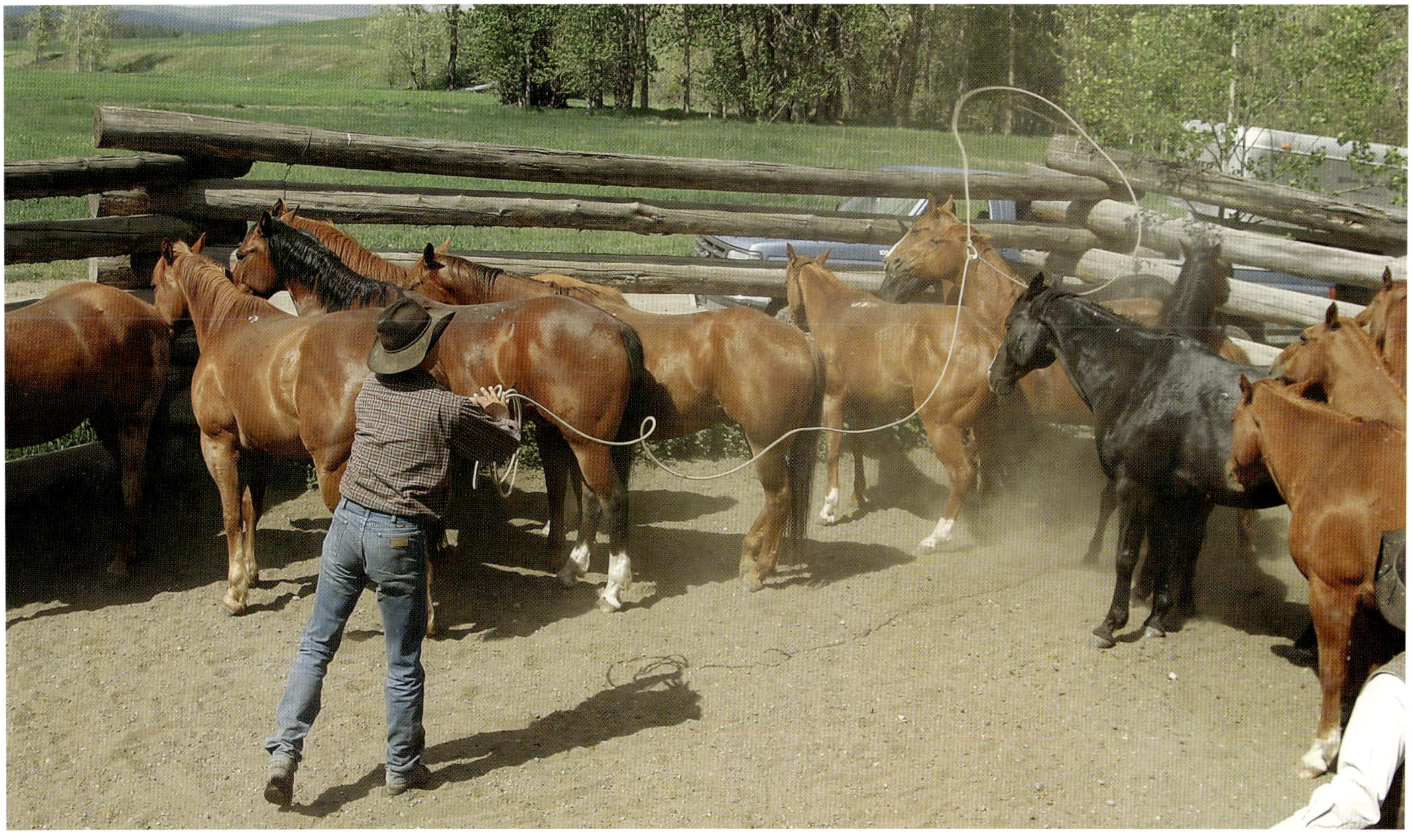

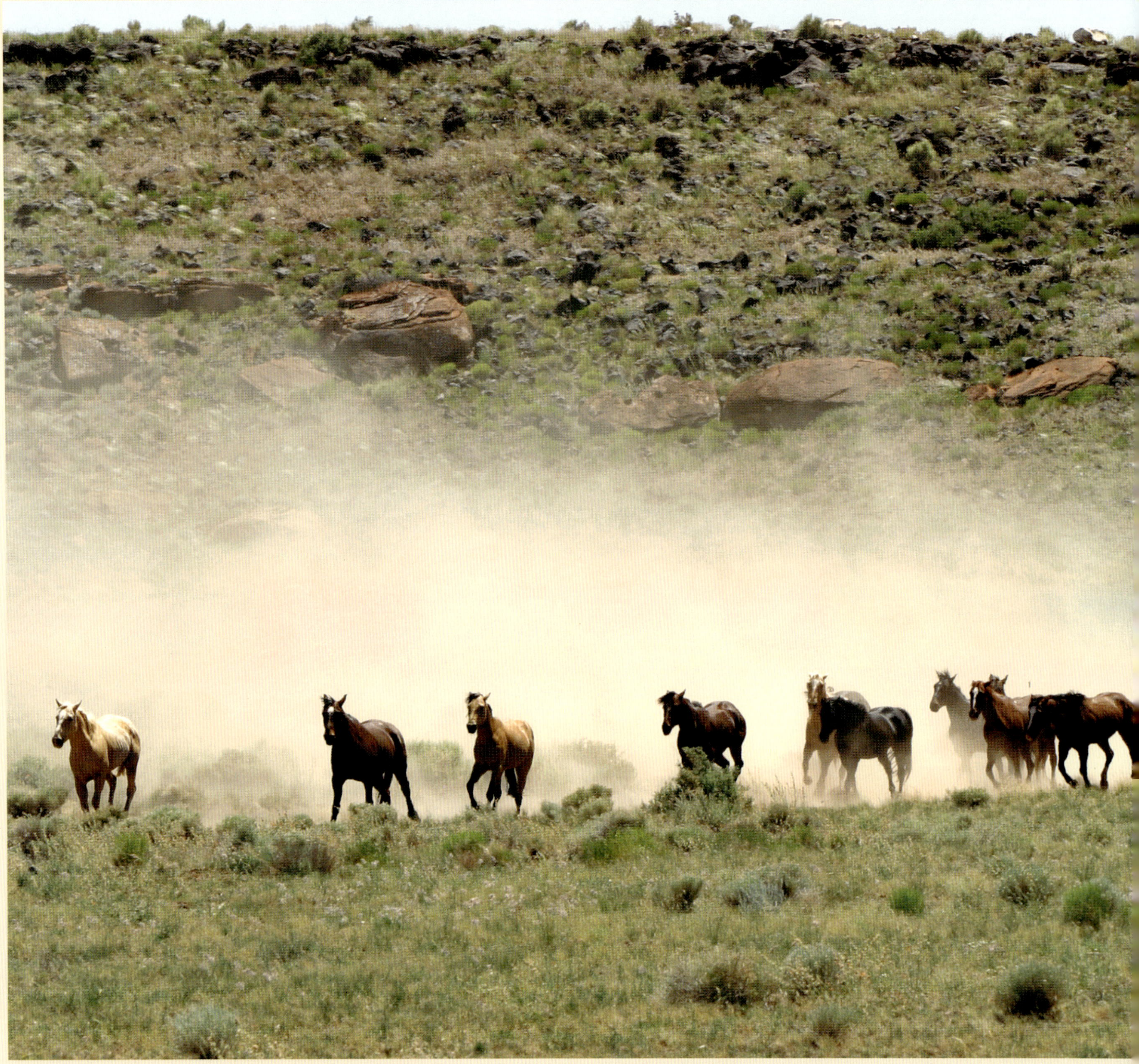

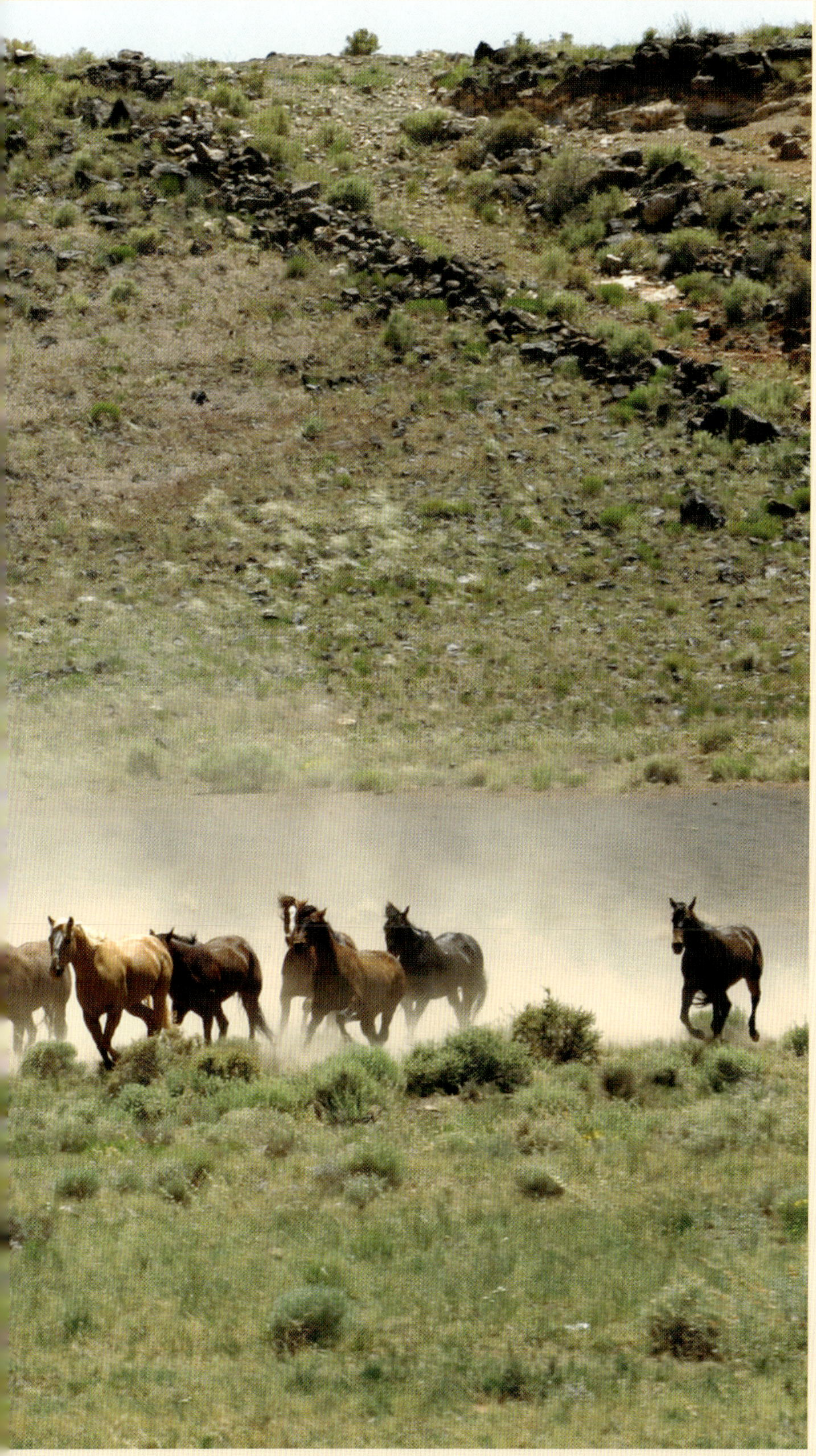

Babbitt Ranches

Flagstaff, Arizona
2005

The Horses of the Babbitts

This year's winner of the Best Remuda Award needs a good cow horse for its northern Arizona ranch land. Babbitt Ranches thinks it has it.

Finished with breakfast, Vic Howell skillfully balances his coffee cup on his plate as he steps from the main room of the bunkhouse into the kitchen. Going to the big sink, he holds his plate and coffee cup under running water until all the remnants of egg and bacon are washed away. One by one, each of the other cowboys files into the kitchen and does the same. Janet, the cook, will wash all the dishes later, but this little bit of courtesy makes her job easier.

The screen door of the bunkhouse then slams as Vic steps into the cool morning air, and it slams again and again as the others follow. Their boots crunch on the gravel parking lot, and the occasional ring of a spur rowel echoes softly between the buildings. It's a few minutes before 6 a.m.—not dark, but not yet daylight either—the moon still hangs over the hills to the west. But it's light enough to see the horses standing in the corrals, and the cowboys walk toward them, bridles in hand. Mounts for the day were roped out of the remuda yesterday afternoon, and the horses selected stand waiting to go to work.

Days on the Babbitt Ranches in northern Arizona start early. More than 700,000 acres constitute the three divisions of the property, and to go from one to another in the bobtail truck that hauls the horses often requires at least an hour. The morning sun can warm the semidesert of the high country quickly, and cattle drive best when it's cool.

Babbitt Ranches lies north of Flagstaff, but the property ranges from beyond Cataract Canyon—near the Grand Canyon—on the west, to the Little Colorado River on the east. The total operation is more than 1,000 square miles.

That's big, but it was once bigger. Established in 1886 by five Babbitt brothers from Cincinnati, Ohio, the original plans were to establish a cattle ranch in New Mexico, but they found the area around Flagstaff more to their liking. The grass was better. The brothers adopted the CO Bar as their brand—in recognition of their hometown, Cincinnati—and by 1919, they owned and operated almost every ranch in northern Arizona. At the height of their operation, they ran cattle and sheep on 100,000 square miles of range in three states.

But the era between World Wars I and II, which included the Depression and a generally depressed cattle market, along with one of the worst droughts the Southwest has ever known, hit the Babbitt brothers hard. In order to survive, much of the land was sold. Finally, in the 1950s, what was left was consolidated into the three divisions that make up the property today—the Espee, the Cataract and the CO Bar.

The terrain of the ranches varies from the sparse grassland of the CO Bar, where the mountains are all former volcanoes and small chunks of lava litter the ground, to the better and stronger grass of the Cataract and Espee. But it's all cow country, and the Babbitts make good use of it. The three divisions are stocked with approximately 4,000 cows and almost as many replacement heifers and yearlings.

That, according to Babbitt Ranches president Bill Cordasco, is why they have horses.

"If there was a better way to handle these cattle, then we would use it. But there's not. With the diversity of our ranch, whether it is mountains, trees, rock or sand, and with the size of

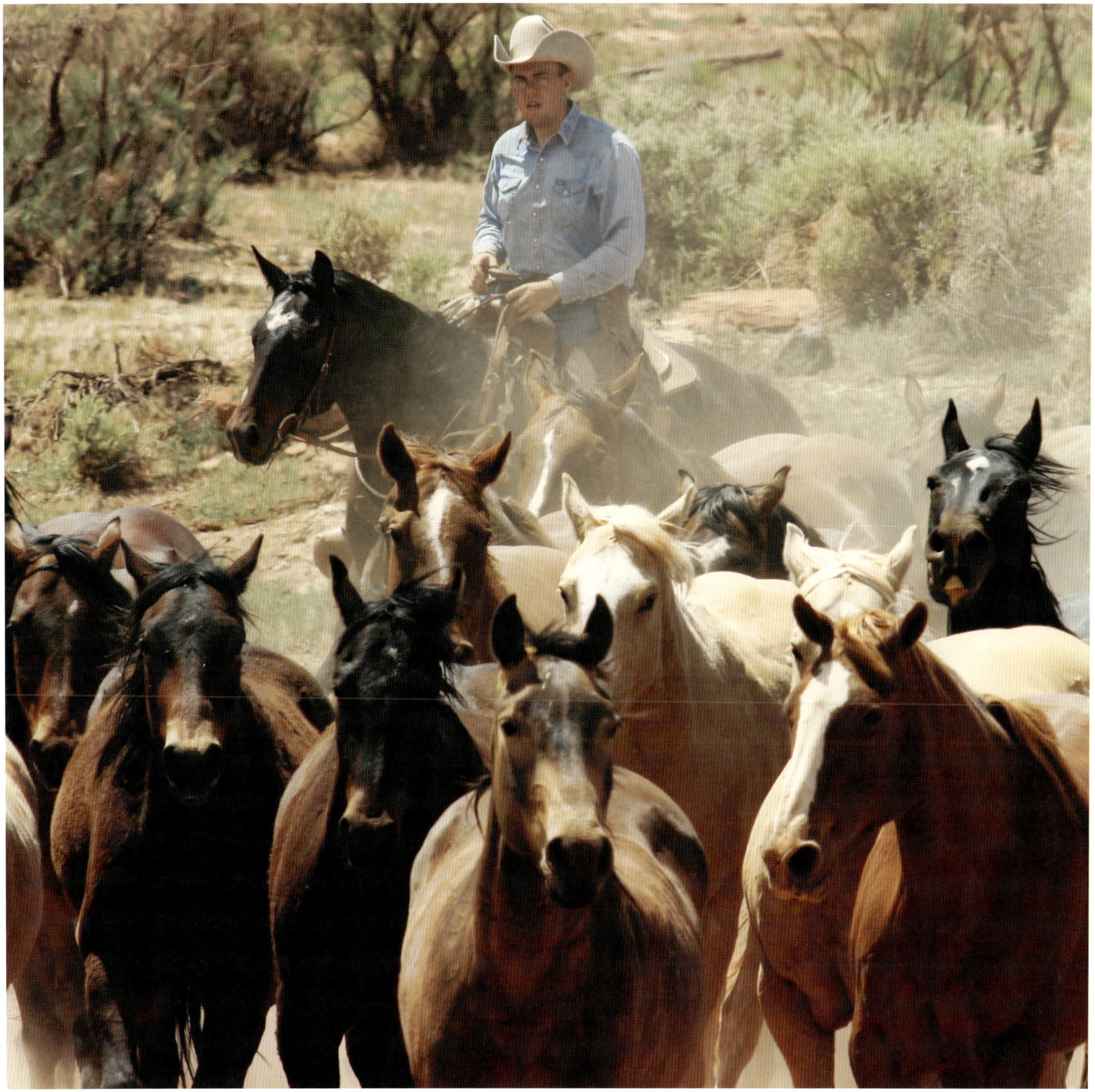

some of our pastures, nothing has come close to the ability of our horses to gather our cattle and work them."

Bill grew up in the shadow of the ranches. Raised by his grandfather, John Babbitt, Bill's heritage goes straight to the founders. John Babbitt was a son of one of the original brothers, C.J. Babbitt. John succeeded his father, and ran the ranches himself for more than 30 years. Bill, as a youngster, rode many miles in the pickup with his grandfather, and today he is applying what he learned, following in his grandfather's footsteps. Not necessarily a cowboy, Bill typically shuns a hat in favor of a ball cap, and he seldom gets horseback. However, he is rarely caught in the home office downtown. He represents the ranches in dealing with any governmental agencies, and he is active with the state livestock associations. And, although he may not help gather the pastures, he is often at the ranches, assisting with the branding, the horse sale and in other areas.

Bill is quick to give credit for much of the current success of Babbitt Ranches to a couple of men named Howell. Bill Howell, retired now, was with the ranches fulltime for about 33 years and managed them for 22. He was succeeded in 1991 by his son Vic, who manages the ranches today. The elder Howell now has his own place and his own cows to look after, but it's not uncommon to find him back on the Babbitts, day working, riding out with the cowboys during branding time.

Upgrading the Horses

Bill Howell, face tanned and weathered by more than 40 years of exposure to the Arizona sun and wind, grins as he tells how he came to work for Babbitts. He says he started as a cowboy up on the Cataract, living in a remote camp with his wife and two small sons. Three years later, the manager at the time, Frank Banks, sent him to the CO Bar to run that division, and when Frank retired two years later, he recommended Bill for his job.

Bill is credited with the decision to upgrade the Babbitts' horse herd. In the early 1970s he bought for the ranches a grandson of Joe Hancock to cross on their mares, and then a grandson of Driftwood, Deck Of Wood. The Hancock and Driftwood lines had already made names for themselves on ranches across the Southwest, and Bill instinctively knew they would be good for the Babbitts. He then followed those with Hanks Chargin Bar, a palomino son of Hanks Star by Tonto Bars Hank, which he crossed on the Deck Of Wood daughters. It was a good start, and when Bill retired from his ranch manager position, Vic stayed with the program.

More Horses

Vic Howell is his father's son. He grew up working on the various divisions of the Babbitt Ranches and learned from both his father and his uncle, Harvey Howell, also a longtime employee. Vic left the ranches long enough to graduate from college, but then returned to work with his dad for several years. When Bill retired as manager, Vic was named to replace him, and the father worked for the son for the next six years.

Cut from the same cloth, Vic sits a saddle in the same easygoing manner as his father. Like Bill, he wears his hat pulled low to shade his face, and his legs are covered with the same type of heavy leather chaps. A couple of giveaway pens stick out the top of one shirt pocket, and, like most ranch managers, he has a well-worn tally book peeking over the edge of the other pocket.

"I'm as proud as I could ever be of my dad," Vic says. "Everybody says that Dad was hard, and he was. He was strict. If you were in the wrong place, you knew it right away. But he made a lot of good cowboys.

"And Dad is the most versatile cowboy I've ever known," Vic adds. "He can design a set of shipping pens, ride a bronc, trip a steer, cut cattle out of a herd, set up a roundup or make a drive, just an all-around cowboy.

"He taught me everything, and we work well together. When we're working cattle, before we ever make a move, we know where the other one is going to be."

The father-son team also thinks alike when it comes to the horses. When Vic took over, he stayed with his father's plan. "The horses we got out of that Hanks Chargin Bar/Deck Of Wood cross were really nice," Vic says. "They were tough for this country, and they could pull. We decided we liked that Driftwood blood, so we bought another Driftwood horse, Cowboy Drift by Orphan Drift, and crossed him back on the mares we had. We really liked the way that went."

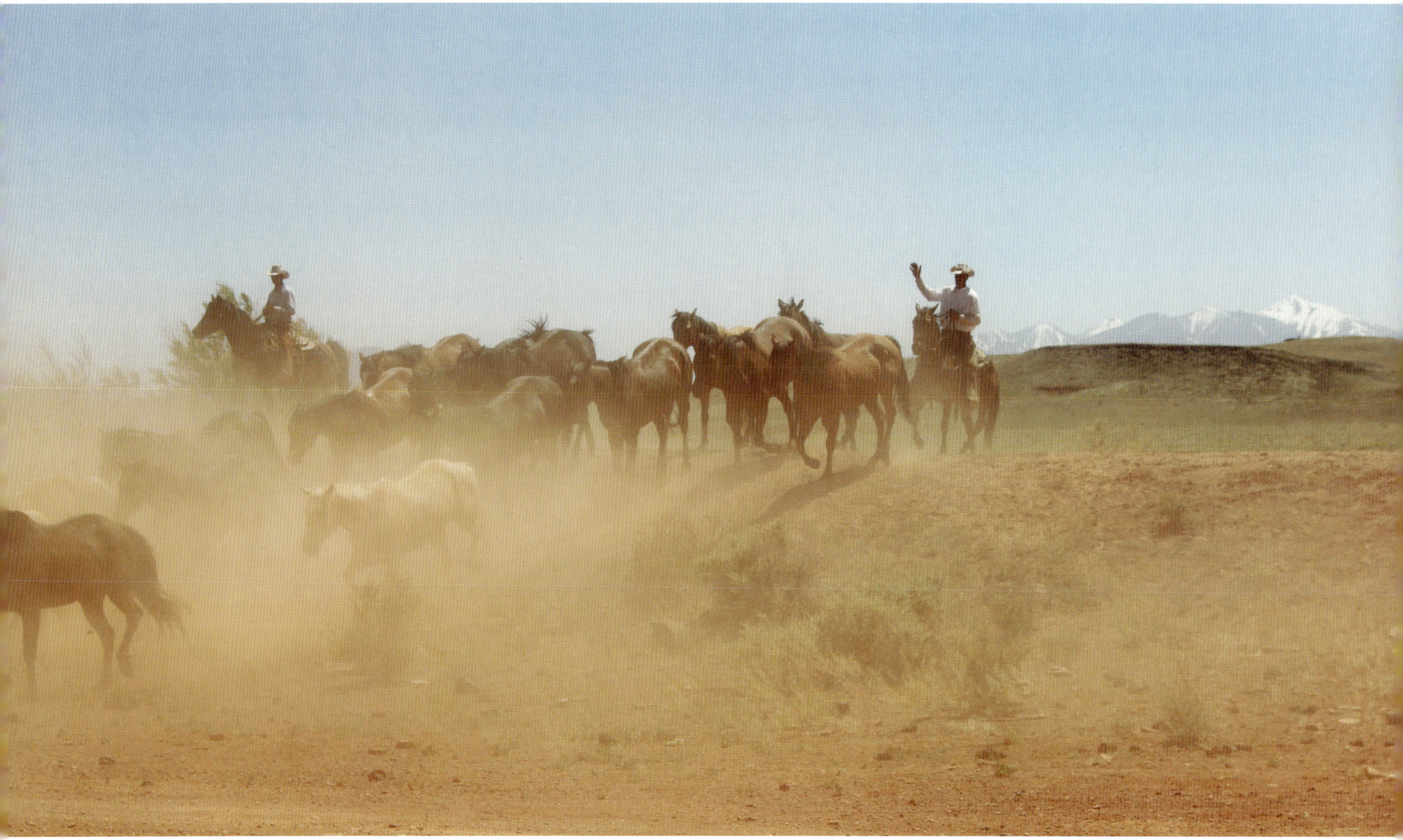

Today the ranch has about 80 broodmares it is breeding to seven stallions—five of which have some Driftwood blood.

Included in the broodmares are 25 Cowboy Drift daughters that are being crossed on Frosty Gold Knight, a palomino stallion whose lineage goes to Doc's Jack Frost, and on Proudgun, a son of Playgun out of a Doc Fitz mare. Other stallions include Cowboy Drift and two sons of his, Cowboy Ben Driftin, out of a Hanks Chargin Bar mare, and The Double Cowboy, out of a Deck Of Wood mare. There is also Cowboywood, by Dynaflow Drift by Double Drift, and a horse they have leased, PC Wades Frost, that is bred similarly to Frosty Gold Knight.

Everything is pasture-bred, and each stallion is turned out with a band of mares about April 20. "A first of April colt is plenty early enough here," Vic says. "We don't expect green grass until then." The stallions are not picked up until the second week of July, when all the mares and foals are gathered for the annual sale.

The Hashknife Horse Sale, named for the brand applied to all horses, is held every year on the second Saturday in July. Only weanlings are sold. Each is auctioned to the highest bidder, who puts down a deposit, but then has to return to the ranch the following March to pick up his purchase. Following the sale, the foals are put back with their mommas. They are weaned in January and February.

All the foals are haltered and branded in late June or early July, just prior to the sale. Those that are kept as possible replacements are worked with again the next month, and then again when they are weaned in February.

Roughly 20 percent of the foals raised each year are kept for the remuda. When the colts turn 2, a couple of the younger cowboys get them started, and then turn them out again until their 3-year-old years. When they are 3, these same cowboys ride them for a month or two and then they are issued to the other cowboys for continued education.

About 80 horses make up the saddle horse remuda, with each cowboy on the ranch being assigned six to eight. The cowboys ride mostly geldings, although all but the older stallions are used just like the geldings from the end of breeding season through the fall works. Also, any fillies destined for the broodmare band are started and ridden 10 to 20 times to be sure they are suitable as replacements.

Babbitt Ranches has had great success with its remuda, both the geldings and the stallions, and not just on the ranch. A number of the horses have been to the National High School Finals Rodeo, and several have seen professional rodeo arenas. And in 2003, a Babbitt Ranches team qualified for the World Championship Ranch Rodeo in Amarillo.

"We are really proud of our remuda," says Bill Cordasco. "Our horses have evolved with our operation. We had horses in the very beginning, so we have nearly 120 years of history where horses have been the primary tool in our cattle operation. In recent years, we have been blessed with some very special people who have a unique passion and understanding of horses and how to fit them together to best work for us.

"I call it the sense of art," he says, "to be able to blend it all and produce the horses we have today."

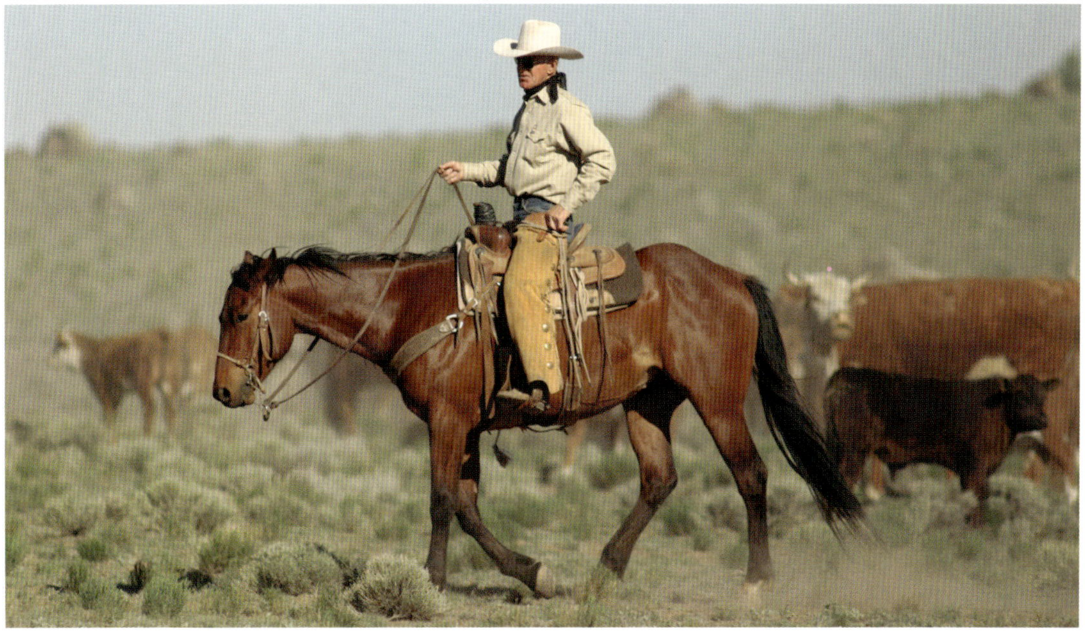

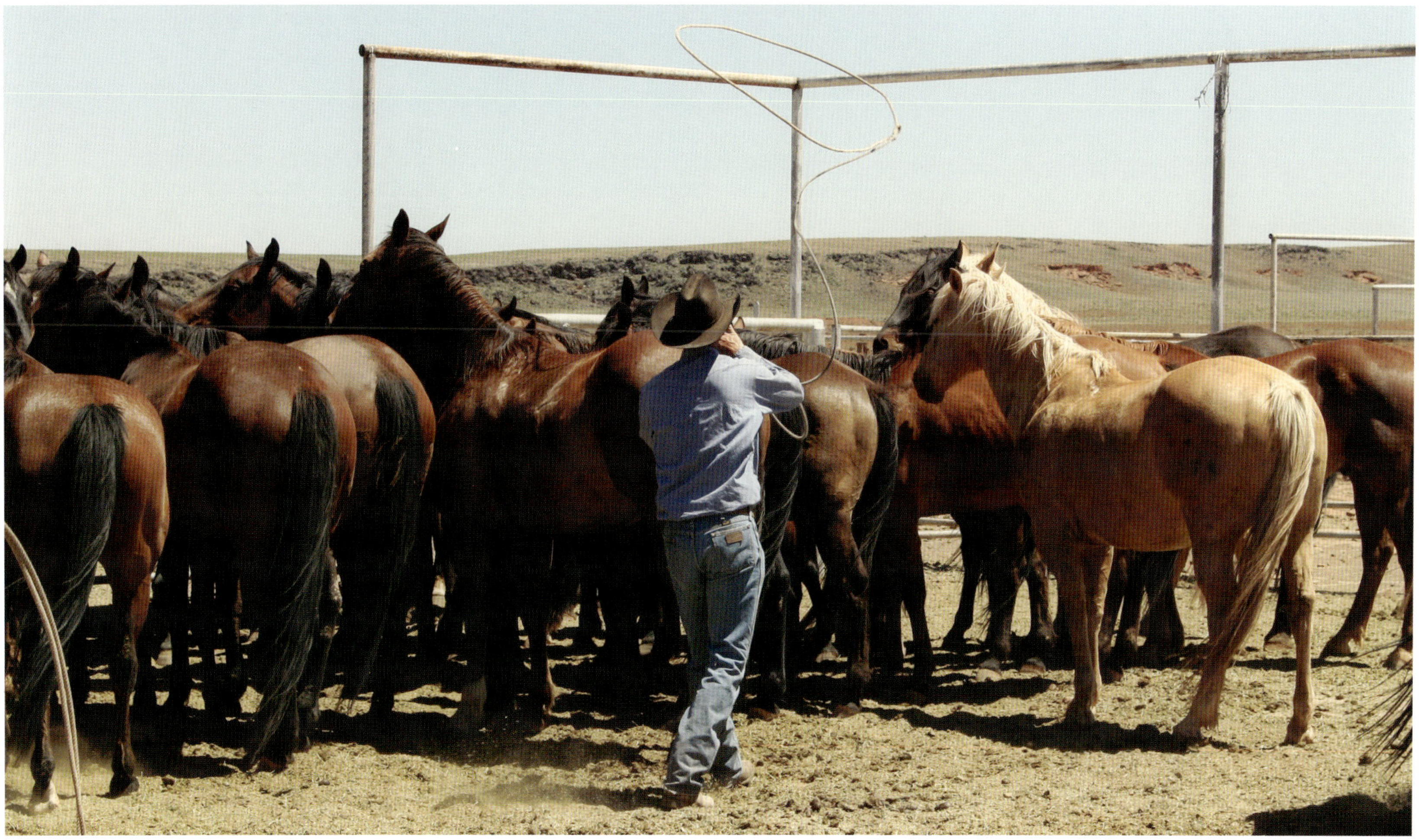

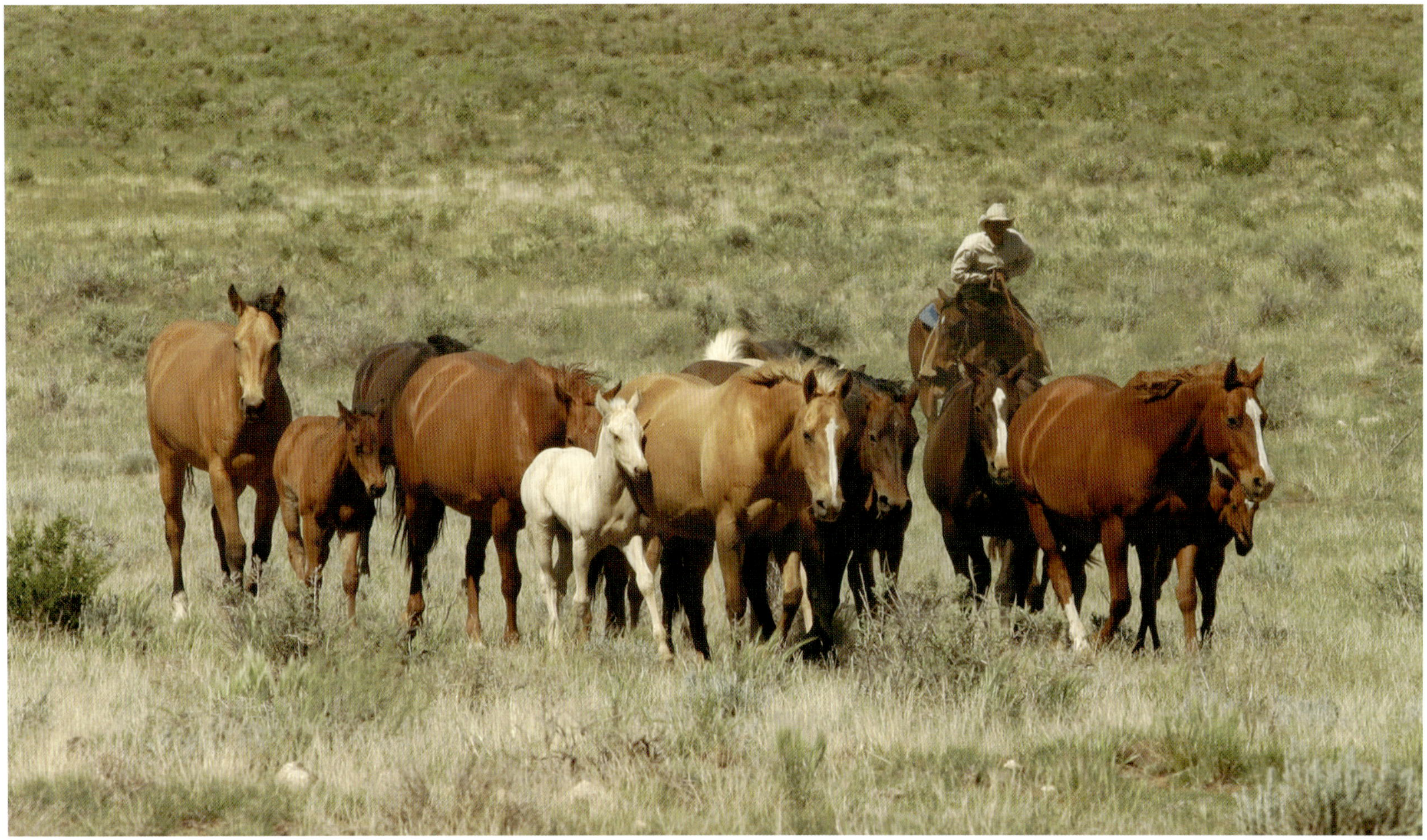

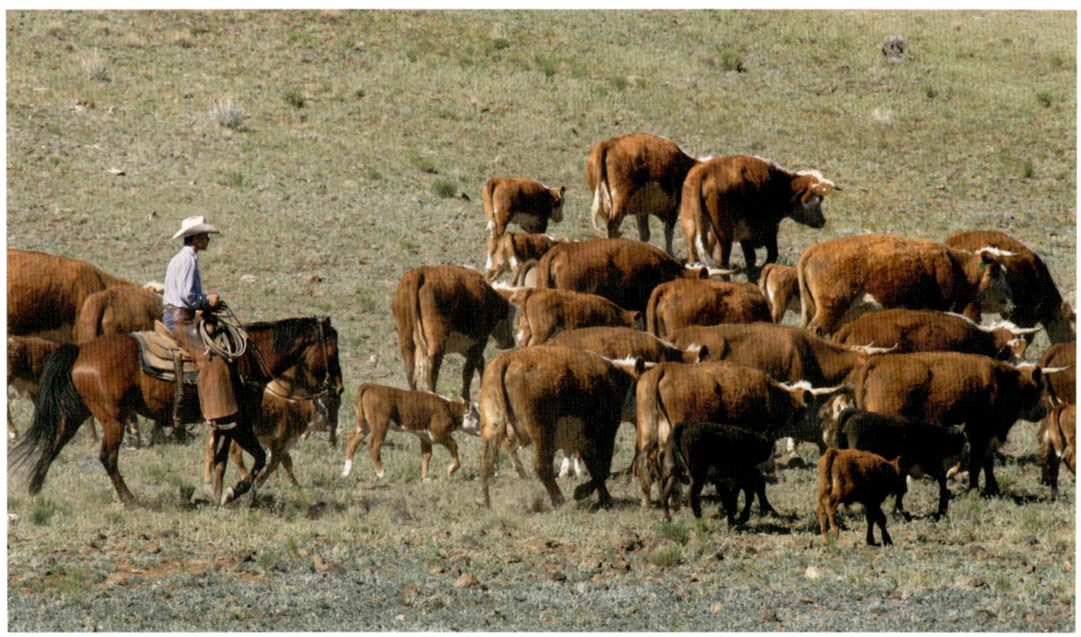

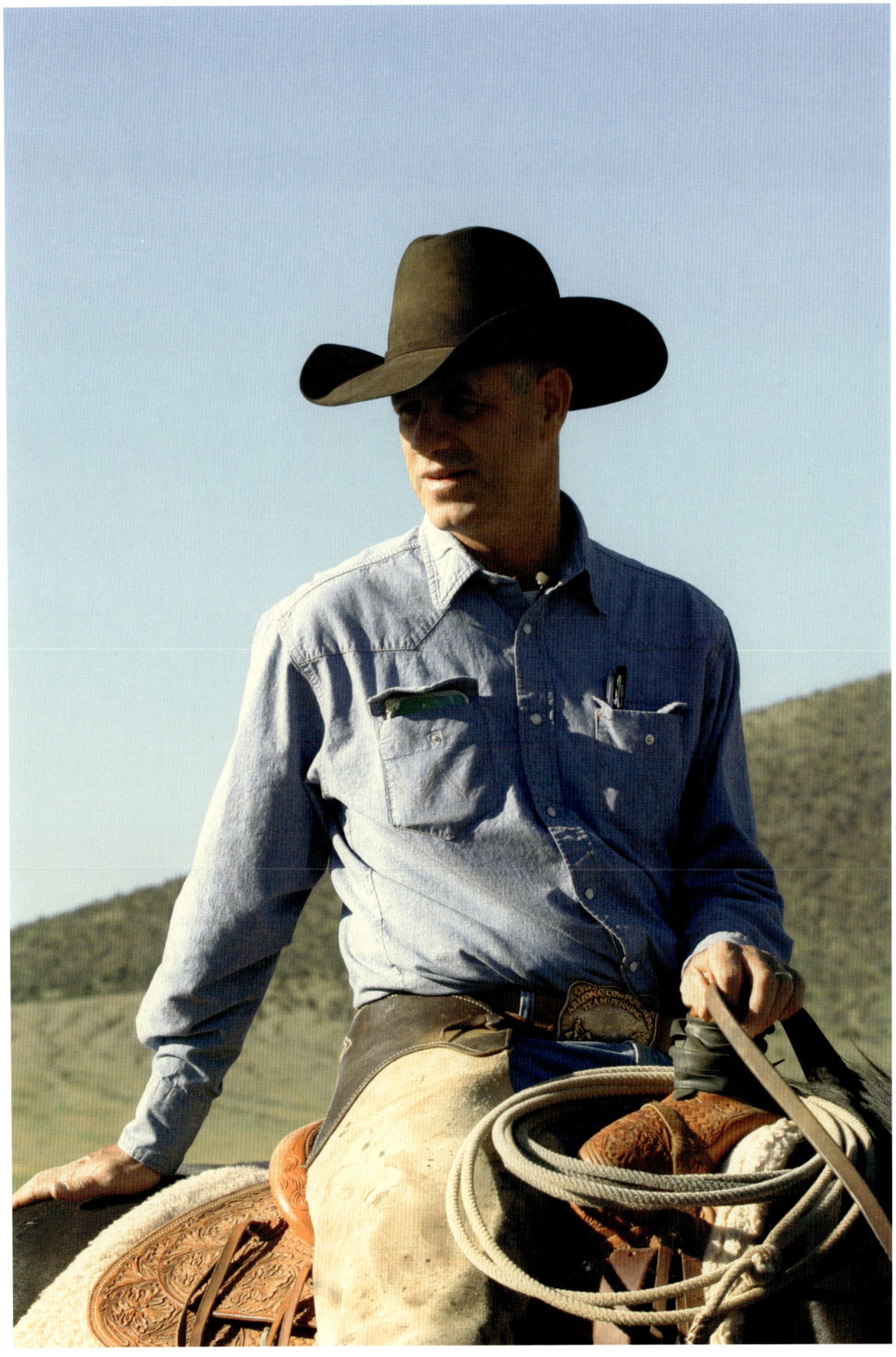

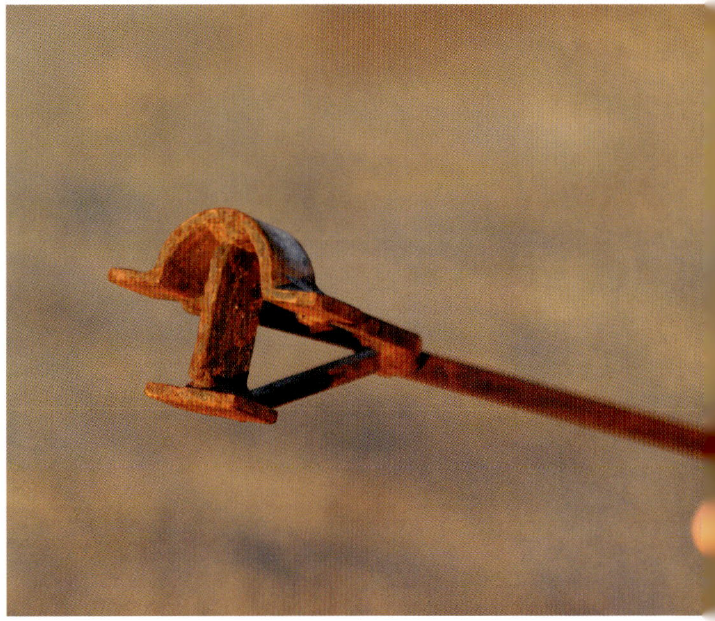

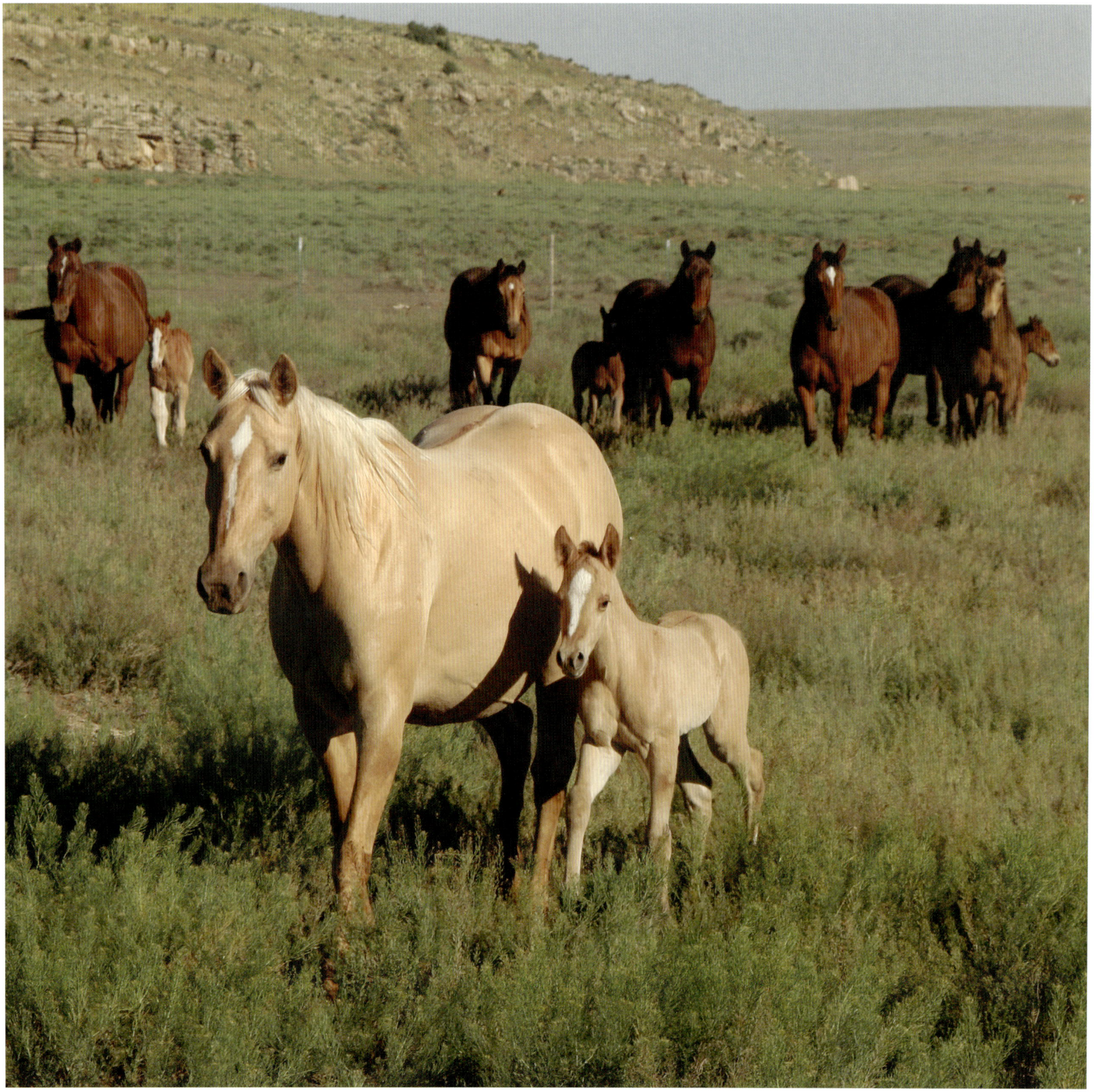

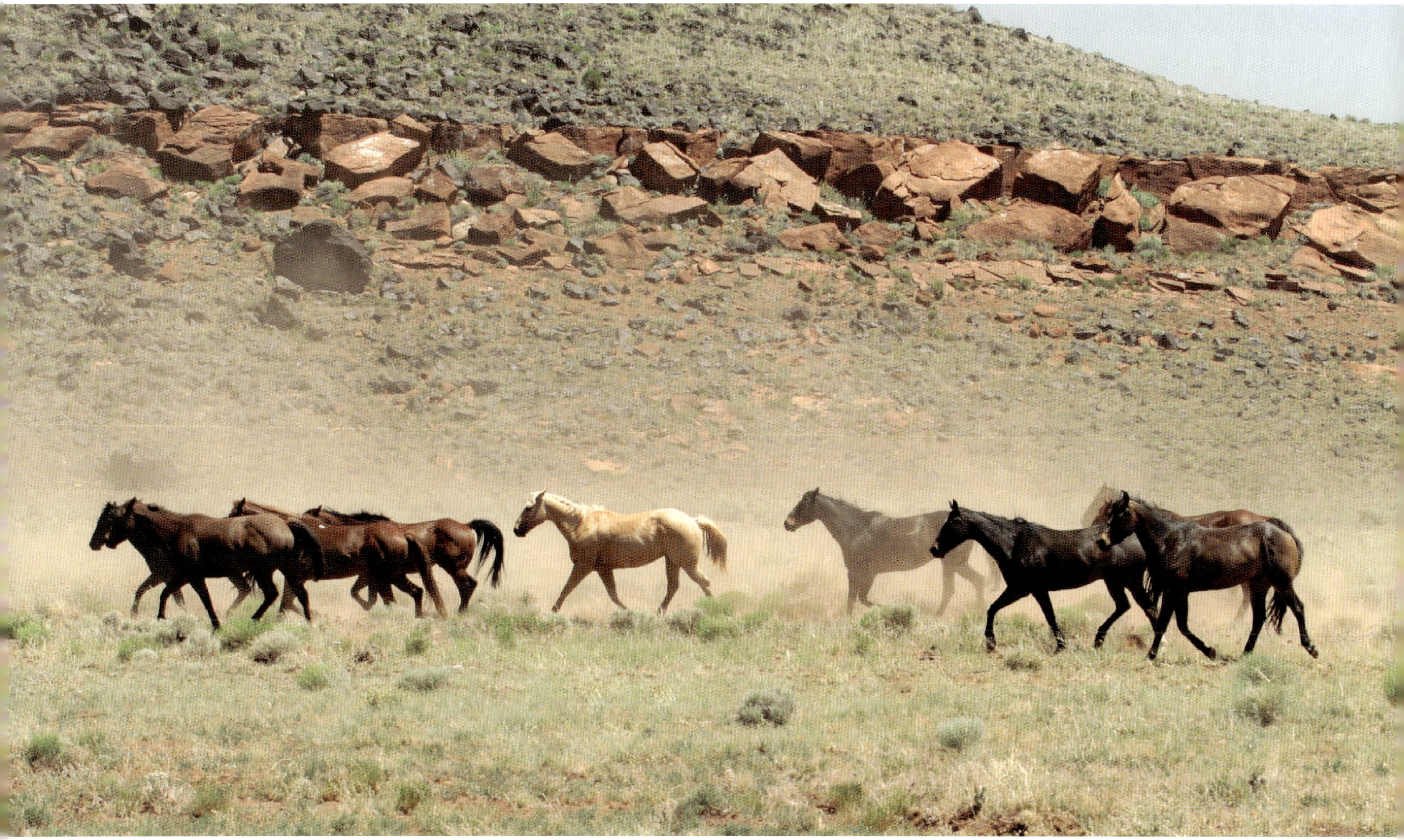

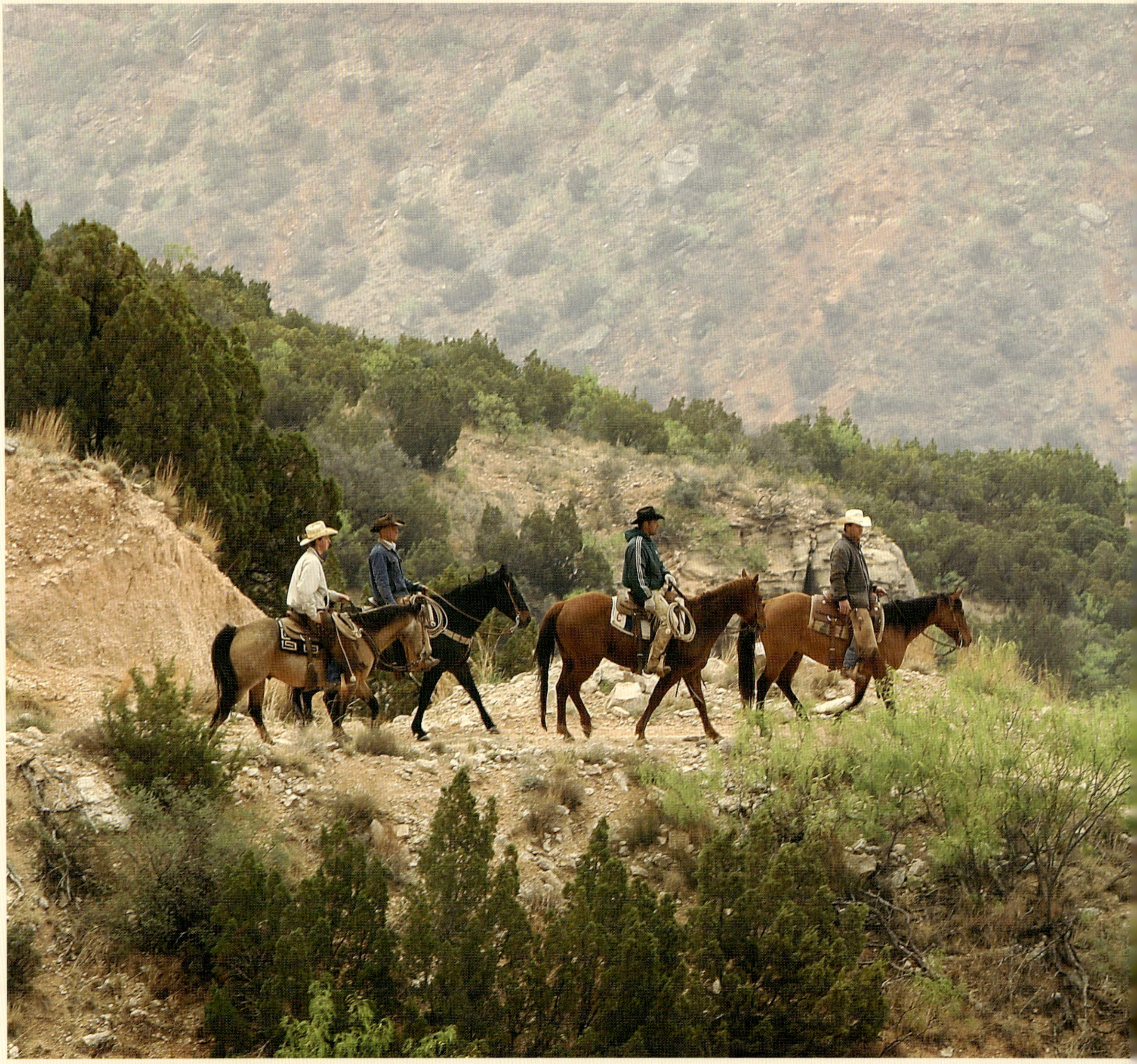

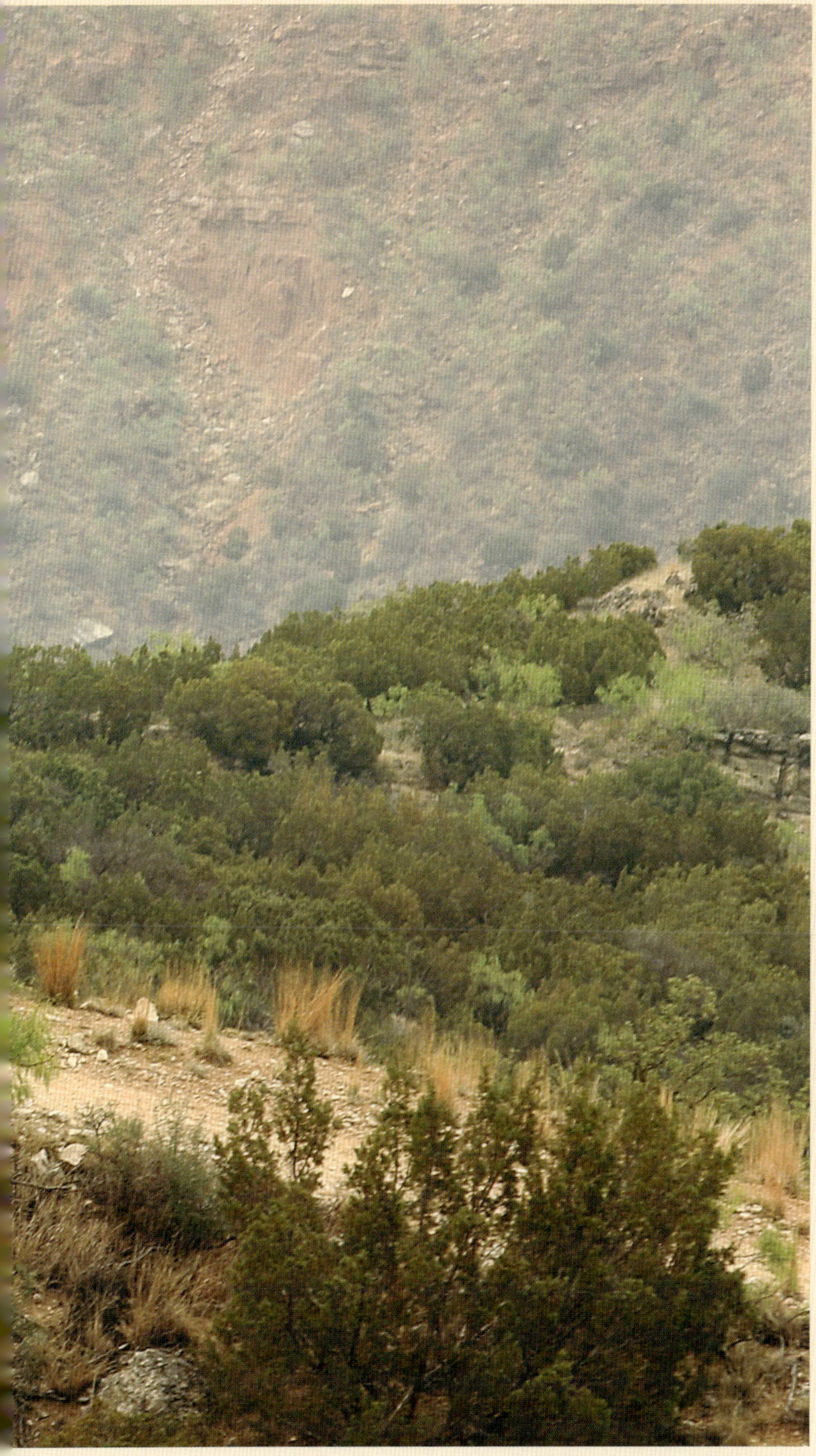

Tule Ranch

Tulia, Texas

2006

The Cutting Cogdells

The Cogdell family has made a name for itself in two areas—ranch horses and cutting horses.

Billy Cogdell was born and raised on his family's ranch north of Snyder, Texas. He grew up horseback, showed his first cutting horse when he was 12, and won the first class he ever entered. It's debatable whether that was good or bad. He found himself hooked on the sport, and for the next eight years he started taking his ranch horses to every cutting he could get to. But that all came to an end shortly after Billy turned 20.

Polio was rampant across the country in the early to mid 1950s, and Billy contracted the disease. It left him 98 percent paralyzed. As a result, the image of Billy Cogdell that most people remember is Billy sitting in his wheelchair at a cutting or a ranch rodeo. Billy died in 2003, but he left a legacy in both the ranching and cutting horse industries that continues today. His three sons and a daughter are carrying on the Cogdell name in both of those industries.

The Beginning

In 1954, Texas was in the fifth year of a terrible drought. It was especially bad in West Texas, where day after day the sky turned brown with dirt picked up from the bare fields and pastures. But as D.M. Cogdell Sr. watched his ranch near Snyder suffer under the unrelenting wind and sun, he kept remembering when, as a youth, he had ridden across the Texas Panhandle plains. And, he remembered how lush the grass was up there. Of course, the Panhandle was suffering from the drought, too, but not as badly as some of the other areas of the state, so D.M. bought the 27,000-acre Tule Ranch, east of the small town of Tulia, and moved his cattle there.

Tule Ranch was named for Tule Creek, a small tributary that cut across the ranch through Tule Canyon and down into the Prairie Dog Town Fork of the Red River and Palo Duro Canyon. The canyon provided both water and protection from the weather, as well as some good grazing, while the flat land up on top offered the hearty buffalo and grama grasses. It was good cow country, and as soon as Billy recovered enough from his bout with polio, D.M. sent him to the Panhandle to take over that ranch.

D.M. died in 1964, and Billy and his brother, D.M. Jr., took over the ranching interests in the Panhandle. Their sister, Jonisue, operated the ranch at Snyder. Then, in the mid 1970s, Billy and D.M. Jr., both with growing families, split their inheritance, and each went out on his own. Billy and his wife, Bette, kept the original Tule Ranch, and were able to add more and more land. At the time of Billy's death, Tule Ranch encompassed more than 160,000 acres with approximately 3,000 mother cows.

Billy was not able to get horseback, but there was virtually nowhere on the ranch he couldn't get with his pickup. And as his children began to grow, he directed their efforts as each of them became a part of the ranch. Sons Rank, Dick and Jim, as well as daughter Penny, all learned to ride on one of the last horses Billy broke before the polio grounded him, and with direction from Billy on the sidelines, they all became good horsemen. But, they also had good horses.

The Horses

D.M. Cogdell Sr. had known the value of a good horse, and he planted the foundation for the Cogdell horses in 1951, when he

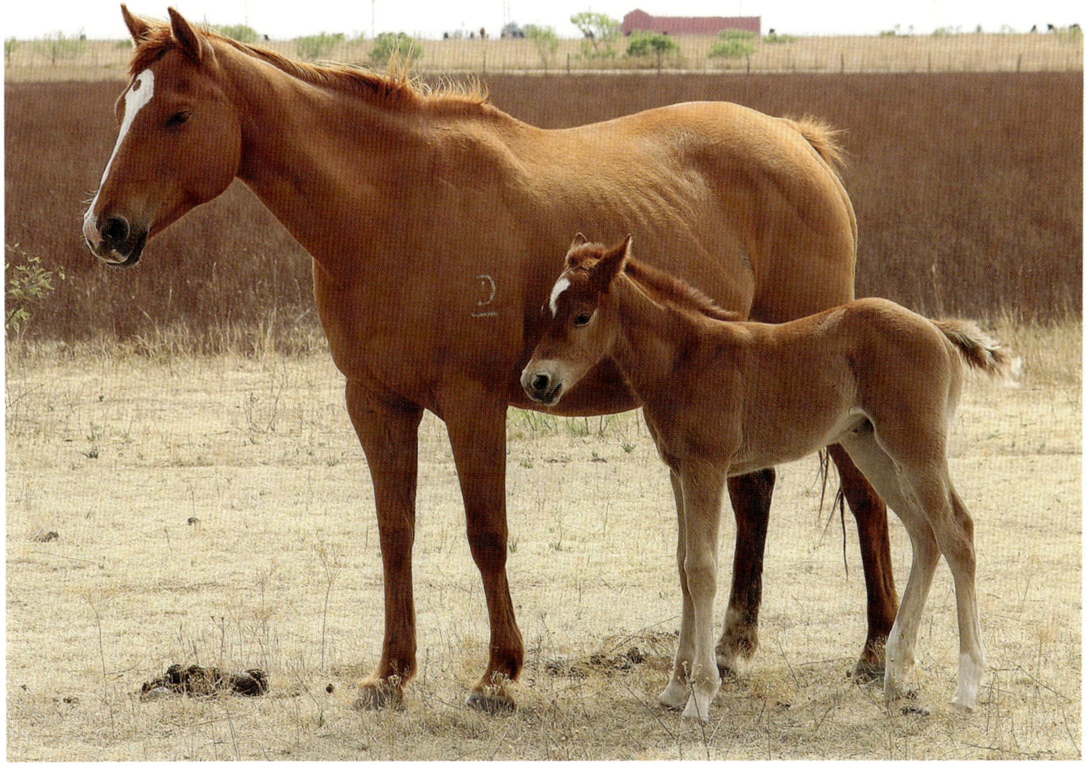

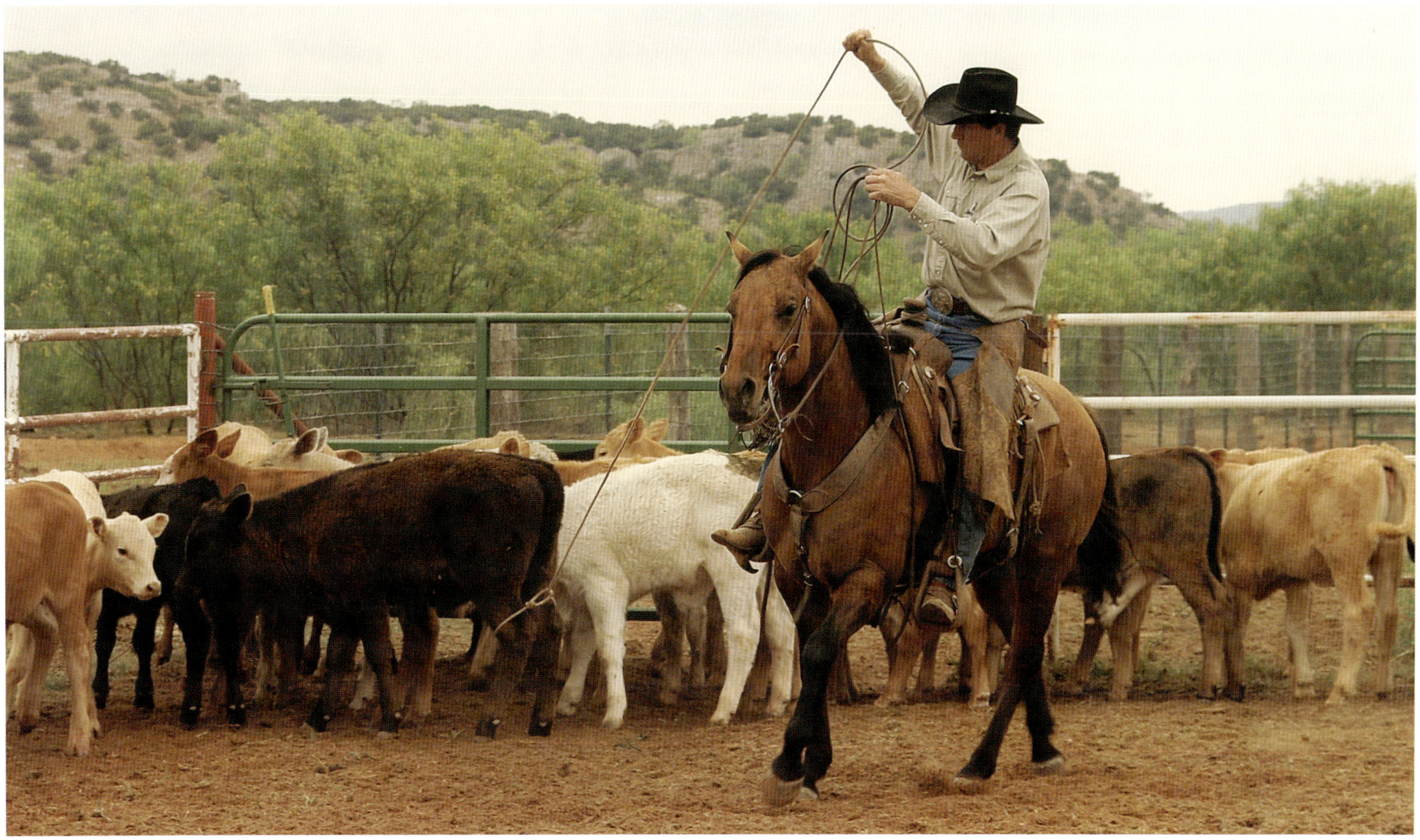

made a trip to the King Ranch and bought 10 yearling fillies and a yearling stallion. The yearling stallion was by Wimpy, and D.M. Sr. named him Solis Cogdell. Later that year, D.M. Sr. went to the Waggoner Ranch at Vernon, Texas, and bought a son of Pretty Buck named Peppy Buck, and then he went down to see the Hankins brothers at Rock Springs, Texas. There he bought a daughter of King. That group of horses provided the basis for a top remuda, one that Billy and his children have continued to develop, even until today.

Even though he was unable to ride, Billy never lost his love for a good cutting horse, and it was instilled in all four of his kids at an early age. They all began to show almost as soon as they learned to ride, and all make their presence known in the cutting arena yet today. But it was probably back in 1978 that the Cogdell name rose to prominence in National Cutting Horse Association competition.

In 1977, while at the NCHA Futurity sale, Billy saw a little 13.2-hand mare by Doc's Lynx come through the ring. "No one wanted her because of her size," Billy was quoted as saying in a February 2000 *Quarter Horse Journal* article. "But I like little horses."

Billy purchased Lynx Melody and sent her to trainer Larry Reeder to ride. Reeder won the Futurity with a performance that is talked about yet today. His stirrups were in the dirt several times as the little mare got down low and worked each cow. "It was a sight," Billy said in the *Journal* article. "Larry is 6'2", and he was on this little bitty mare."

Although unknown at the time, Billy, when he bought Lynx Melody, had purchased a two-in-one package—Lynx Melody was in foal to Doc's Stormy Leo. Billy named the foal Docs Accident and immediately fostered it onto a Shetland mare so that nothing would interfere with Lynx Melody's training. However, Docs Accident also made a name for himself. He placed eighth in the 1981 NCHA Futurity and, during his career, earned almost $43,000.

Lynx Melody went on to win the 1979 NCHA Derby and was the 1980 NCHA World Champion Mare. Her lifetime earnings were more than $113,000.

Retired from the arena, Lynx Melody was brought home and placed in the broodmare band, where she produced foal after foal that earned thousands of dollars in the cutting arena. Then, in 1996, she produced a filly by Peppys Boy 895, a stallion Billy owned by Vibrant Peppy by Peppy San Badger. Billy named the mare Shania Cee. In 1999, Shania Cee won the NCHA Futurity, making Lynx Melody the only mare ever to win the Futurity and produce a Futurity winner, and making Billy Cogdell only the second person to ever own two Futurity winners. Lynx Melody's foals have now earned more than $1 million in cutting competition.

Shania Cee finished her career with more than $400,000 earned, and then she joined her mother in the Cogdells' broodmare band.

That broodmare band reflects the Cogdells' interest in the cutting arena, as well as in ranch horses, because to the Cogdells, they are one and the same. "We all love to ride cutting horses," says Billy's daughter Penny Carpenter. "But cutting is a sport that came from the ranch, and we are continually cutting cattle in our ranching operations. We cut out dry cows, sort pairs, separate steer and heifer calves and cut out sick cattle. We are all trying to breed cutting horses, but many of our cutting horses are ranch horses."

An example of that is former ranch sire Maxi Lena. A son of Doc O'Lena out of a daughter of Gay Bar King, Maxi Lena's foals have earned more than $850,000 in NCHA competition, but he has also sired six ranch horse and ranch rodeo champions. Maxi Lena died in 2002, but a number of his daughters are in the broodmare band. There are also several daughters of Peppys Boy 895, the sire of Shania Cee. Peppys Boy 895 is an all-around sire, with his get earning money in cutting, reining and working cow horse competition, as well as in the ranch horse arena.

Stallions used on the ranch today include Shadowcatchin Cattin by Cattin, Shiver Shake by Doc's Hickory, and Lucky Bottom Tom, a Doc Bar-bred horse that goes back to Gold King Bailey on the bottom.

After Billy died, the ranch was split among the four kids. Each of them continues to operate a share of Tule Ranch today, and, as before, all of them work together. As an example, Dick happens to own all three of the stallions to which his siblings are breeding their mares.

All of the stallions are pasture bred, with each turned out with a band of mares. All the foals are evaluated as to their abilities,

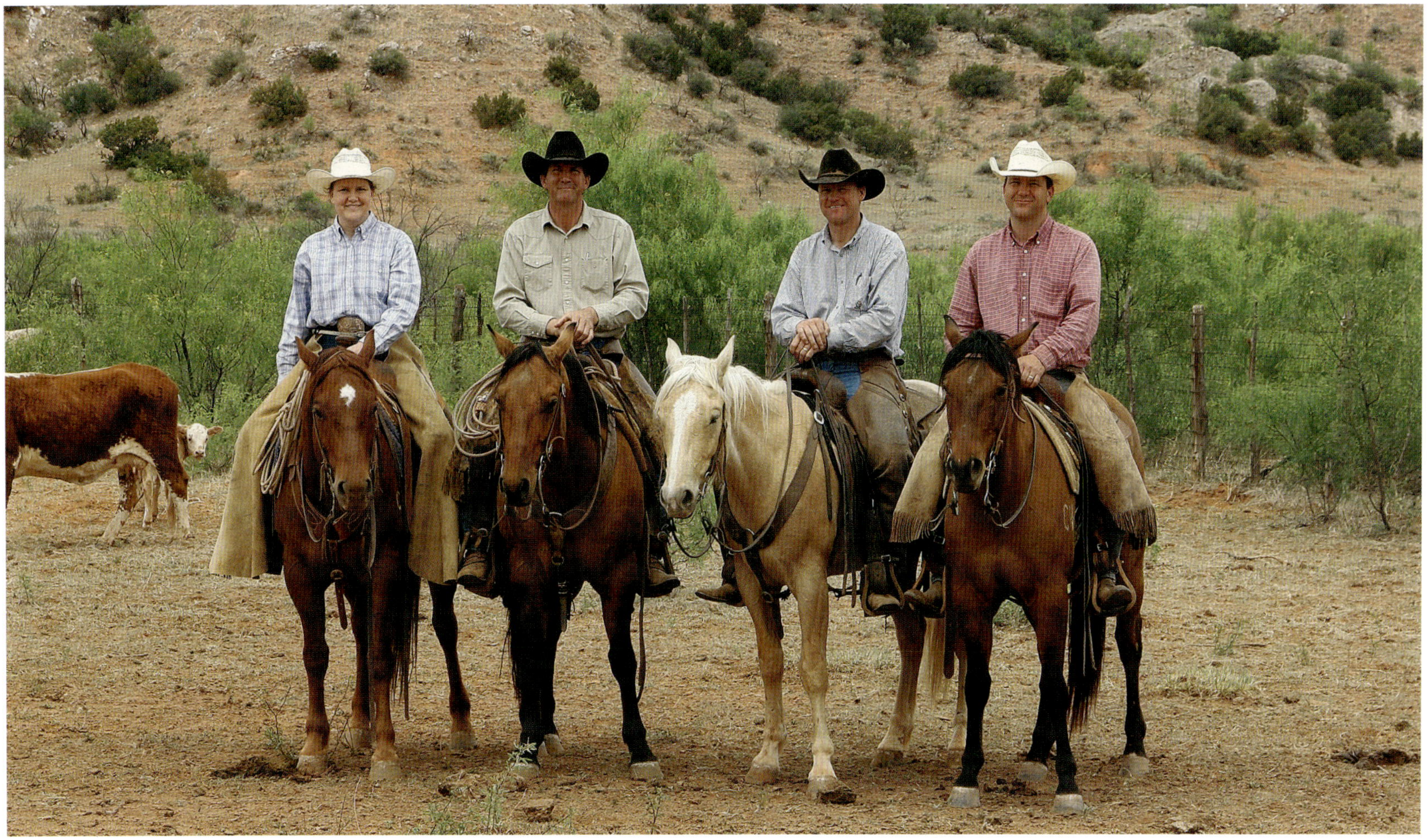

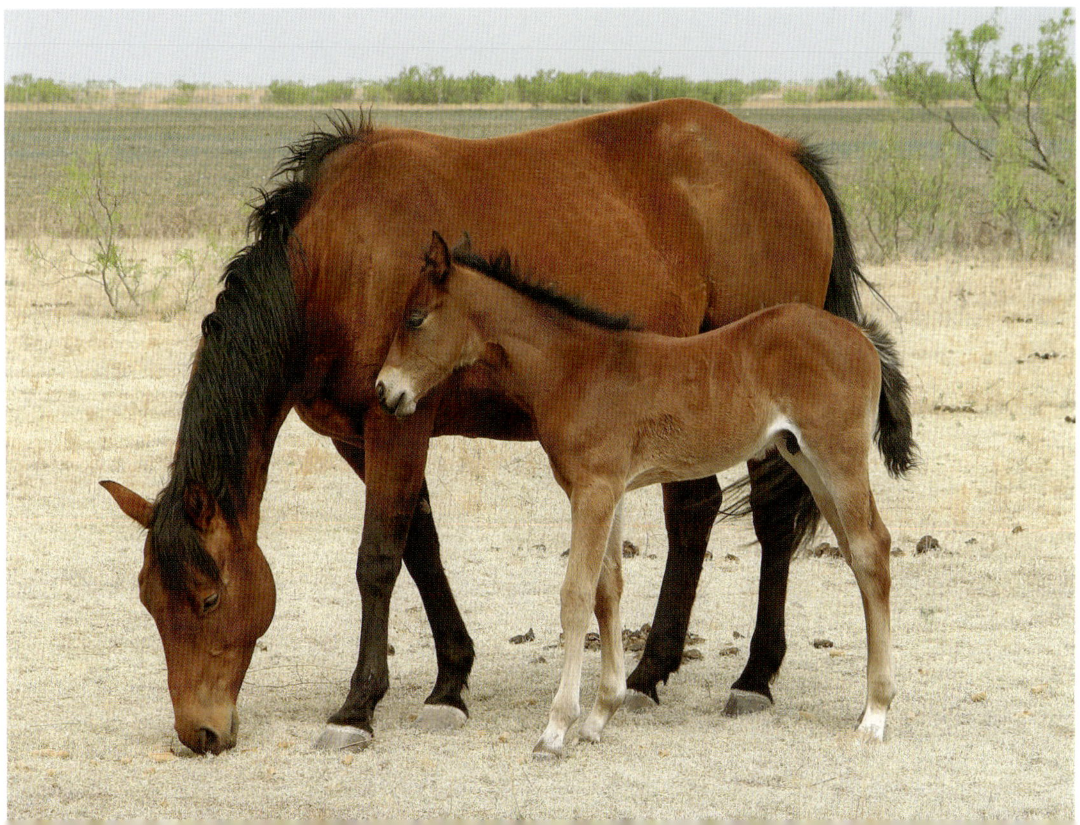

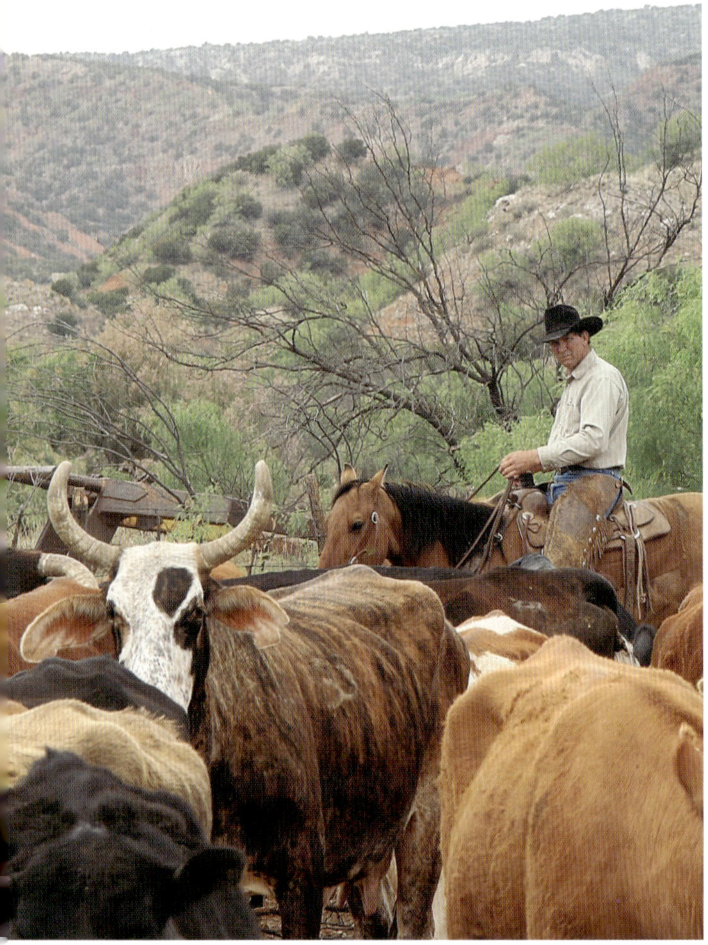

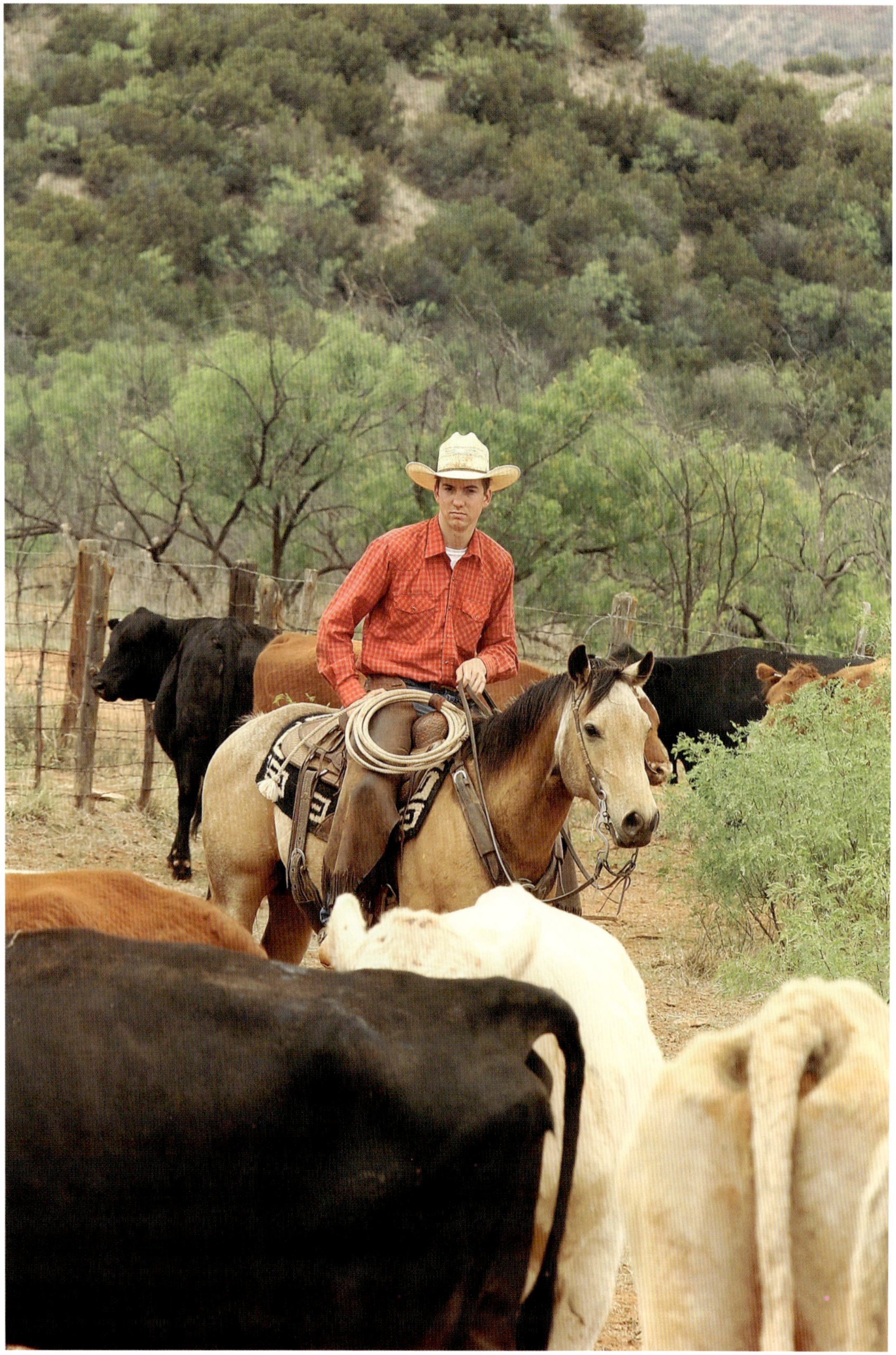

and then a determination is made as to which ones to keep and which to sell. Some of the foals are at least tried in the cutting arena, while others are destined to a future as a ranch horse. Fillies are also broken and evaluated. Some will go into training, some will be returned to the broodmare band and some are sold.

"The primary goals of the Tule Ranch remuda program are to produce good quality all-around ranch horses with the ability and breeding to also compete as cutting horses," says Penny. "Ranch horses and cutting horses are all chosen from the same group of colts. There is no separate group of horses designated for cutting horse prospects."

The Cogdells today operate their ranches just as their father did. In the rough Tule Canyon, there is no substitute for a good horse. Cattle are gathered and calves branded, all with horses, and in the fall, during shipping time, some cattle are driven close to 15 miles up and out of the canyon, where trucks can pick them up.

"Daddy always wanted to win the Best Remuda Award," Penny says, "because it encompassed everything that he spent his life doing—breeding and raising good horses. It's through his leadership that Tule Ranch has won it, and he certainly gets all the credit."

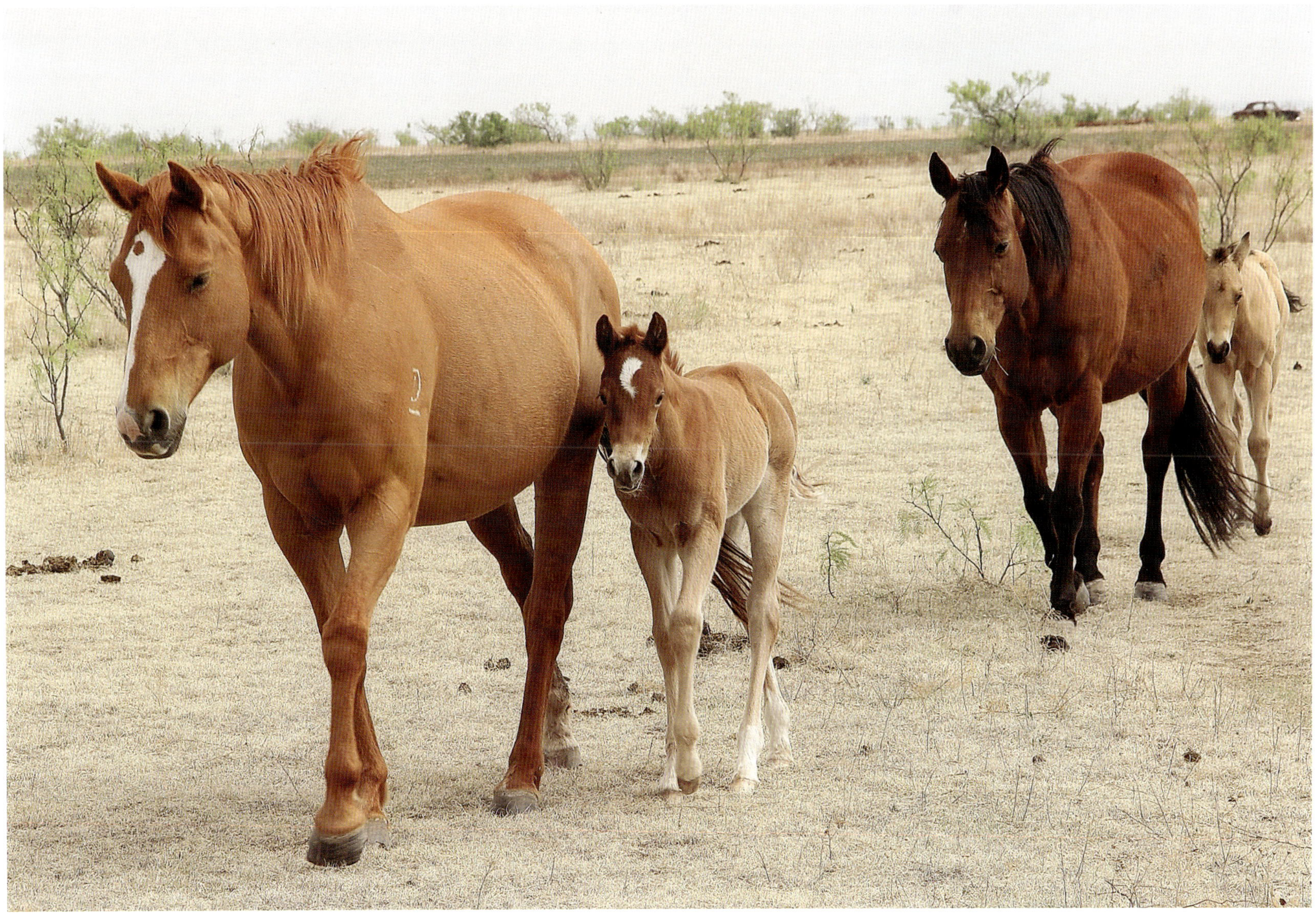

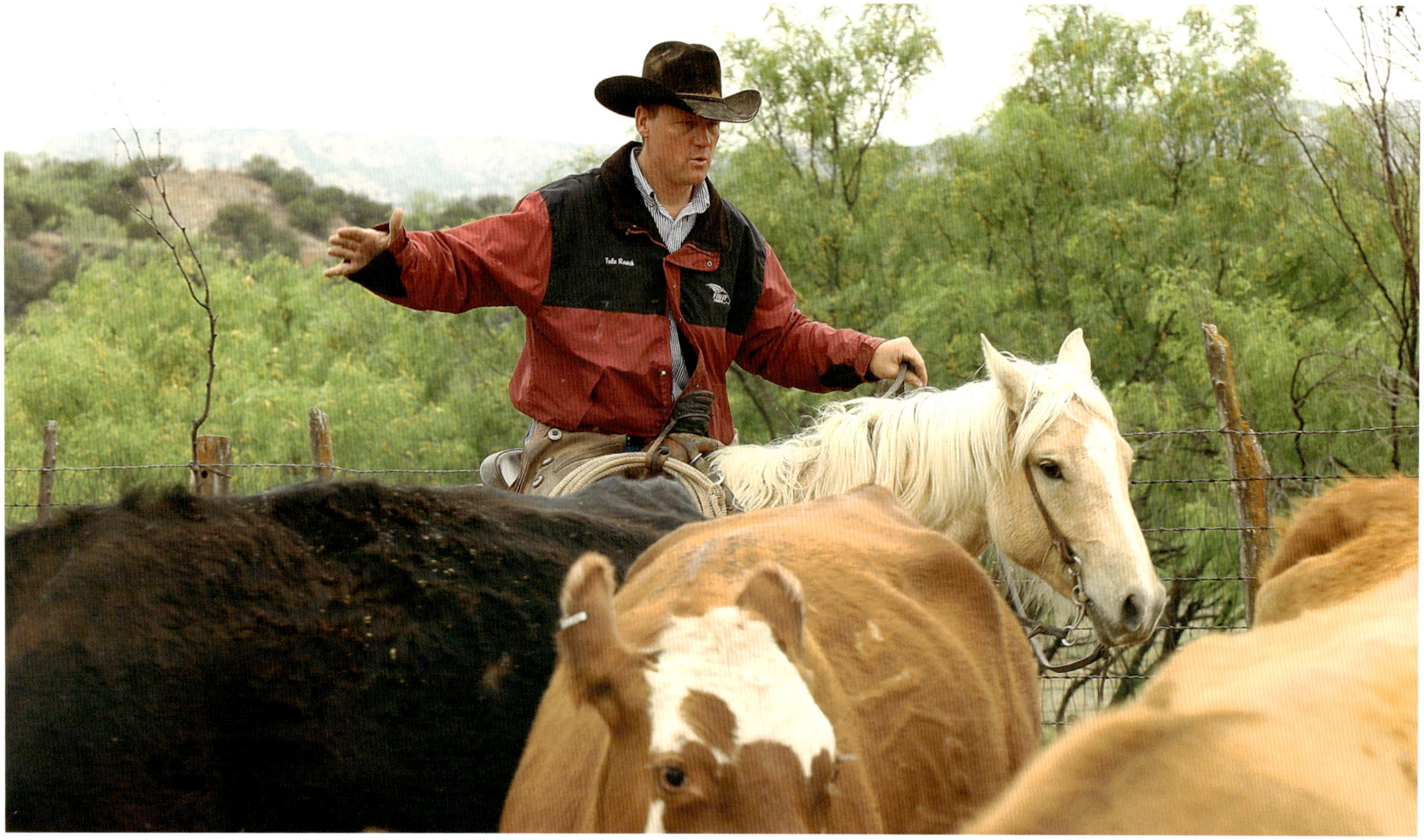

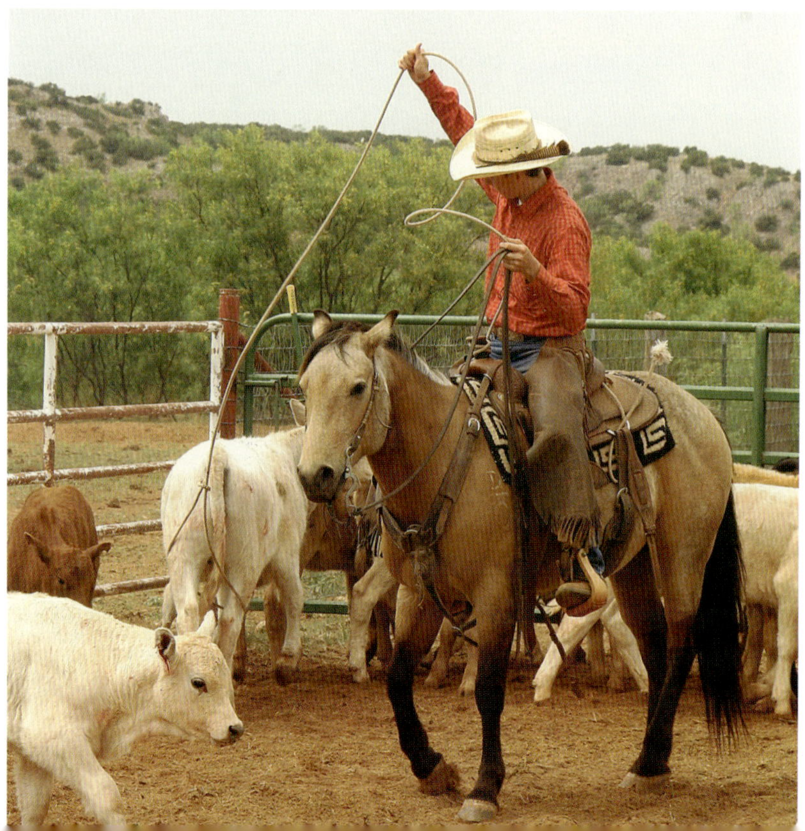

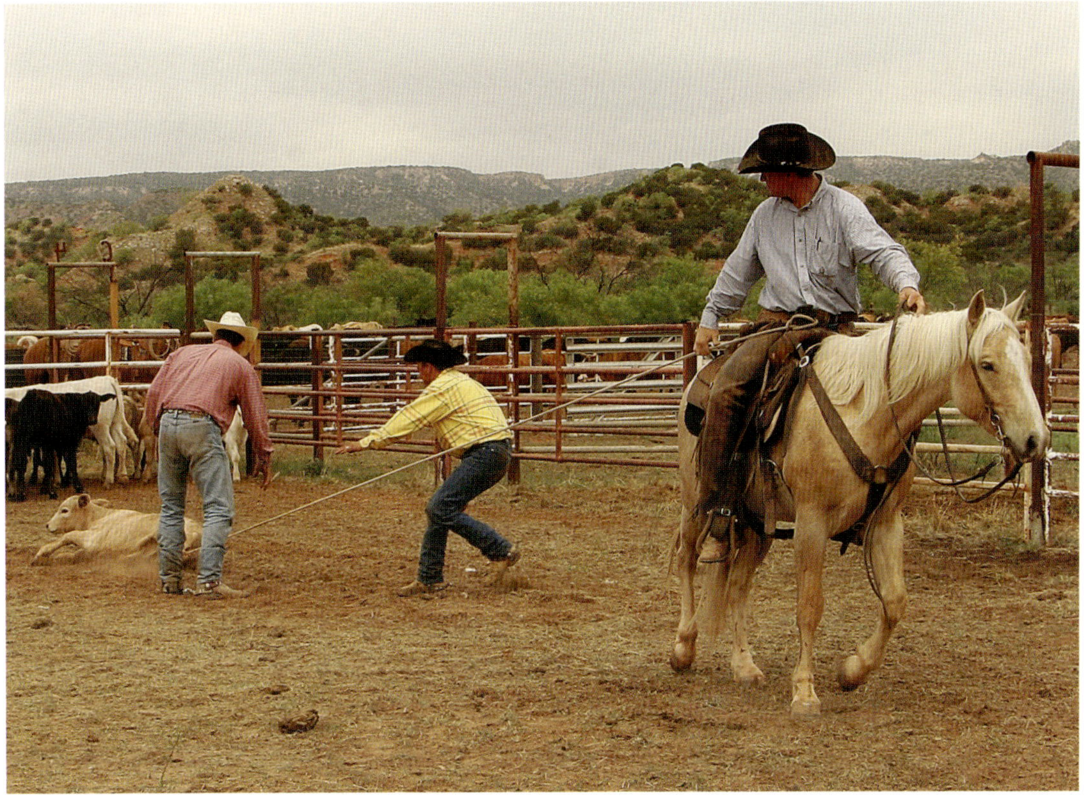

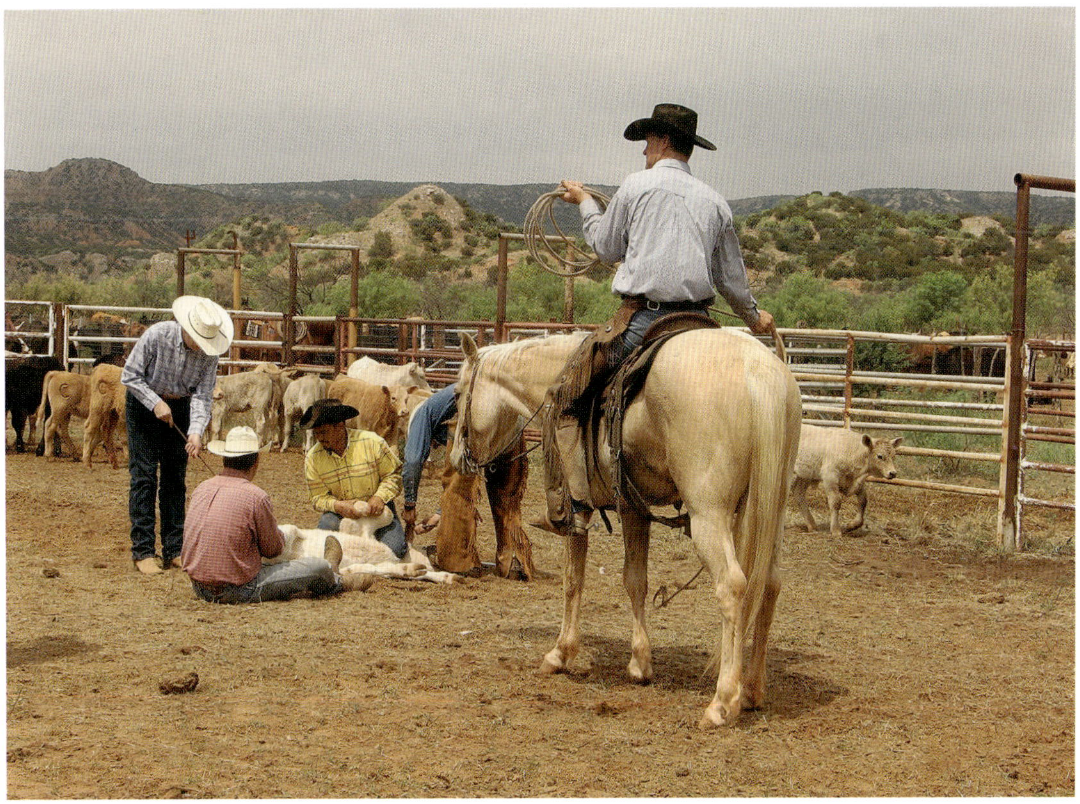

Pedigrees

Cowboy Drift 1932436 1982 bay stallion

- Orphan Drift
  - Driftwood Ike
    - Driftwood
      - Miller Boy
      - The Comer Mare
    - Hancock Belle
      - Buck 8
      - Miss Hancock 1
  - Orphan Annie C
    - Tailor Made C
      - Joe Reed II
      - Betty Wood
    - Twilight C
      - Clabber
      - Phoebe C
- Poco Judy Sue
  - Poco Speedy
    - Poco Bueno
      - King
      - Miss Taylor
    - Nellie D
      - Blackburn
      - Rainy Day Mare
  - Judy Sue
    - Driftwood
      - Miller Boy
      - The Comer Mare
    - Nita H
      - Bonite
      - Honey D

Frosty Gold Knight 3788770 1999 palomino stallion

- AP Frosty Knight
  - Sun Frost
    - Doc's Jack Frost
      - Doc Bar
      - Chantella
    - Prissy Cline
      - Driftwood Ike
      - Josephine El
  - AP Night
    - Rapidland
      - Mr Midland
      - Roy's Rapid Miss
    - Mary Dokota
      - Laughing Boy
      - Scotch Mary
- Conchos Lady 034
  - War Concho
    - War Chief
      - Joe Hancock
      - Big Mountain 1
    - Brown Beulah
      - Driftwood
      - Queen Ann
  - Speedy Beth
    - Driftwood Ike
      - Driftwood
      - Hancock Belle
    - Squeeky Beth
      - Squeeky
      - Johnson's Smokey Sue

Proudgun 3630985 1998 sorrel stallion

- Playgun
  - Freckles Playboy
    - Jewel's Leo Bars
      - Sugar Bars
      - Leo Pan
    - Gay Jay
      - Rey Jay
      - Georgia Cody
  - Miss Silver Pistol
    - Doc's Hickory
      - Doc Bar
      - Miss Chickasha
    - Pistol Lady 2 Be
      - King's Pistol
      - Miss Bailey 24
- Betty Otoe Fitz
  - Doc Fitz
    - Doc Bar
      - Lightning Bar
      - Dandy Doll
    - Jean Chance
      - Beau Chance
      - Jean Dexter
  - Otoes Chip
    - War Chip's
      - War Leo
      - Brady Lady
    - Betty Otoe
      - Otoe
      - Betty Rebo

The Double Cowboy 3293140 1994 bay stallion

- Cowboy Drift
  - Orphan Drift
    - Driftwood Ike
      - Driftwood
      - Hancock Belle
    - Orphan Annie C
      - Tailor Made C
      - Twilight C
  - Poco Judy Sue
    - Poco Speedy
      - Poco Bueno
      - Nellie D
    - Judy Sue
      - Driftwood
      - Nita H
- CO Bar Four One
  - Deck Of Wood
    - Speedy II
      - Driftwood
      - Dusky Ruth
    - Deck Maid
      - Yendis
      - Pima Maid
  - Babbitt 42
    - Ginger's Barhank
      - Sugar Bars' Hank
      - Wilson's Ginger
    - Babbitt Nineteen
      - Clabber Boy
      - Babbitt Mare

Cowboy Ben Driftin 3274628 1994 buckskin stallion

- Cowboy Drift
  - Orphan Drift
    - Driftwood Ike
      - Driftwood
      - Hancock Belle
    - Orphan Annie C
      - Tailor Made C
      - Twilight C
  - Poco Judy Sue
    - Poco Speedy
      - Poco Bueno
      - Nellie D
    - Judy Sue
      - Driftwood
      - Nita H
- Cataract Twenty One
  - Hanks Chargin Bar
    - Hanks Star
      - Tonto Bars Hank
      - Pollet
    - Miss Chargin Bar
      - Bar Breaker
      - Miss Bobby King
  - CO Bar Seven Three
    - Deck Of Wood
      - Speedy II
      - Deck Maid
    - Babbitt 42
      - Ginger's Barhank
      - Babbitt Nineteen

Cowboywood 3830227 1999 bay stallion

- Dynaflow Drift
  - Double Drift
    - Gray Chip
      - Driftwood
      - Sugar Tit
    - Rose Wood
      - Driftwood
      - Miss Hondo
  - King Bunny King
    - Country Boy Lauro
      - Lauro
      - Clyde Sis
    - Bunny McDuff
      - Tony Boy
      - Johnie Belle
- Sons Foxy Lady
  - Son O Sugar
    - Sugar Bars
      - Three Bars (TB)
      - Frontera Sugar
    - Leo Pan
      - Leo
      - Panchita
  - Caraboo Kitten
    - Poco Husky
      - Poco Bueno
      - Wimpy's Ojos
    - Mary D
      - Blackburn
      - Mare by Bailey

PC Wades Frost 4064173 2001 bay stallion

- Sun Frost
  - Doc's Jack Frost
    - Doc Bar
      - Lightning Bar
      - Dandy Doll
    - Chantella
      - War Chant
      - Bardella
  - Prissy Cline
    - Driftwood Ike
      - Driftwood
      - Hancock Belle
    - Josephine El
      - Super Charge
      - Webb's Demon
- PC Wades World
  - Docs Oaks Sugar
    - Doc's Oak
      - Doc Bar
      - Susie's Bay
    - Miss Sugar Bingo
      - Mr Sugar Boy
      - Cutie Bingo
  - Cowan Kitten
    - Orphan Drift
      - Driftwood Ike
      - Orphan Annie C
    - Poco Cokey
      - Poco Speedy
      - Rivbear Coke

Bar B Top Gun 4237531 2002 bay stallion

- Real Gun
  - Playgun
    - Freckles Playboy
      - Jewel's Leo Bars
      - Gay Jay
    - Miss Silver Pistol
      - Doc's Hickory
      - Pistol Lady 2 Be
  - Miss War Chips
    - War Chip's
      - War Leo
      - Brady Lady
    - Patsy Chip
      - Tona's Streak
      - Patsy Mahoney
- Seven S Bliss
  - Zans Misty Gold
    - Zan Parr Bar
      - Par Three
      - Terry's Pal
    - Skips Misty Gold
      - Silver Skip
      - Dainty King
  - Seven S Ruffles
    - Son O Leo
      - Leo
      - Sue Sugar
    - Tammie Do
      - Beggar Toots
      - Nebraska Lady

Bar B Red Kiowa 4091197 2001 bay stallion

- Tiveos Roan Sugar
  - Sparta Tiveo
    - Poco Tivio
      - Poco Bueno
      - Sheilwin
    - Sorrel Pokey
      - Pokey Red Smoke
      - Orovada Rose
  - Sons Roan Sugar
    - Son O Sugar
      - Sugar Bars
      - Leo Pan
    - Goldy Roan
      - Don Pedro
      - Goldy Ritchey's
- Kiowa Diamond Lady
  - Kiowa Thad
    - Mr Thad
      - Zantanon H
      - Etta's Birthday
    - Ruby's Liza
      - Kiowa Dusty
      - Duchess Ruby
  - Beenolady
    - Mr Nothing
      - Sapp Head
      - Poco Five Below
    - Bee Stella
      - Bee Line
      - Stella Straw

JR Driftwood Spice 4524389 2004 buckskin stallion

- MJH Rosewood Ike
  - Wayward Ike
    - Driftwood Ike
      - Driftwood
      - Hancock Belle
    - Katy Mas
      - Poquito Mas
      - Wayward Katy
  - Lady Roanie Bay
    - Roan Prairie
      - Cibecue Roan
      - Tom's Rosewood
    - Travelin Tivo
      - Traveler Sam
      - Tivolee
- Wilywoods Spice
  - Wilywood
    - Orphan Drift
      - Driftwood Ike
      - Orphan Annie C
    - Oui Oui
      - Poco Speedy
      - Woodfern
  - Miss Senorita Chick
    - Nebraska Boy
      - Diamond Charge
      - Spanish Senorita
    - Miss Spanish Face
      - Spanish Fiddle
      - Chicks Face

Jaysons Dunit 4059664 2001 bay stallion

- Definitly A Dunit
  - Hollywood Dun It
    - Hollywood Jac 86
      - Easter King
      - Miss Hollywood
    - Blossom Berry
      - Dun Berry
      - Regina Bella
  - Zans Diamond Squaw
    - Zans Diamond Sun
      - Zan Parr Bar
      - Diamonds Sparkle
    - Major Classy Squaw
      - Major's Manana
      - Miss Maysville
- Ronella Lyn
  - Scott Will Win
    - Scottish Will
      - Scottish
      - Elm's Siesta
    - Miss Suzie Dunn
      - Kiowa Dusty
      - Laverne Miss
  - Ronella Bar
    - Rattler's Bar
      - Cold Wave
      - Lula Dee
    - Miss Ronweiser
      - Budweiser
      - Rondy Dun

Lonestars Pistolero 4065121 2001 gray stallion

- Playgun
  - Freckles Playboy
    - Jewel's Leo Bars
      - Sugar Bars
      - Leo Pan
    - Gay Jay
      - Rey Jay
      - Georgia Cody
  - Miss Silver Pistol
    - Doc's Hickory
      - Doc Bar
      - Miss Chickasha
    - Pistol Lady 2 Be
      - King's Pistol
      - Miss Bailey 24
- Lone Star Holly
  - Texas Kicker
    - Colonel Freckles
      - Jewel's Leo Bars
      - Christy Jay
    - Angels Rosie
      - Mr Gold 95
      - Bill's Shu Baby
  - Holly Debra
    - Hollywood Bill
      - Hollywood Gold
      - Miss Jo Kenney
    - Poco Debra
      - Poco Bueno
      - Serenade Girl

Four Figure 3881810 1999 dun stallion

- WD Five
  - War Drift
    - War Chief
      - Joe Hancock
      - Big Mountain 1
    - Brown Beulah
      - Driftwood
      - Queen Ann
  - My Beaver 31
    - My Beaver
      - Beaver Creek
      - FL Pee Wee
    - Haythorn Gal 6
      - Eddie 40
      - Corbett's Gal
- Figure Four 266
  - Lone Drifter
    - Driftwood Ike
      - Driftwood
      - Hancock Belle
    - Moore Yen
      - Yendis
      - Katy Mas
  - Miss Setta
    - Eddie 40
      - Eddie
      - Sports Roan
    - Setta Hancock 66
      - Buck Hancock
      - Gilda Hancock

Joes Country Tune 4357711 2003 buckskin stallion

- Mr Joes Song
  - Mr Baron Red
    - Red Baron Bell
      - Baron Bell
      - Queen Vallerina
    - Two Eyed Patti
      - Two Eyed Jack
      - Patty's Queen
  - Watch Joes Song
    - Watch Joe Jack
      - Two Eyed Jack
      - Watch Jo Moore
    - Pat's Last
      - Pat Star Jr
      - Little Music
- Red Bucks Babe
  - Two Eyed Red Buck
    - Mr Baron Red
      - Red Baron Bell
      - Two Eyed Patti
    - Ima Tyree
      - Harlan's Tyree
      - Ima McKee
  - Fancy Jenny Bark
    - Fancy Roan Bar
      - Roan Bar
      - Fancy Baker 2
    - Jenny Lee Bark
      - Cedar Bark
      - Jenny Lee Bar

Rattlin Ron 3364157 1995 bay stallion

- Gray Badgers Tip
  - San Tip
    - Peppy San
      - Leo San
      - Peppy Belle
    - Tippy Tivio
      - Poco Tivio
      - Lucille M
  - Gray Badgers Gin
    - Shot O'Gin
      - Doc O'Lena
      - Gin Echols
    - Hattie's Badge
      - Grey Badger II
      - Hattie O'Neill
- My Ronella Lady
  - Scott Will Win
    - Scottish Will
      - Scottish
      - Elm's Siesta
    - Miss Suzie Dunn
      - Kiowa Dusty
      - Laverne Miss
  - Ronella Bar
    - Rattler's Bar
      - Cold Wave
      - Lula Dee
    - Miss Ronweiser
      - Budweiser
      - Rondy Dun

Lil Peppys Sugar Bar 3024969 1991 sorrel stallion

- Peppy San Badger
  - Mr San Peppy
    - Leo San
      - Leo
      - San Sue Darks
    - Peppy Belle
      - Pep Up
      - Belle Burnett
  - Sugar Badger
    - Grey Badger III
      - Grey Badger II
      - Mary Greenock (TB)
    - Sugar Townley
      - Lucky Jim
      - R J Clark Mare
- Sugar's Lady Bar
  - Rube Williams
    - Dee Gee's King
      - King
      - Dee Gee
    - Eddie's Bill
      - Eddie
      - Bellbyjohntreadwell
  - Sugar C Lady
    - Sugar Bars
      - Three Bars (TB)
      - Frontera Sugar
    - Lady Raw Hide
      - Dodger Crocker
      - Raw Hide

Ray Ray Leo 1215625 1976 gray stallion

- War Leo
  - Leo's Question
    - Leo
      - Joe Reed II
      - Little Fanny
    - Questionaire's Miss
      - Question Mark
      - Gipsy Rose 1
  - War Bird
    - War Star
      - Oklahoma Star Jr
      - Sheridina
    - Star Bird
      - Oklahoma Star
      - Moore's Fleet
- Tee Jay Sue
  - Jackie Bee
    - Jimmy Mac Bee
      - Sonny Day Bee
      - Miss Hackberry
    - Jackie Diane
      - Jack R
      - My Jackie Waggoner
  - Ronnie's Sue
    - My Little Joe
      - My Red Money
      - Lady Querida
    - Kenny Linn
      - Boys' Buck
      - Masie's Pride

CJ Goldrush 4042159 2001 buckskin stallion

- CJ Sugar
  - Son O Sugar
    - Sugar Bars
      - Three Bars (TB)
      - Frontera Sugar
    - Leo Pan
      - Leo
      - Panchita
  - Christy Jay
    - Rey Jay
      - Rey Del Rancho
      - Calandria K
    - Christy Carol
      - Leo Bob
      - Mike's Dilly
- Smooth Dot
  - Smooth Herman
    - Jet Smooth
      - Jet Deck
      - Lena's Bar (TB)
    - Carol's Ethel
      - King
      - Evans Mare
  - NU Creampuff
    - NU Bar
      - Doc Bar
      - Teresa Tivio
    - Holly Music
      - Sir Raleigh
      - De Oro Bonita

Play To The Max 4359146 2003 sorrel stallion

- Playgun
  - Freckles Playboy
    - Jewel's Leo Bars
      - Sugar Bars
      - Leo Pan
    - Gay Jay
      - Rey Jay
      - Georgia Cody
  - Miss Silver Pistol
    - Doc's Hickory
      - Doc Bar
      - Miss Chickasha
    - Pistol Lady 2 Be
      - King's Pistol
      - Miss Bailey 24
- Miss Special Quixote
  - Especial
    - Doc Bar
      - Lightning Bar
      - Dandy Doll
    - War Leo's Lady
      - War Leo
      - Brady Lady
  - Jesse Quixote
    - Doc Quixote
      - Doc Bar
      - Magnolia Gal
    - Jessica King
      - Captain Jess
      - Blondie R

LH Red Hot Affair 3714812 1998 black stallion

- LH King Star
  - Colonel Fandel
    - Colonel Hankins
      - Colonel Freckles
      - Queenie Star
    - Helen Fandel
      - Doc's Zimfandel
      - Sister Schuhart
  - Docs Lily
    - Doc's Zimfandel
      - Doc Bar
      - Pura Fina
    - Lily Hankins
      - Hankins' Bars
      - King's Lily
- Blues Red Hot Affair
  - Red Hot Hancock
    - Hancock's Dude
      - Joe Queen
      - Flicka Hancock
    - Mira Bay Hancock 3
      - Lowry Hancock Jr
      - Bay Ella
  - Blue Affair
    - Sonny's Nightshade
      - Sonny Dee Bar
      - Kim's Bonanza
    - Gold Deline
      - Gold Leo
      - Jay's Roan

Mr Remind 3621396 1997 brown stallion

- Letters Reminder
  - Doc's J Jay
    - Doc Bar
      - Lightning Bar
      - Dandy Doll
    - Paula Jumpup
      - Jimeny Jumpup
      - Stormy Rose
  - Eighteen Letters
    - Againu
      - Parker's Trouble
      - Irene 26
    - Miss Berty Abner
      - Sonny Bert
      - Silk Pool
- Snoopers Sassy
  - Mr Snooper
    - Snooper Bid Hill
      - Snooper Bid
      - Brenda Lee
    - Falcon Girl 24
      - Barney Whiskers
      - Bennett Star Bar
  - Sagey Bar
    - Mr Blue Bar
      - Bright Bar
      - Blue Bonnet Moore
    - Hula Skipper
      - Skipper M
      - Brush Slippers

Roosters Shorty 3978347 2000 bay stallion

- Gallo Del Cielo
  - Peppy San Badger
    - Mr San Peppy
      - Leo San
      - Peppy Belle
    - Sugar Badger
      - Grey Badger III
      - Sugar Townley
  - Doc's Starlight
    - Doc Bar
      - Lightning Bar
      - Dandy Doll
    - Tasa Tivio
      - Poco Tivio
      - Chowchilla Pee Wee
- Shorty Lena Six
  - Shorty Lena
    - Doc O'Lena
      - Doc Bar
      - Poco Lena
    - Moira Girl
      - Mora Leo
      - Sonoita Queen
  - Miss Quixote Six
    - Doc Quixote
      - Doc Bar
      - Magnolia Gal
    - Chick Duster
      - Six Chick
      - Dusty Josephine

Fritzi Badger 3096485 1992 sorrel stallion

Parents	Grandparents	Great-grandparents	Great-great-grandparents
Peppy San Badger	Mr San Peppy	Leo San	Leo
			San Sue Darks
		Peppy Belle	Pep Up
			Belle Burnett
	Sugar Badger	Grey Badger III	Grey Badger II
			Mary Greenock (TB)
		Sugar Townley	Lucky Jim
			R J Clark Mare
Shirley's Doc Bar	Doc Bar	Lightning Bar	Three Bars (TB)
			Della P
		Dandy Doll	Texas Dandy
			Bar Maid F
	Shirley Chex	King Fritz	Power Command
			Poco Jane
		Ottilie	Billy Lenart
			Sandra S

Massasuta 3400056 1995 sorrel stallion

Parents	Grandparents	Great-grandparents	Great-great-grandparents
Dash For Cash	Rocket Wrangler	Rocket Bar (TB)	Three Bars (TB)
			Golden Rocket (TB)
		Go Galla Go	Go Man Go
			LA Galla Win
	Find A Buyer (TB)	To Market (TB)	Market Wise (TB)
			Pretty Does (TB)
		Hide and Seek (TB)	Alibhai (TB)
			Scattered (TB)
Tara Named Me	Beduino (TB)	Romany Royal (TB)	Grey Sovereign (TB)
			Romany Belle (TB)
		Jo-Ann-Cat (TB)	Rejected (TB)
			Quick Eye (TB)
	I Owe You Me	Merridoc	Tiny's Gay
			Sparkling Tip
		Armadette	Armsmear
			Model Tee

Drifts Chip 1495390 1979 black stallion

Parents	Grandparents	Great-grandparents	Great-great-grandparents
Double Drift	Gray Chip	Driftwood	Miller Boy
			The Comer Mare
		Sugar Tit	Kelly
			Louse Schuyler
	Rose Wood	Driftwood	Miller Boy
			The Comer Mare
		Miss Hondo	Hank 1
			Nellie 21
Diamond Isle	Diamond Chip	Silver King	Old Sorrel
			Clegg Mare No 3
		Scar Face S	Bailey 1
			Mare By Midnight
	Dusty Babe	Dusty Hancock	Roan Hancock
			Triangle Lady 12
		803 Babe	Red Joe Of Arizona
			Dreamy 1

Sixes Pick 3744428 1998 sorrel stallion

Parents	Grandparents	Great-grandparents	Great-great-grandparents
Tanquery Gin	Doc O'Lena	Doc Bar	Lightning Bar
			Dandy Doll
		Poco Lena	Poco Bueno
			Sheilwin
	Gin Echols	Ed Echols	Zantanon
			Dorothy E (TB)
		Gin Squirt (TB)	Grog (TB)
			Breakfast Time (TB)
Natural Pick	Tenino Badger	Peppy San Badger	Mr San Peppy
			Sugar Badger
		Tenino Fair	Doc Bar
			Fairway's Gal (TB)
	Natural	Coe Badger Two	Grey Badger II
			Buck's Lass
		Miss Cee Bars 99	Cee Bars
			Holly Bugg 99

Paddys Irish Whiskey 2983308 1991 bay stallion

- Peppy San Badger
  - Mr San Peppy
    - Leo San
      - Leo
      - San Sue Darks
    - Peppy Belle
      - Pep Up
      - Belle Burnett
  - Sugar Badger
    - Grey Badger III
      - Grey Badger II
      - Mary Greenock (TB)
    - Sugar Townley
      - Lucky Jim
      - R J Clark Mare
- Doc's Starlight
  - Doc Bar
    - Lightning Bar
      - Three Bars (TB)
      - Della P
    - Dandy Doll
      - Texas Dandy
      - Bar Maid F
  - Tasa Tivio
    - Poco Tivio
      - Poco Bueno
      - Sheilwin
    - Chowchilla Pee Wee
      - Candy Kid
      - Dora

Playin Stylish 3391651 1995 sorrel stallion

- Docs Stylish Oak
  - Doc's Oak
    - Doc Bar
      - Lightning Bar
      - Dandy Doll
    - Susie's Bay
      - Poco Tivio
      - Susie L
  - Docs Stylish
    - Doc's Prescription
      - Doc Bar
      - Jameen Tivio
    - Holleo
      - Heleo
      - Holly Badger
- Playboys Mom
  - Freckles Playboy
    - Jewel's Leo Bars
      - Sugar Bars
      - Leo Pan
    - Gay Jay
      - Rey Jay
      - Georgia Cody
  - Lenaette
    - Doc O'Lena
      - Doc Bar
      - Poco Lena
    - Bar Socks Babe
      - Bar El Do
      - Dusty Socks

Seven From Heaven 3793189 1999 gray stallion

- Playgun
  - Freckles Playboy
    - Jewel's Leo Bars
      - Sugar Bars
      - Leo Pan
    - Gay Jay
      - Rey Jay
      - Georgia Cody
  - Miss Silver Pistol
    - Doc's Hickory
      - Doc Bar
      - Miss Chickasha
    - Pistol Lady 2 Be
      - King's Pistol
      - Miss Bailey 24
- Peppys From Heaven
  - Peppy San Badger
    - Mr San Peppy
      - Leo San
      - Peppy Belle
    - Sugar Badger
      - Grey Badger III
      - Sugar Townley
  - Royal Blue Boon
    - Boon Bar
      - Doc Bar
      - Teresa Tivio
    - Royal Tincie
      - Royal King
      - Texas Dottie

Four Roan Fly 3455576 1996 red roan stallion

- Continental Fly
  - The Continental
    - Torino
      - Ledo Bars
      - Rita Blue
    - Frosty Money
      - Jimmy Mac Bee
      - Red Dee Money
  - My Beaver 6
    - My Beaver
      - Beaver Creek
      - FL Pee Wee
    - Brown's Firefly
      - Black Hancock 1
      - Miss Yellow Wolf
- Happy Hancock 23
  - Happy Hancock
    - Pepper Hancock
      - Roan Hancock
      - Triangle Lady 8
    - Betty Buster
      - Chester B
      - Betty Hall
  - Last Pick 7
    - Last Pick
      - Old Man
      - First Pick
    - Lady Triangle
      - Little Sport
      - Lady Beverley

Royal Fletch 3542512 1997 sorrel stallion

- Jae Bar Fletch
  - Doc's Jack Sprat
    - Doc Bar
      - Lightning Bar
      - Dandy Doll
    - Pura Fina
      - Puro Tivio
      - Handy Annie
  - Jae Bar Lena
    - Leo's Question
      - Leo
      - Questionaire's Miss
    - Royal Bonita
      - Royal King
      - Jal
- Royal Blue Dually
  - Dual Pep
    - Peppy San Badger
      - Mr San Peppy
      - Sugar Badger
    - Miss Dual Doc
      - Doc's Remedy
      - Miss Brooks Bar
  - Royal Blue Boon
    - Boon Bar
      - Doc Bar
      - Teresa Tivio
    - Royal Tincie
      - Royal King
      - Texas Dottie

Hickorys Lowenbrau 3231700 1992 sorrel stallion

- High Brow Hickory
  - Doc's Hickory
    - Doc Bar
      - Lightning Bar
      - Dandy Doll
    - Miss Chickasha
      - Chickasha Mike
      - Maggie Cowden
  - Grulla San
    - Leo San Hank
      - Leo San
      - Hank's Sue
    - Blackburn 36
      - Poco Nino
      - Lady Black 62
- Lemac Jan
  - Dan's Sugar Bars
    - Sugar Bars
      - Three Bars (TB)
      - Frontera Sugar
    - Lena Leo
      - Leo
      - Lena Horn
  - Simla Sue
    - Hollywood Gold
      - Gold Rush
      - Triangle Lady 17
    - Miss Ollie
      - Tom (Scooter)
      - Joe Graham Mare

Ikes Double Drift 3696423 1998 buckskin stallion

- White Lighting Ike
  - Driftwood Ike
    - Driftwood
      - Miller Boy
      - The Comer Mare
    - Hancock Belle
      - Buck 8
      - Miss Hancock 1
  - Katy Was A Lady
    - K4 Hickory Skip
      - Skipity Scoot
      - Hickory Ann Hill
    - Katy Mas
      - Poquito Mas
      - Wayward Katy
- Dianes Cold Nose
  - Drifts Chip
    - Double Drift
      - Gray Chip
      - Rose Wood
    - Diamond Isle
      - Diamond Chip
      - Dusty Babe
  - West 301
    - Double Measure
      - Double Deck Bar
      - Tea Time Music
    - Skullberri
      - Skull Valley
      - Blue Angel Baby

Snickelfritz Chex 1158399 1975 black stallion

- King Fritz
  - Power Command
    - King
      - Zantanon
      - Jabalina
    - Crickett McCue
      - Barney McCue
      - Fanny McCue
  - Poco Jane
    - Poco Bueno
      - King
      - Miss Taylor
    - Mary Jane W
      - Pretty Boy
      - Waggoner Mare
- Poco Brickhouse
  - Zub
    - Buzzie Bell H
      - Daybreak
      - Luckystrikebell Mare
    - Goldie III
      - Jiggs II
      - Goldie 15
  - Poco Bonita
    - Poco Turp
      - Poco Bueno
      - Cowgirl M
    - Bonita Browning
      - Browningsbillythekid
      - Funny Face

Eddie Eighty 1293625 1977 sorrel stallion

- Eddie 40
  - Eddie
    - Raffles (TB)
      - Luke McLuke (TB)
      - Phebe G. (TB)
    - Greta 1
      - Bonnie Jack (TB)
      - Chessie E
  - Sports Roan
    - Sport
      - Zandy
      - Lowry's Mabel
    - Honey Lady
      - Pep Up
      - Lady Blackburn VI
- Tootie Toot 71
  - Buck Hancock
    - Joe Hancock
      - John Wilkins
      - Range Mare
    - Triangle Lady 40
      - Red Buck
      - Burnett Mare
  - Tootie Rock 71
    - Brown Rock
      - Blue Rock
      - Lady Black
    - Triangle Susan
      - Grey Badger II
      - Triangle Lady 71

Happy Hancock 0097947 1956 roan stallion

- Pepper Hancock
  - Roan Hancock
    - Joe Hancock
      - John Wilkins
      - Range Mare
    - Riding Type Mare
      - Unknown Sire
      - Unknown Mare
  - Triangle Lady 8
    - Tom (Scooter)
      - Midnight
      - Trixie 8
    - Mare by Mike Beetch
      - Mike Beetch 1
      - Unknown Mare
- Betty Buster
  - Chester B
    - Chubby
      - Midnight
      - Fourth Of July 1
    - Old Vaughn Mare
      - John Ellard Horse
      - Mare by Ivey Roan
  - Betty Hall
    - Yahootie
      - Silk McCue
      - Golden Wolf
    - Avent Mare 1
      - Unknown Sire
      - Unknown Mare

Eddie 40 0103667 1958 sorrel stallion

- Eddie
  - Raffles (TB)
    - Luke McLuke (TB)
      - Ultimus (TB)
      - Midge (TB)
    - Phebe G. (TB)
      - Mazagan (TB)
      - Unknown Mare (TB)
  - Greta 1
    - Bonnie Jack (TB)
      - Wrack (TB)
      - Mary's Aunt (TB)
    - Chessie E
      - Red Bug 2
      - Mare by Master Gould
- Sports Roan
  - Sport
    - Zandy
      - Zantanon
      - Mare by Strait Horse
    - Lowry's Mabel
      - Master Gould
      - Steel Dust Mare
  - Honey Lady
    - Pep Up
      - Macanudo
      - Petra R2
    - Lady Blackburn VI
      - Blackburn
      - Mare by Bailey

Continental Fly 1733408 1981 chestnut stallion

- The Continental
  - Torino
    - Ledo Bars
      - Aledo Bar
      - Queen Paula
    - Rita Blue
      - Todo Bueno
      - Jessie Blue
  - Frosty Money
    - Jimmy Mac Bee
      - Sonny Day Bee
      - Miss Hackberry
    - Red Dee Money
      - My Red Money
      - Irvin's Babe
- My Beaver 6
  - My Beaver
    - Beaver Creek
      - King
      - Stifle
    - FL Pee Wee
      - Peter McCue Jr
      - Prairie Dog L
  - Brown's Firefly
    - Black Hancock
      - Joe Hancock
      - Burnett Mare
    - Miss Yellow Wolf
      - Skeet 1
      - Old Yellow 1

RSI Doctor Pepper 3520044 1996 gray stallion

- Peppy Taito
  - Peppys Pavo
    - Mr San Peppy
      - Leo San
      - Peppy Belle
    - Pava De Bobby
      - Hiredhand's Cardinal
      - Coquena De Bobby
  - LA Taita
    - Alexander's Hand
      - El Vaiven
      - LA Alejandra
    - Taito Lass 5
      - Poco Taito
      - Lady Nifty 5
- RS Poco Too
  - Poco Sassy Doc
    - Sassy Doc
      - Doc Bar
      - Sassy Bars
    - Rack Lady 23
      - Poco Rack
      - Nifty Miss 23
  - Spade Sancho Too
    - Gray Dee Too
      - Doc's Dee Bar
      - Rick's Leola
    - Miss Premo Proctor
      - Sancho Lit
      - Quien Sabe 18

Pay Forty Four 2819393 1989 chestnut stallion

- Preferred Pay
  - Dash For Cash
    - Rocket Wrangler
      - Rocket Bar (TB)
      - Go Galla Go
    - Find A Buyer (TB)
      - To Market (TB)
      - Hide and Seek (TB)
  - Fancy Three
    - Three Bars (TB)
      - Percentage (TB)
      - Myrtle Dee (TB)
    - Annie Echols
      - Ed Echols
      - Orphan Annie
- Docs Monkey
  - Genuine Doc
    - Doc Bar
      - Lightning Bar
      - Dandy Doll
    - Gay Bar's Gen
      - Gay Bar King
      - Princess Piper
  - Monkey Dori
    - Chunky's Monkey
      - Peppy San
      - Stardust Desire
    - Hands Off
      - Parker's Trouble
      - Miss Mae

Wranglers Starlight 3770947 1998 bay stallion

- Grays Starlight
  - Peppy San Badger
    - Mr San Peppy
      - Leo San
      - Peppy Belle
    - Sugar Badger
      - Grey Badger III
      - Sugar Townley
  - Doc's Starlight
    - Doc Bar
      - Lightning Bar
      - Dandy Doll
    - Tasa Tivio
      - Poco Tivio
      - Chowchilla Pee Wee
- Wranglers Connie
  - Doc's Wrangler
    - Doc Bar
      - Lightning Bar
      - Dandy Doll
    - Miss Bar 89
      - Hollywood Gold
      - Jiggs' Mother
  - Sugar Leo Kit
    - Sugar Cro Bar
      - Sugar Bars
      - Sissie Croton
    - Leo Kit
      - Mucho Leo
      - Short Kit

Tejons New Star 3588720 1997 sorrel stallion

- Grays Starlight
  - Peppy San Badger
    - Mr San Peppy
      - Leo San
      - Peppy Belle
    - Sugar Badger
      - Grey Badger III
      - Sugar Townley
  - Doc's Starlight
    - Doc Bar
      - Lightning Bar
      - Dandy Doll
    - Tasa Tivio
      - Poco Tivio
      - Chowchilla Pee Wee
- New Orlena
  - Doc O'Lena
    - Doc Bar
      - Lightning Bar
      - Dandy Doll
    - Poco Lena
      - Poco Bueno
      - Sheilwin
  - Tisa McCoy
    - Bob McCoy
      - Sugar Bars
      - Bare's Pretty Lady
    - Bonnie Pistol
      - King's Pistol
      - Duchess Bonnie

Special Gun 3737774 1998 gray stallion

- Playgun
  - Freckles Playboy
    - Jewel's Leo Bars
      - Sugar Bars
      - Leo Pan
    - Gay Jay
      - Rey Jay
      - Georgia Cody
  - Miss Silver Pistol
    - Doc's Hickory
      - Doc Bar
      - Miss Chickasha
    - Pistol Lady 2 Be
      - King's Pistol
      - Miss Bailey 24
- Special Medicine
  - Doc's Prescription
    - Doc Bar
      - Lightning Bar
      - Dandy Doll
    - Jameen Tivio
      - Poco Tivio
      - Jameen
  - Bridgette Bar
    - Two Joe
      - Double Joe
      - Bobbita
    - Bar Bomb
      - Cue Reed
      - Beaver Bomb

Cat Silver 3650009 1997 gray stallion

- High Brow Cat
  - High Brow Hickory
    - Doc's Hickory
      - Doc Bar
      - Miss Chickasha
    - Grulla San
      - Leo San Hank
      - Blackburn 36
  - Smart Little Kitty
    - Smart Little Lena
      - Doc O'Lena
      - Smart Peppy
    - Doc's Kitty
      - Doc Bar
      - Kitty Buck
- Purdy Pistol
  - Peppy San Badger
    - Mr San Peppy
      - Leo San
      - Peppy Belle
    - Sugar Badger
      - Grey Badger III
      - Sugar Townley
  - Miss Silver Pistol
    - Doc's Hickory
      - Doc Bar
      - Miss Chickasha
    - Pistol Lady 2 Be
      - King's Pistol
      - Miss Bailey 24

MH Dyno Peppy 4430320 2003 sorrel stallion

- Little Dyno
  - Smart Little Lena
    - Doc O'Lena
      - Doc Bar
      - Poco Lena
    - Smart Peppy
      - Peppy San
      - Royal Smart
  - A Lenaette
    - Freckles Playboy
      - Jewel's Leo Bars
      - Gay Jay
    - Lenaette
      - Doc O'Lena
      - Bar Socks Babe
- Rosalina Peppy
  - Peppy San Badger
    - Mr San Peppy
      - Leo San
      - Peppy Belle
    - Sugar Badger
      - Grey Badger III
      - Sugar Townley
  - Rocinante Quixote
    - Doc Quixote
      - Doc Bar
      - Magnolia Gal
    - Sooty Doc
      - Doc's Yakoon
      - Sooty Sue

Real Gun 3660150 1997 gray stallion

- Playgun
  - Freckles Playboy
    - Jewel's Leo Bars
      - Sugar Bars
      - Leo Pan
    - Gay Jay
      - Rey Jay
      - Georgia Cody
  - Miss Silver Pistol
    - Doc's Hickory
      - Doc Bar
      - Miss Chickasha
    - Pistol Lady 2 Be
      - King's Pistol
      - Miss Bailey 24
- Miss War Chips
  - War Chip's
    - War Leo
      - Leo's Question
      - War Bird
    - Brady Lady
      - Cuellar
      - Estrella 1
  - Patsy Chip
    - Tona's Streak
      - Scotty's Billy Star
      - Tona La Negra II
    - Patsy Mahoney
      - Smoky H
      - Mare by Louis Farr

Shadowcatchin Cattin 4071128 2001 red roan stallion

- Cattin
  - Smart Little Lena
    - Doc O'Lena
      - Doc Bar
      - Poco Lena
    - Smart Peppy
      - Peppy San
      - Royal Smart
  - Lynx Melody
    - Doc's Lynx
      - Doc Bar
      - Jameen Tivio
    - Trona
      - Leon Bars
      - Miss Royal Fleet
- San Mans Dottie
  - Freckles San Man
    - Jewel's Leo Bars
      - Sugar Bars
      - Leo Pan
    - San Benito Belle
      - Doc Bar
      - Ben's Loxie C
  - Texas Dottie
    - Royal Texas
      - Royal King
      - Texas Kitty
    - Dottie Black
      - King Black
      - Fifty Rosy

Lucky Bottom Tom 3283244 1994 palomino stallion

- Lucky Bottom Rufus
  - Doc Bars Boy 2
    - Doc Bar's Boy
      - Doc Bar
      - Water Lady
    - Lucky Bottom 7
      - Lucky Star Mac
      - Nellie Gray H
  - Lucky Little Baby
    - Lucky Star Mac
      - Earl M's Mac
      - V's Quarter Ladystar
    - Easy Baby
      - Lowrance Dun
      - Bingo Belle
- Cindee Ann Bailey
  - Gold Bert Bailey
    - Gold King Bailey
      - Hank H
      - Beauty Bailey
    - Maudie Bert
      - Bert Baker
      - Maud Star
  - Fan's Cindee
    - Cactus Comet
      - Lightning Bar
      - Black Beauty Pollard
    - Fannie James
      - Jessie James
      - Rene 2

Shiver Shake 3768194 1999 bay stallion

- Doc's Hickory
  - Doc Bar
    - Lightning Bar
      - Three Bars (TB)
      - Della P
    - Dandy Doll
      - Texas Dandy
      - Bar Maid F
  - Miss Chickasha
    - Chickasha Mike
      - Billy Clegg
      - Millie M
    - Maggie Cowden
      - Tucumcari
      - Grey Eagle 3
- Loverboys Nurse
  - Freckles Loverboy
    - Jewel's Leo Bars
      - Sugar Bars
      - Leo Pan
    - Miss Reyco
      - Rey Jay
      - Copoco
  - Nurse Rey
    - Wyoming Doc
      - Doc O'Lena
      - Magnolia Holly
    - Jay Moss
      - Rey Jay
      - Mos Tony Top

Colonel Zippo Pine 2423139 1985 sorrel stallion

- Colonel Freckles
  - Jewel's Leo Bars
    - Sugar Bars
      - Three Bars (TB)
      - Frontera Sugar
    - Leo Pan
      - Leo
      - Panchita
  - Christy Jay
    - Rey Jay
      - Rey Del Rancho
      - Calandria K
    - Christy Carol
      - Leo Bob
      - Mike's Dilly
- Annie Pines Zip
  - Zippo Pat Bars
    - Three Bars (TB)
      - Percentage (TB)
      - Myrtle Dee (TB)
    - Leo Pat
      - Leo
      - Dunny Girl
  - Pine's Annie Lee
    - Poco Pine
      - Poco Bueno
      - Pretty Rosalie
    - Annie Lee Cody
      - Bill Cody
      - Annie Lee S

Madonnas Blue Bee 3155942 1992 red roan stallion

- Flying X 6
  - Salty Blue
    - Blue Rock
      - Lost Cause (TB)
      - Yellow Wolf Mare
    - Salty Susan
      - Hollywood Gold
      - Salty Mama
  - Marlene Kid
    - Martin's Kid
      - Arizona Yellowjacket
      - Ellanita
    - Eddie 33
      - Eddie
      - Susie S Hancock
- Tee J Madonna Bee
  - Jackie Bee
    - Jimmy Mac Bee
      - Sonny Day Bee
      - Miss Hackberry
    - Jackie Diane
      - Jack R
      - My Jackie Waggoner
  - Go Diamond Deck
    - Diamond Tender
      - Diamond Signal
      - Brandetta
    - Go Mandy Lady
      - Sir Bongo
      - Twisty's Priss

Showstoppin Boon 3684164 1998 chestnut stallion

- Peptoboonsmal
  - Peppy San Badger
    - Mr San Peppy
      - Leo San
      - Peppy Belle
    - Sugar Badger
      - Grey Badger III
      - Sugar Townley
  - Royal Blue Boon
    - Boon Bar
      - Doc Bar
      - Teresa Tivio
    - Royal Tincie
      - Royal King
      - Texas Dottie
- Doc Olenas Showgirl
  - Doc O'Lena
    - Doc Bar
      - Lightning Bar
      - Dandy Doll
    - Poco Lena
      - Poco Bueno
      - Sheilwin
  - Smokes Showgirl
    - Freckles Smoke
      - Jewel's Leo Bars
      - Croark
    - Gay Jay
      - Rey Jay
      - Georgia Cody

Little Elmer Fudd 3532890 1996 bay stallion

- Little Lena Doc
  - Smart Little Lena
    - Doc O'Lena
      - Doc Bar
      - Poco Lena
    - Smart Peppy
      - Peppy San
      - Royal Smart
  - Doc's Chico Girl
    - Doc Bar
      - Lightning Bar
      - Dandy Doll
    - Chico Baby
      - Top Breeze
      - Chico Boom
- Mandy Oak
  - Doc's Oak
    - Doc Bar
      - Lightning Bar
      - Dandy Doll
    - Susie's Bay
      - Poco Tivio
      - Susie L
  - Lite Up Peppy
    - Peppy San Badger
      - Mr San Peppy
      - Sugar Badger
    - Cigarrera
      - El Rey Rojo
      - Tacha Martin

Greyt Whiz 4455906 2002 gray stallion

- Topsail Whiz
  - Topsail Cody
    - Joe Cody
      - Bill Cody
      - Taboo
    - Doc Bar Linda
      - Doc Bar
      - Betty's Mount
  - Jeanie Whiz Bar
    - Cee Red
      - Cee Bars
      - Miss Jo Holly
    - Jeanie Whiz
      - Billy's Whizzer
      - Jeanie Bugle
- Easter Cabin
  - Easter Gentleman
    - Easter King
      - King
      - Gocha H
    - Bobbin Badger
      - Grey Badger II
      - Buckann
  - Forest Cabin Moon
    - Hobby Doc
      - Hobby Horse
      - Doc Bertie
    - Moon Child
      - Moon Deck
      - First Bid

Poco Tuck Zero 2708010 1988 bay stallion

- Poco King Tuck
  - Poco Bueno
    - King
      - Zantanon
      - Jabalina
    - Miss Taylor
      - Poco Bueno 1
      - Mare by Hickory Bill
  - Lady Illini 25
    - Illini King
      - King Hand
      - Peppy Mack Medias
    - Miss Bee 25
      - Bee Line
      - Waggoner Mare
- Senorita Zero 127
  - Zero Trouble
    - Taco Bar
      - Parker's Trouble
      - Susie Lewis L
    - Debra Zero
      - Johnny Zero
      - Peggy O'Grady
  - Dividen Miss 127
    - Mr Dividend
      - Dividend
      - Lady Beaver 65
    - Miss Nifty 127
      - Nifty Pep
      - Pretty Lady 127

Doc O Boots 3478820 1996 sorrel stallion

- Grays Starlight
  - Peppy San Badger
    - Mr San Peppy
      - Leo San
      - Peppy Belle
    - Sugar Badger
      - Grey Badger III
      - Sugar Townley
  - Doc's Starlight
    - Doc Bar
      - Lightning Bar
      - Dandy Doll
    - Tasa Tivio
      - Poco Tivio
      - Chowchilla Pee Wee
- Doc'O Lenita
  - Doc O'Lena
    - Doc Bar
      - Lightning Bar
      - Dandy Doll
    - Poco Lena
      - Poco Bueno
      - Sheilwin
  - Bonita Tetana
    - Barney Boots
      - Barney Blue
      - Lady Boots
    - Lulu Plunkett
      - Brown Chubby
      - Little Dora 1

Mr Sorrel Bo 3480399 1996 sorrel stallion

- Peeping Bo Badger
  - Peppy San Badger
    - Mr San Peppy
      - Leo San
      - Peppy Belle
    - Sugar Badger
      - Grey Badger III
      - Sugar Townley
  - Doc's Bo Peep
    - Doc Bar
      - Lightning Bar
      - Dandy Doll
    - Buxom Ella
      - Moolah Bux (TB)
      - Chantella
- Docs Pavo Lady
  - Doc O Rack
    - Doc's Prescription
      - Doc Bar
      - Jameen Tivio
    - Miss Rack 10
      - Poco Rack
      - Pep's Lady 10
  - Miss Peppy Petra
    - Peppys Pavo
      - Mr San Peppy
      - Pava De Bobby
    - Miss Petra 61
      - Nifty Pep
      - Lady Tony 61

Index

W

Y

Z

ACKNOWLEDGMENTS

I have so many people to thank, but I first have to start with Rue Judd, publisher of Bright Sky Press. Rue led me through the process, from beginning to end, and all of her suggestions made for an even better book.

Isabel Lasater Hernandez did an outstanding job on the design. She made the photographs leap off the page, and I can never thank her enough for what she's done with the presentation of the book.

Dixie Nixon is the world's greatest editor. Although I have been editing magazine articles for many years, she pointed out errors that not only I had missed, but that also were missed by others who had proofed my stories before Dixie got them.

The owners, managers and cowboys on the various ranches were the biggest help of all. Without exception, I was welcomed into their homes and on their ranches and provided with anything I asked for. When it was necessary to photograph from horseback, I was furnished a good horse, and, by the way, not once did one of those horses buck. I especially thank them for that.

Bayer Animal Health, of course, deserves a tremendous amount of thanks for co-sponsoring, with the American Quarter Horse Association, the Best Remuda Award. Bayer has been an AQHA corporate partner longer than any other company.

I have to thank those in my office who helped me with the photo scanning and the editing, but I especially have to thank my secretary, Tawanna Walker. She was with me every step of the way, from transcribing my taped interviews to helping me get everything sorted out and sent to the publisher.

And finally, I have to thank my wife, Mavis. Although she was able to accompany me on a couple of the trips, for the most part, she stayed at home and kept the home fires burning—something she's been doing for 36 years now—and I appreciate that very much. Without her support, none of it would have been possible.

Jim Jennings